PLAIN MEANING, PRECEDENT, AND METAPHYSICS: INTERPRETING THE ELEMENTS OF THE CLEAN WATER ACT OFFENSE

Jeffrey G. Miller

Professor Emeritus, Pace Law School

ENVIRONMENTAL LAW INSTITUTE
Washington, D.C.

1730 M Street NW, Washington, DC 20036

Published October 2017.

Printed in the United States of America
ISBN 978-1-58576-189-0

Dedicated to my wife and love,

Karen Mignone

Contents

About the Author

Jeffrey G. Miller served as a Professor of Law at Pace Law School for 25 years before his retirement in 2013. He served as its Vice Dean for Academic Affairs, Associate Dean of Graduate Studies, and Associate Dean of Environmental Programs. He has focused his teaching and scholarship primarily on water pollution, hazardous waste management, enforcement of environmental law, and citizen suits. He has written three books, several chapters, and numerous articles on these and related subjects. He was the faculty advisor to Pace's National Environmental Moot Court Competition at its inception, has continued in that capacity most years since, and has written the majority of the problems for the Competition. He serves on the board of directors of the Pace Environmental Law Clinic. Before he joined the Pace faculty in 1987, Professor Miller spent ten years with the U.S. Environmental Protection Agency and ten years in private practice. He began his EPA career as Enforcement Branch Chief and Enforcement Division Director in EPA's Boston office and subsequently joined its Washington, D.C., office where he served as Deputy Assistant Administrator for Water Enforcement and then acted as its Assistant Administrator for Enforcement from 1979 to 1981. He practiced environmental law as a partner in Verner, Liipfert, Bernhard, McPherson & Hand and in Bergson, Borkland, Margolis & Adler, both in Washington, D.C. He was subsequently of counsel to Perkins Coie for several years. He began his legal career as an associate in Bingham, Dana & Gould in Boston. Professor Miller has served as Treasurer and on the board of directors of the Environmental Law Institute and as an advisor to its *Environmental Law Reporter*. He served on the boards of directors of two corporations engaged in the management of hazardous waste. He has testified on issues of environmental law before subcommittees of the U.S. Senate, the House of Representatives, and state legislatures, as well as an expert witness in state and federal courts and private arbitrations. He has lectured on and taught environmental law throughout the United States and in Europe, South America, and Asia. He is a graduate of Princeton University and Harvard Law School, where he was a research associate for a year after his graduation.

Prologue and Acknowledgments

When I was a professor of environmental law at Pace Law School, academic colleagues from Pace and other law schools often asked me to suggest a topic for writing a law review article. One of my perennial suggestions was the interpretation of the elements of the Clean Water Act (CWA) offense. Over time, I had noticed increasing numbers of judicial decisions and EPA interpretations of the elements, many of them interesting, some of them important, and a few of them puzzling. As my academic career drew to a close, no one had followed my suggestion, giving me no excuse not to do so. I soon recognized that I should have addressed the task far earlier, for the number of judicial decisions and the number of analytical issues had grown exponentially over the years, making the task far more formidable than I had expected, particularly as an academic swan song. There was one benefit from waiting, however. I had matured as an academic. I now recognized that the task I undertook was not just determining meanings of the elements of the CWA offense, but also examining the process of statutory interpretation itself. This, of course, led to my five recently published articles in the *Environmental Law Reporter* on interpreting the elements of the CWA offense and to this volume, based on those articles.

Although I have written on many issues of environmental law and environmental law statutes, I have had a love affair with the Clean Water Act from the beginning of my environmental law career. That started when I joined EPA's newly formed Region I office in Boston in September of 1971, four months after the Agency was formed by an Executive Order. At that time, water pollution was addressed at the federal level by the Federal Water Pollution Control Act (FWPCA). The FWPCA, however, was cumbersome, incomplete and ineffective, including only minimal federal authority. Our major water pollution efforts, therefore, were directed at developing a permit program for industries under the Refuse Act of 1899, a statute aimed at preventing obstructions to navigation. The Refuse Act Permit program featured permits issued by the U.S. Army Corps of Engineers with pollution reduction requirements developed by EPA. The pervasive amendments to the FWPCA in 1972, now known as the CWA, adopted, subsumed, and expanded the Refuse Act permit program. As an EPA lawyer, I negotiated the terms of a permit for a pulp and paper mill in New Hampshire, one of the first CWA permits issued in New England. The mill was one of the largest industrial pollution sources in the region and wiped out much of the fish life in the Androscoggin River downstream in New Hampshire and Maine, as well as the lower Kennebec River into which the Androscoggin flowed at Merry Meeting Bay. The pollution reduction at the mill required by the permit cleaned up both rivers downstream. Those waters, which flow past my office as I write this, today support a healthy and diverse fishery. I later administered the national water pollution permit and enforcement program at EPA headquarters in Washington.

While with EPA in Washington, I worked with ELI to develop and conduct negotiation training for EPA and state enforcement personnel. A change of administration prompted me to enter the private practice of environmental law in Washington. During this period, my involvement with ELI strengthened. I participated as a faculty member in many educational programs, as well as in several multi-day workshops on negotiation and other topics, including several thinly disguised "rule of law" exercises in newly democratic countries in eastern Europe and South America, workshops overseen by Elisa Parker, Jay Pendergrass, and Jay Austin. Bill Futrell asked me to join the Board and later to become an officer. On the Board, I was privileged to serve with Ed Muskie, father of the CWA.

Fascinated by the possibility of private enforcement of the federal environmental statutes at a time when federal enforcement waned, I undertook a few such cases, but, more importantly, I published three pioneering articles in the *Environmental Law Reporter*, analyzing the legal issues involved in the statutes' citizen suit provisions. Because citizen suits were most easily prosecuted at the time under the CWA, this effort brought me back to my favorite statute. I also helped oversee an ELI study, commissioned by EPA, to determine whether citizen enforcement cases were "legitimate," ultimately concluding that most citizen suits were brought under the same circumstances that EPA suits were brought. ELI ultimately republished the articles and a summary of the study in a book with Wiley Publishing Company. Those publications propelled me into a 25-year career teaching environmental law at Pace Law School.

The publication of this volume, therefore, is a perfect bookend to my career as an environmental law academic enamored with the CWA. I am deeply grateful to ELI for publishing the articles and the books at the beginning and end of my career. I am particularly pleased that Jay Pendergrass and Jay Austin are involved in this project and that Elisa Parker is still active with

ELI. I thank Scott Schang for initially agreeing to publish the five recently published articles forming the basis of this book. I am also pleased to work with a newer ELI team member, Rachel Jean-Baptiste, the able and gracious editor of those articles and of this book. My heartfelt thanks to ELI for its role over the years in the development and practice of environmental law, as well in its role in my own environmental law career.

My thanks go to Laura Young, Pace Law 2015, for assisting me in completing much of the research for this endeavor and, in particular, for much of the analysis in Chapter 4, and to Christine Swatzell, Pace Law 2012, for assisting me with much of the initial research and analysis for the endeavor. I am also thankful for the Pace Law School Library staff, who helped me greatly with research, both while I was at Pace and after I retired and moved to Maine. The Pace IT staff has provided great long-distance service, helping me keep my laptop functioning.

My most profound gratitude goes to my wife, Karen Mignone, who has graciously tolerated weeks, months, and years of my slogging away at these and other research and writing endeavors, too often in blithe disregard of her needs and sometimes even of her presence. She has served as my valuable first-round editor, alerting me when my writing makes no sense at all or suggesting just the right word when I am having difficulty with a passage. Finally, she cooks wicked Italian dinners for a weary writer.

Jeff Miller, Arrowsic, Maine. 2016

Introduction

This volume examines the six most critical words in the Clean Water Act[1] (CWA). Those six words frame the elements of the water pollution offense: the "discharge of any pollutant by any person," unless in compliance with several listed sections of the statute, CWA §301(a).[2] The listed sections authorize the issuance of two types of CWA permits[3] and specify their substantive requirements. The statute defines "discharge of a pollutant" in the offense to mean: "any addition of any pollutant to navigable waters from any point source," CWA §502(12).[4] In sum, the subsection prohibits: (1) any addition (2) of any pollutant (3) to navigable waters (4) from any point source (5) by any person, except in compliance with a CWA permit. Justice Ruth Bader Ginsburg called this the "core command" of the CWA.[5]

The most critical words in the statute occur in the elements of the primary offense[6] as single words or as two word phrases: addition, pollutant, navigable waters, and point source. Those six words give meaning to the first four of the five elements of the water pollution offense. The fifth element, whose key word is "person," is important, but has not led to judicial or administrative disputes.[7] Without meeting all five of the elements, water pollution does

---

1 33 U.S.C. §§1251 et seq.

2 33 U.S.C. §1311(a).

3 Permits issued pursuant to §402, 33 U.S.C. §1342, regulate water pollution, and permits issued pursuant to §404, 33 U.S.C. §1344, regulate filling wetlands.

4 33 U.S.C. §1362(12). Because the term defined in §502(12), "discharge of *a* pollutant," is not exactly the same as the term used in §301(a), "the discharge of *any* pollutant," it could be argued that the definition in §502(12) does not apply to the phrase used in §301(a). However, courts routinely refer to §502(12) as defining "discharge of *a* pollutant" in §301(a), without noting the difference. (Emphasis added throughout.) Committee to Save Mokelumne River v. East Bay Mun. Dist., 13 F.3d 305, 307 (9th Cir. 1993); Apalachicola Riverkeeper v. Taylor Energy Co., LLC, 2013 WL 3779166 (E.D. La.); United States v. Bailey, 516 F. Supp. 2d 998 (D. Minn. 2007); Leslie Salt Co. v. Froehlke, 403 F. Supp. 1292, 1295 (N.D. Cal. 1974), *rev'd in part & mod. in part on other grounds*, Sierra Club v. Leslie Salt Co., 412 F. Supp. 1096 (N.D. Cal. 1976), *aff'd*, Leslie Salt Co. v. Froehlke, 578 F.2d 741 (9th Cir. 1978).

5 Coeur Alaska, Inc. v. Southeast Alaska Conservation Council, 557 U.S. 261, 298 (2009). The author has called it elsewhere "the basic prohibition" of the CWA. *See* Ann Powers et al., Introduction to Environmental Law: Cases and Materials on Water Pollution Control 141 (2008).

6 There are secondary offenses in the statute, such as a civil offense for spills of oil and hazardous substances in CWA §311(b)(6), 33 U.S.C. §1321(b)(6).

7 "By any person" is not included in or addressed by the §502(12) definition. It is the last of the elements because it follows "discharge of a pollutant" in §301(a). This volume does not focus on this element because there is little controversy about the interpretation of "person," which is defined in §502(5). The element is important because it confines violations of the CWA to the consequences of human activity. Of course, it is difficult to conceive of a civil or criminal offense without a human action.

not require compliance with a permit under either of the CWA's two permit programs, the engines of the statute's regulation of water pollution. Without meeting all five of the elements, water pollution is not subject to enforcement under the statute. These elements are jurisdictional requirements for the water pollution offense, for the requirement of compliance with a permit, and for enforcement. They are not the only important provisions in the statute, but they frame the jurisdiction and extent of the statute's regulatory programs.

One measure of the importance of the first four elements of the water pollution offense is that disputes over their meaning have led to 258 reported judicial decisions, including 11 U.S. Supreme Court decisions between 1974 and 2013.[8] Because many of the decisions interpret more than one element of the offense, they include 347 separate interpretations of the elements.[9] These judicial decisions and interpretations, however, are not the only materials studied for the analyses in this volume. Significant legislative history[10] and administrative interpretations supplement the judicial material.

This volume has two primary purposes. The first is to provide definitive and comprehensive analyses and understandings of each of the first four elements of the offense. The second is to determine what, if anything, can be learned about the process of statutory interpretation itself from studying the interpretations of the elements.

Each of the first four chapters of this volume examines a single element of the CWA offense. The elements are very different. Neither the U.S. Congress nor EPA has defined the first element, "addition." It's most obvious meaning is mathematical, which does not fit its context in the CWA. The common meaning of "pollutant" is material that causes pollution. Congress, however, defined "pollutant" and "pollution" without cross-referencing them. It defined "pollutant" as an exclusive list of 18 materials and categories of materials, with two exclusions. The list includes some materials that are usually thought of as polluting, e.g., sand and rock.[11] The Supreme Court has developed the meaning of "navigable water" for over a century to describe

8 See the decisions listed in *infra*, Table A. These constitute all of the reported decisions the author could locate within the time frame studied. There may be others, particularly in unofficial reporters. If others exist, they are infrequently cited and hence not important for the substantive issues analyzed here.

9 Count the numbers of judicial interpretations for each element listed *infra*, in Table B: 63 for Addition, 70 for Pollutant, 142 for Navigable Waters, and 72 for Point Source. The sum of interpretations for the four elements is 347 in the 258 decisions examined.

10 Attacks on the efficacy of legislative history by Justice Antonin Scalia and others caused judicial use of legislative history as a statutory interpretation device to decline significantly over the four decades of this study. The author examines this decline *infra*, in Chapter 5.

11 33 U.S.C. §1362(6).

the extent of federal Commerce Clause jurisdiction to improve and regulate waterways and their use. In that context, it means waters presently used, used in the past, or potentially usable in the future, with reasonable improvements, for waterborne transportation.[12] Congress defined the term, however, as "the waters of the United States," a broad but vague concept having nothing to do with waterborne transportation. "Point source" is an artificial legislative construct. Congress defined it as a "discernable, confined and discrete conveyance," followed by lists of 11 examples and two exclusions.[13] Congress defined "person" in the normal legal sense,[14] and the term has led to virtually no litigation. Disputes over the interpretations of the first four statutory terms have produced a steady stream of reported decisions since the initial implementation of the statute. Even after four decades, many of these issues are unresolved and new issues continue to arise.

The natures of the key words and phrases in these elements are very different. Those differences suggest that courts may use different approaches and methods to interpret the elements. For instance, the Supreme Court's earlier interpretation of "navigable waters" as a Commerce Clause jurisdictional term suggests that courts initially may be more inclined to interpret "navigable waters" as an element of the CWA offense based on precedent more than to interpret the other elements based on precedent, which lack such historical Supreme Court attention. Similarly, Congress' use of lists in its definitions of "pollutant" and "point source" suggests that courts may interpret them using canons of statutory interpretation focusing on the natures of lists, canons that are simply inapplicable to the other elements.[15] These differences in interpretive approach could have other consequences, such as the frequency of judicial challenges to agency actions and encouraging or discouraging environmentally positive results. While these questions and issues may be addressed when examining the meaning or interpretation of any one element, some may become apparent only when examining them simultaneously. This volume does so, particularly in the fifth chapter.

The decisions examined in this analysis were handed down over a 40-year period. This presents the opportunity to determine whether their interpretations have changed over the years, either as a result of the maturing of the statute or of developing jurisprudence. This long-term view of interpreting

---

12 33 U.S.C. §1362(7). The touchstone of "navigable water" for Commerce Clause jurisdictional purposes is use in waterborne transportation. The Danial Ball, 77 U.S. 557 (1870).

13 33 U.S.C. §1362(14).

14 33 U.S.C. §1362(5).

15 The canons that apply to lists include: *ejusdem generis*, *noscitur a sociis*, and the distinction between exclusive and inclusive lists.

the elements of the CWA offense also provides the opportunity to observe the continuing interaction between Congress, EPA, and the courts in giving meaning to the elements.

Chapter 1. "Addition"

This chapter examines the meaning of the simple word "addition" in the first element of the Clean Water Act (CWA)[1] prohibition against any addition of any pollutant to navigable waters from any point source by any person, unless in compliance with a permit. Neither the U.S. Congress nor the U.S. Environmental Protection Agency (EPA) has defined "addition" in this context. EPA and the courts have interpreted the element broadly to expand the offense. Some of their broad interpretations threaten to make felons of innocent persons owning a particular class of point sources: those conveying pollutants added by others to navigable waters. EPA's conflation of "addition" with "navigable waters," including its theory that all navigable waters are one entity, threatens to eviscerate half of the CWA's regulatory strategy programs, water quality standards, half of its permit programs, and the §404 program protecting wetlands. This chapter examines administrative and judicial interpretations of "addition" as an element of the CWA. It suggests a definition that fits all appropriate fact situations, while avoiding both threats to innocent point source owners and to the viability of the water quality standards and wetlands protection programs. It rejects the need for and legality of EPA's theory of unitary navigable water, EPA's water transfer rule based on that theory, and much of EPA's "outside world" theory of "addition."

I. Legislative Use of "Addition"

Section §301(1) prohibits "the discharge of any pollutant by any person,"[2] unless in compliance with several listed sections. The listed sections authorize the issuance of two types of CWA permits[3] and specify their substantive requirements. Section 502(12) defines "discharge of a pollutant" to mean "any addition of any pollutant to navigable waters from any point source."[4]

1. Clean Water Act (CWA), 33 U.S.C. §§1251-1387, ELR Stat. FWPCA §§101-607.
2. 33 U.S.C. §1311(1).
3. Permits issued pursuant to §402, 33 U.S.C. §1342, regulate water pollution; permits issued pursuant to §404, 33 U.S.C. §1344, regulate filling wetlands.
4. 33 U.S.C. §1362(12). Because the term defined in §502(12), "discharge of *a* pollutant," is not exactly the same as the term used in §301(a), "the discharge of *any* pollutant," the definition in §502(12) arguably does not apply to the phrase used in §301(a). However, courts routinely refer to §502(12) as defining "discharge of *a* pollutant" in §301(a), without noting the difference. (Emphasis added

In sum, the subsection prohibits: (1) any addition (2) of any pollutant (3) to navigable waters (4) from any point source (5) by any person,[5] except in compliance with a CWA permit. Justice Ruth Bader Ginsburg has called this the "core command "of the CWA.[6]

CWA §502, the general definitional section of the statute, does not define "addition," the statute does not define it elsewhere, and there is no legislative history directly suggesting its meaning. Congress used "addition" in the CWA only in §§301(a) and 502(12),[7] except in the irrelevant phrase "in addition to" or a variant.[8] This should make interpretation of "addition" easy in most cases, because there is no congressional interference with the ordinary meaning of this straightforward, almost mathematical concept.[9] And as we will see below, some cases are easy. On the other hand, with no indication of what Congress had in mind by using "addition," beyond the word itself and the purpose of the statute to control water pollution, some "addition" decisions may not be easy. Indeed, some are difficult, with courts almost resorting to metaphysics.[10]

---

throughout.) *See, e.g.*, Committee to Save Mokelumne River v. East Bay Mun. Dist., 13 F.3d 305, 307, 24 ELR 20225 (9th Cir. 1993); Apalachicola Riverkeeper v. Taylor Energy Co., LLC, 2013 WL 3779166 (E.D. La.); United States v. Bailey, 516 F. Supp. 2d 998 (D. Minn. 2007); Leslie Salt Co. v. Froehlke, 403 F. Supp. 1292, 1295, 5 ELR 20039 (N.D. Cal. 1974), *rev'd in part & modified in part on other grounds*, Sierra Club v. Leslie Salt Co, 412 F. Supp. 1096, 6 ELR 20363 (N.D. Cal. 1976), *aff'd*, Leslie Salt Co. v. Froehlke, 578 F.2d 742, 8 ELR 20480 (9th Cir. 1978).

5. "By any person" is not included in the §502(12) definition. It is the last of the elements because it follows "discharge of a pollutant" in §301(a). There is little controversy about the interpretation of "person" as defined in §502(5). The element is important because it confines violations of the CWA to the consequences of human activities. Of course, it is difficult to conceive of a civil or criminal offense without a human action.
6. Coeur Alaska, Inc. v. Southeast Alaska Conservation Council, 557 U.S. 261, 298 (2009). The author has called it elsewhere "the basic prohibition" of the CWA. *See* Jeffrey G. Miller et al., Introduction to Environmental Law: Cases and Materials on Water Pollution Control 141 (2008).
7. Sierra Club v. El Paso Gold Mines, Inc., 421 F.3d 1133, 1143 (10th Cir. 2005); Catskill Mountain Chapter of Trout Unlimited v. City of New York, 273 F.3d 481, 493 (2d Cir. 2001).
8. For example, §212(2)(B), 33 U.S.C. §1292(2)(B), "*In addition to* the definition contained in subparagraph (A) of this paragraph, 'treatment works' means . . ." Section319(a)(1)(A), 33 U.S.C. §1329(A)(1)(A), also provides that "without *additional* action to control nonpoint sources . . . ." (Emphases added.)
9. The length to which Congress sometimes goes to interfere with straightforward scientific concepts is illustrated by its definition of "solid wastes" as "solid, liquid, semisolid or contained gaseous material" in the Resource Conservation and Recovery Act (RCRA), 42 U.S.C. §§6901-6992k, ELR Stat. RCRA §§1001-11011. *See* RCRA §1004(27), 42 U.S.C. §6903(27).
10. For instance, consider the idea that soil and vegetation withdrawn from a wetland in a landclearing operation are metamorphosed into pollutants the instant they are returned to the same wetland and that the pollutant is added to the wetland because no pollutant was there before, even though the same soil and vegetation were in the same wetland before and after the operation. Borden Ranch v. U.S. Army Corps of Eng'rs, 261 F.3d 810, 814, 32 ELR 20011 (9th Cir. 2001); United States v. Deaton, 209 F.3d 331, 335-36, 30 ELR 20508 (4th Cir. 2000). Consider also the bizarre idea that all the disparate bodies of navigable water in the country are one and therefore channeling a heavily polluted navigable water into a pristine navigable water adds no pollutants to the pristine water because the polluted water and the pristine water are the same navigable water. Catskill Mountain Chapter

"Addition," the noun form of the verb "to add," is the action noun in the offense. Because offenses are actions, "addition" is the central element in the offense. This is more easily seen by rearranging the offense as a sentence: A person illegally adds pollutants to navigable water from a point source, unless he is in compliance with a permit. "Person" is the subject of the sentence, "adds" is the verb, and "pollutant" is the object. The rest of the sentence is a series of prepositional phrases: "to navigable water," "from a point source," "in compliance," and "with a permit." All of these prepositional phrases are adverbial phrases, modifying "adds." The last phrase, a defense rather than an element of the offense, is a compound phrase, but is still an adverbial propositional phrase modifying "adds."

It is important to keep the centrality of "addition" in mind when addressing the disputes considered in this chapter. Many of the defendants in these disputes raised the "passive point source" defense. They argued that their point sources added no pollutants to water, but merely conveyed already polluted water to downstream navigable water. The point sources were wholly passive, adding nothing that was not in the water already, and the defendants owning the point sources therefore did not violate the statute. This argument subtly rearranges the structure of the offense, making "point source" rather than "person" the subject of the sentence. It would make the sentence read: A point source illegally adds pollutants to navigable water by any person except in compliance with a permit. That, of course, makes no sense. The offense, like all other offenses, prohibits human conduct and activity, not point source conduct and activity.

Typical CWA cases involve industries or municipalities adding their pollutants to navigable waters through their own point sources (usually outfall pipes) from their own operations. The §402 permit program is designed with these cases in mind. Atypical cases involve persons adding their pollutants to navigable waters through point sources owned or operated by others. The statute explicitly deals with many but not all of these atypical cases. For instance, the pretreatment program in §307(b) controls industrial discharges of pollutants into municipal sewers leading to municipal treatment systems, from which they are added to navigable

---

of Trout Unlimited, Inc. v. City of New York, 273 F.3d 481 (2d Cir. 2001). Equally metaphysical is the notion that water withdrawn with live fish from navigable water, run through a power plant to generate energy, and returned as fish puree to the same navigable water body adds nothing to the navigable water because it never lost its status as navigable water, while water withdrawn from same navigable water to cool the same power plant and returned hot to the same navigable water body adds heat and fish puree to the navigable water because the water lost its status as navigable water during its journey cooling the generator. National Wildlife Fed'n v. Consumers Power Co., 862 F.2d 580, 589, 19 ELR 20235 (6th Cir. 1988).

waters by municipal outfalls, a large part of this atypical universe. These "indirect" industrial dischargers are not required to have §402 permits,[11] but are directly regulated by technology-based effluent limitations for toxic pollutants and other limitations designed to prevent interference with the municipal treatment systems into which they are discharged or from passing through those systems untreated. The municipal treatment systems' subsequent addition of the indirect industrial dischargers' treated pollutants to navigable waters is regulated by §402 permits issued to the municipalities, because the municipal systems are designed and intended to collect, treat, and discharge these industrial wastes. Similarly, both municipal and industrial stormwater systems are required under §402(p) to have permits. That is not remarkable: But for the municipal street and storm sewer systems, polluted stormwater would not be discharged to nearby streams; and but for the industrial operations, polluted stormwater would not be discharged to nearby streams.

However, there is an entirely different category of these atypical cases, in which the pollutants passing through point sources have no other relationship with the point sources or their owners. The following hypothetical illustrates these cases. Water from a spring on A's property flows into a river adjoining that property. A maintains a farm road bordering the river. Because land at the intersection of the spring flow and the river becomes muddy and obstructs the passage of vehicles, A installs a culvert for the spring flow to the river, and elevates the farm road to pass over the culvert. Unknown to A, B is upset with C, his former girlfriend, and her family because she will no longer see him. C and her family live on the river, just downstream from A's property, and use it as their water supply. In revenge for C's renunciation of him, B pours several buckets of a deadly poison into the spring water just before it passes through the culvert, so that the poison enters the river through the culvert, flows into C's water supply, and kills C and her family. Who would the prosecutor charge with homicide, A or B? Of course, she would charge B because B is the actor intending and acting to cause the deaths. Would the prosecutor charge A, because of the role that his culvert played in the action? Of course not. She would not charge A because he did not act toward C and his culvert was not a but-for cause of the deaths. If A had never installed the culvert and B had poured poison in the spring flow, the same deaths would have occurred.

11. See 40 C.F.R. §122.2, which defines "discharge of a pollutant" not to include additions from "indirect sources," which it, in turn, defines as sources adding nondomestic waste to municipal sewage treatment plants.

In the above hypothetical, who would the prosecutor charge with the crime of water pollution, A or B? Should she charge A, because A owned and operated the point source from which the pollutants flowed into the river and the owner or operator of a point source may add pollutants to rivers from a point source only in compliance with a CWA permit? No, because A did not act toward the river with regard to the pollutants and was not a but-for cause of their entry into the river. If A had never installed the culvert and B had poured pollutants into the spring, the same pollutants would have entered the river. A would have had no reason to know he should apply for a CWA permit to add pollutants to the river. On the other hand, B acted toward the river with regard to the pollutants, B was the but-for cause of the pollutants entering the river, and B had every reason to know that he should apply for a CWA permit to add pollutants to the river. B, not A or A's point source, is the "any person" who violated CWA §301(a); B added pollutants to the river from a point source. (Of course, if A installed the culvert knowing that B's poison would enter the river more quickly and more completely, A would be complicit with B's offenses.[12])

Courts, including the U.S. Supreme Court,[13] have generally rejected the "passive point source" owner defense, but only in cases in which the point source owner was not truly passive and without considering the above hypothetical or situations like it. Many decisions interpret "addition" broadly enough to convict the culvert owner in the above hypothetical, even though treating him as a §301(a) violator makes no sense and may even raise constitutional issues.[14] The point source owner in the hypothetical, however, is defended by recognizing that under §301(a), a person must act to add pollutants to navigable water from a point source. In the typical cases and most of the atypical cases, the point source owners are not really passive, they act to add pollutants to water where the pollutants would not otherwise be; but for their actions, the pollutants would not be in the water.[15] In our hypothetical,

12. Violations of the CWA are criminal felonies if they are knowing, criminal misdemeanors if they are negligent, and may be civil offenses with no fault. §309(c), 33 U.S.C. §1319.
13. In *South Florida Water Management District v. Miccosukee Tribe of Indians*, 541 U.S. 95, 105, 34 ELR 20021 (2004), the Court held that the passive point source defense was "untenable."
14. United States v. Apollo Energies, Inc. 611 F.3d 679 (10th Cir. 2010), in which the court treated lack of causation and of mens rea as raising similar constitutional issues. See also the helpful note by Alex Arensberg, *Are Migratory Birds Extending Environmental Criminal Liability*, 38 Ecology L.Q. 427 (2011). Because §309(c) requires mens rea for criminal offenses and A in our hypothetical has no level of mens rea, the constitutionality of criminal prosecution for a CWA offense without causation is unlikely to be an issue. However, because civil liability under §309(a) or (b) is strict liability, requiring no mens rea, a civil prosecution against A for violating §301(c) without an act by A causing the violation is possible.
15. In *South Florida Water Management District*, 541 U.S. 95, for instance, the defendant pumped polluted donor water into unpolluted receiving water and hence was a but-for cause of adding pollutants to

however, A is a purely passive point source owner and is not a but-for cause of adding poison to the river. If the culvert had not existed, the poisons would still have entered the river and killed C and her family. B is the only but-for cause of the pollutants entering the river and the resulting deaths, and B added pollutants to navigable water from a point source, even if he did not own the point source. The distinction is more clear if the question is whether the point source owner is a but for cause of adding a pollutant rather than whether the point source owner is passive.

In the absence of a statutory or regulatory definition of "addition" or controlling precedent, courts "are left to construe it 'in accordance with its ordinary or natural meaning.'"[16] Indeed, courts interpreting "addition" in the CWA use plain meaning and precedent more frequently than other interpretive devices.[17] To determine the plain meaning of a word, courts may consult a dictionary,[18] although there is no unanimity which dictionary is preferred.[19] A dictionary definition particularly helpful to understanding "addition" in §301(a) is "the act . . . of adding something to something else,"[20] which can

---

the receiving water. If the defendant had not acted, the pollutants would not have reached the receiving water.

16. S.D. Warren Co. v. Maine Bd. of Envtl. Prot., 547 U.S. 370, 376 (2006), quoting FDIC v. Meyer, 510 U.S. 471, 476 (1994).

17. Courts used the precedent and plain meaning canons of construction 51 and 52 times, respectively, in the 63 decisions interpreting "addition," or in 81% and 83% of those decisions, respectively. *See infra* Table B for Addition. Count the number of times devices 22, plain meaning, and 23, precedent, are cited in Column E. Courts employed the next most commonly used canon, deference or device 4, in 19 of those decisions. Initially courts used plain meaning almost to the exclusion of precedent. In the 11 decisions in 1983 and earlier, they used plain meaning 10 times and precedent 5 times. Subsequently, they used both of the canons almost equally. In the 52 decisions after 1983, they used plain meaning 42 times and precedent 46 times. Count the number of times that devices 22 and 23 are cited in Column E of Table B for Addition, *infra*, in decisions dated 1983 and earlier and in decisions dated after 1983 in Column B.

18. *See* William N. Eskridge Jr., Dynamic Statutory Interpretation 275 (1987) (identifying one of the Chief Justice William Rehnquist Court's canons of statutory construction as "follow dictionary definitions of terms unless Congress has provided a specific definition," and noting that the dictionary should be "of the era in which the statute was enacted." See also Antonin Scalia & Bryan A. Garner, Reading Law: The Interpretation of Legal Texts 415-24 (2012), for a detailed discussion of the nuances of using dictionaries in determining the "plain meaning" of statutes.

19. According to Justice Antonin Scalia and Bryan Garner, *supra* note 18, only a "scholarly" or "weighty" dictionary will do, and they list several dictionaries from different eras as possibilities. *See* Scalia & Garner at 415-24. They also suggest using a dictionary contemporary with the use of the word being interpreted, the year 1972 in the case of the CWA, or later because dictionaries' definitions usually trail actual usage. *Id.* at 419.

20. *See* Oxford English Dictionary Online. Courts have consulted other dictionaries. For instance, in *Sierra Club v. El Paso Gold Mines*, 421 F.3d 1133, 1143 (10th Cir. 2005), the court referred to the definition of "addition" in *Webster's New International Dictionary* (2002): "the act or process of adding." In *Friends of the Everglades v. South Fla. Water Mgmt. Dist.*, 2006 WL 3635465 (S.D. Fla. 2006), the court referred to the definition in *Webster's 3rd International Dictionary Unabridged* (1993): "joining of one thing to another." See also *S.D. Warren Co*, 547 U.S. at 376, in which the Court looked to the dictionary definition of "discharge" in *Webster's New International Dictionary* at 742 (2d ed. 1954).

be restated as "the act of adding something (A) to something else (B)," when (A) would not otherwise be in (B)." In the context of §301(a), it would read: "the act of adding A (any pollutant) to B (navigable waters) from a point source," when A (that pollutant) would not otherwise be in B (those navigable waters)."

The working hypothesis of this chapter is that "addition" means "the act of a person adding a pollutant to navigable waters from a point source, when that pollutant would not otherwise be in those navigable waters." Because the author's definition of "addition" focuses on addition by a person rather than by a point source, it is possible for a point source to convey pollutants to navigable water without adding them to it. This formulation of the definition still covers most passive point source situations, but does not include the passive owner in our hypothetical. This formulation emphasizes the causal concept inherent in the definition; addition does not happen spontaneously, something or some person must cause it.[21] It also incorporates part of what will be discussed below as EPA's "outside world" theory of "addition."[22] This suggested definition should yield appropriate results in the decisions examined below. EPA and the courts should adopt it.

II. Administrative Definition of "Addition"

The general definition section of EPA's §402 regulations, 40 C.F.R. §122.2, does not define "addition." EPA's few attempts to clarify the meaning of "addition" have only muddied the waters. In its initial attempt in *National Wildlife Federation v. Gorsuch*,[23] EPA asserted that "addition from a point source occurs only if the point source itself physically introduces a pollutant into water from the outside world."[24] In *South Florida Water Management District v. Miccosukee Tribe of Indians*, the petitioner did not introduce pollutants into water, but pumped an already polluted water body into a less-polluted water body. The petitioner argued from EPA's "outside world" theory that §402 permits are required only "'when a pollutant originates from a point source,' and not when pollutants originating elsewhere merely pass through the point source."[25] The Supreme Court decisively rejected this

21. The Eleventh Circuit recognized this in *Miccosukee Tribe of Indians of Fla. v. South Fla. Water Mgmt. Dist.*, 280 F.3d 1364, 1368-69, 32 ELR 20475 (11th Cir. 2002), *aff'd on other grounds*, South Fla. Water Mgmt. Dist. v. Miccosukee Tribe of Indians, 541 U.S. 95, 34 ELR 20021 (2004).
22. As discussed below in Section III.A.1.c. of this chapter, the Supreme Court rejected the remainder of EPA's "outside world" theory in *Miccosukee. See* 541 U.S. at 105.
23. 693 F.2d 156, 13 ELR 20015 (D.C. Cir. 1982).
24. 693 F.2d at 175.
25. 541 U.S. 95, 104, 34 ELR 20021 (2004).

argument. "A point source need not be the original source of the pollutant: it need only convey the pollutant to navigable water."[26] In doing so, however, the Court interpreted "point source" rather than "addition." This chapter addresses the "outside world" concept in several contexts in detail below, rejects it in part, and questions the usefulness of the remainder.

EPA's second attempt to clarify "addition" was in a policy and later in the preambles to a proposed and final rule exempting water transfers from the requirement of obtaining §402 permits. While the rule does not mention "addition," the preambles state that EPA's interpretation of "addition" is the basis of the rule. However, on close reading of the preambles, they did not define "addition," but instead interpreted the statute as a whole or interpreted "addition" as conflated with "navigable waters" to mean all navigable waters as a unitary national entity, rather than individual navigable water bodies as singular entities. The chapter addresses the policy and the rule in detail below in Section A.3. and concludes they are erroneous interpretations of the CWA.

III. Judicial Interpretation of "Addition"

As with most statutory interpretation, the meaning of "addition" is uncontroversial in most cases. It is usually fairly clear whether a person is adding pollutants to navigable water. You can often see a factory discharging discolored or foaming water into a river, visually adding pollutants. Indeed, most industrial and municipal dischargers admit they are adding pollutants to navigable water, by applying for CWA permits to do so. With that in mind, it is no surprise that the "addition" element is not contested in many litigated cases.[27] Even in decisions in which the "addition" element is challenged, courts often summarily hold that an addition has taken place.[28]

Most of these cases fall into three broad categories. The first, and most numerous, are situations in which point source owner defendants claim they are not liable because other persons add pollutants to water upstream from the point sources, and the pollutants merely flow through the defendants' passive point sources. The second, and least numerous, are situations

26. 541 U.S. at 105.

27. Sixty three decisions interpret "addition," 70 interpret "pollutant," 72 interpret "point source," and 142 interpret "navigable waters." Many of the decisions interpret more than one element. Count the number of decisions for each element, *see infra* Table B. Many of the decisions interpret more than one element, *see infra* Table A, Element Interpreted column.

28. In 14 of the 63 decisions considered in this chapter, courts decided "addition" was or was not satisfied based solely on a recitation of either plain meaning or precedent, *infra* Table B for Addition, Columns E & F.

in which the additions are indirect, often from the air. The final category consists of §404 cases involving additions in the process of clearing wetlands.

A. *Passive Point Sources*

Before examining the patterns in which defendants may raise the passive point source defense, it is useful initially to develop the concept of the discharge of pollutants and its supporting vocabulary. A point source, for instance a pipe, has two ends, one into which water enters and another from which water is discharged. Water entering the pipe is withdrawn from donor water and is referred to as influent. Water discharged from the pipe flows through the point source into the receiving water and is referred to as effluent. The donor water may be non-navigable (for instance, stormwater runoff or groundwater) or navigable (the same navigable water body as the receiving water or a different navigable water body). Passive point source owner or operator defendants argue that they are liable only for pollutants they add to water and that they are not liable for pollutants added by other persons to donor water before it reaches the defendants' passive point sources. They argue that those others are the but-for causes of adding the pollutants to the receiving water; without the addition of pollutants *by others* to the donor water, pollutants would not enter the receiving navigable water. A significant flaw in this argument is that there may be more than one but-for cause of a violation.

Causation is an element of negligence and nuisance, the common-law precedents of basic pollution control legislation, including the CWA.[29] It might be assumed that causation has no place in pollution control legislation, because it creates strict liability offenses.[30] But strict liability at common law means liability without fault or intent, not liability without cause.[31]

The U.S. Court of Appeals for the Eleventh Circuit rejected the passive point source defense in *Miccosukee Tribe of Indians of Florida v. South Florida Water Management District*,[32] because the passive source was the but-for cause of adding pollutants to navigable water. In *Miccosukee*, plaintiffs challenged the transfer through pumps and pipes of a polluted navigable donor water into a less-polluted navigable receiving water without a §402 permit. Defendants argued they did not add the pollutants to the less-polluted receiving

29. Miller et al., *supra* note 6, at 11-48.
30. The CWA's civil offenses are strict liability; they have no mens rea component. *See* §309(a) & (b); American Canoe Ass'n v. Murphy Farms, 412 F.3d 536, 540 (4th Cir. 2005). The CWA criminal offenses, however, are not strict liability; the misdemeanor offense requires negligence, and the felony offense requires knowledge. *See* §309(c).
31. Dan B. Dobbs et al., The Law of Torts 959-61 (2011).
32. 570 F.3d 1364 (11th Cir. 2002).

water because others added them earlier to the more-polluted donor navigable water. The court concluded:

> When a point source changes the natural flow of a body of water which contains pollutants and causes that water to flow into another distinct body of navigable water into which it would not have otherwise flowed, that point source is the cause-in-fact of the discharge of pollutants. And, because the pollutants would not have entered the second body of water but for the change in flow caused by the point source, an addition of pollutants from a point source occurs.[33]

In other words, the point source owner defendant in this case was not truly passive.

On appeal from the Eleventh Circuit's decision, the Supreme Court unanimously[34] held that the passive point source defense in that case was "untenable," putting the defense to rest, at least under the fact pattern at issue in the case. However, the Court did not base its holding on interpreting "addition," as the Eleventh Circuit had, but on interpreting "point source," for the definition of that element "makes it plain that a point source need not be the original source of the pollutant; it need only convey the pollutant to navigable waters."[35] The definition of point source is a "discernible, confined and discrete *conveyance*." (Emphasis added.) The Court's failure to address causation was not a rejection of causation's role in "addition," because the Court simply did not address the meaning of "addition." The Eleventh Circuit's and Supreme Court's *Miccosukee* decisions will be discussed in greater detail below.

Owners of point sources accused of violating the CWA have raised the passive owner defense when others initially introduced pollutants into donor

33. *Id.* at 1368-69. Courts commonly use but-for analysis to determine causation for tort purposes. *See* Dobbs et al., *supra* note 31 at §168. The defendant's action is the cause-in-fact of a harm if, but for the defendant's conduct, the harm would not have occurred. This necessarily requires a comparison with what would have occurred if the defendant had not acted. *Id.* at §§168-169. Of course, it is possible to have two defendants whose actions both are but-for causes of a harm. *Id.* at §171. A but-for analysis of the passive point source owner and the person adding pollutants to the donor water will come to different results depending on the facts. If the pollutant added to the donor water would have entered the receiving water without the point source, the owner or operator of the point source is not the but-for cause of the addition and is not liable, at least not under a but-for causation analysis. In either case, the person adding pollutants to the donor water is a but-for cause of the addition of the pollutants to navigable water from the point source, regardless of the liability of the owner or operator of the point source.
34. Although Justice Scalia dissented in part, he concurred with the majority on the passive point source issue. South Fla. Water Mgmt. Dist. v. Miccosukee Tribe of Indians, 541 U.S. 95, 112-13, 34 ELR 20021 (2004).
35. *Miccosukee*, 541 U.S. at 105. The Court's second conclusion is that Congress clearly intended that sewage treatment plants (publically owned treatment works or POTWs) require §402 permits; POTWs "treat and discharge pollutants added to water by others."

water subsequently flowing through defendants' point sources into navigable receiving water.[36] These cases involved many fact patterns, including runoff from abandoned mining operations, other surface runoff, dam-induced changes in water quality, water recirculated in industrial activity, and transfers of water between watersheds. The analyses in these cases by EPA and the courts,[37] however, fall into three other categories based on whether the pollutants: (1) are recirculated within the same navigable water; (2) originate in non-navigable stormwater runoff; or (3) are transferred from one navigable water to another navigable water. Decisions within these categories sometimes conflate the analysis of two elements, "addition" and "navigable waters." They also routinely neglect to determine whether the defendants are truly "passive" point source owners or whether they act in some way to add pollutants to navigable water.

1. Additions From Circulated or Recirculated Water

In the decisions analyzed here, defendants withdrew polluted navigable donor water, used it, and then discharged it through a point source to the same navigable receiving water from which the donor water was diverted prior to use. If the defendants did no more, under both EPA's "outside world" theory and our suggested definition of "addition," no permit is necessary.

a. Net/Gross Rule

At the outset of the implementation of the §402 permit program, EPA confronted the issue that many dischargers of polluted effluent withdrew pol-

36. Such cases arose under the Refuse Act, the predecessor of the CWA's §402 program, and early in the enforcement of the CWA. In *United States v. Granite State Packing Co.*, 470 F.2d 303, 3 ELR 20074 (1st Cir. 1972), the industrial defendant was liable under the Refuse Act for discharging its waste into water that subsequently flowed through a municipally owned culvert to navigable water. And in an early case under the CWA, *United States v. Velsicol Chem. Corp.*, 438 F. Supp. 945 (W.D. Tenn. 1976), the industrial defendant was liable under the CWA for the discharge of its waste into a municipally owned storm sewer that subsequently flowed through a point source into navigable water. In both cases, the enforcement target was the person initially adding pollutants to donor water, not the passive point source owner. The results were not unexpected; similar results would have occurred under negligence and nuisance law, the common-law antecedents of modern pollution control statutes. *See, e.g.*, *Springer v. Jos. Schlitz Brewing Co.*, 510 F.2d 468 (4th Cir. 1975), in which Schlitz discharged brewing waste into a municipal sewage treatment plant, causing the plant to discharge untreated waste into a river, injuring a downstream landowner's property. The court held that Schlitz could be liable in negligence if in the exercise of reasonable care it could have anticipated that the treatment plant could not adequately treat the brewery waste. We do not know enough facts in these cases to determine whether the point source owner was also a but-for cause of adding pollutants to the receiving water, i.e., whether it discharged donor water into receiving water that the donor water would not otherwise have entered.
37. Courts in these cases typically deferred to one degree or another to EPA's interpretation.

luted influent from and discharged polluted effluent to the same water body; some of the pollutants in their effluents were already in the influent donor water when they withdrew it. The dischargers were passive point source owners for pollutants already in their influent water: other polluters or nature itself initially introduced the pollutants to the donor water and therefore were the but-for causes of the defendants' point sources adding those pollutants to the receiving water.

Thus, the question: Did the statute require point source owners or operators to treat pollutants that other persons or nature itself caused to be in the donor water before the dischargers first withdrew influent from it, used the influent, and then discharged it back to the same receiving water? The point source owners were not the but-for causes of the pollutants from the influent; the same pollutants would be in the receiving water if the point source never existed.[38] Industries challenged technology-based effluent standards because they perceived EPA would apply them in permits to require treatment of all pollutants in an industry's discharge, including pollutants present in the water before the industry withdrew it for use.[39] Before courts dealt with the issue, EPA proposed and thereafter adopted a regulation allowing permit applicants to seek credit toward meeting their permits' effluent limitations for pollutants already in their influents, the so-called net/gross regulation.[40] It should be noted that EPA's rule applies only when the "intake water is drawn from the same body of water into which the discharge is made," excluding water transfers later authorized by EPA's water transfer rule.[41]

The net/gross rule is consistent with EPA's concept that "addition" must come from the "outside world." When an industry withdraws donor water

---

38. The issue is more complicated in some circumstances. Industries often require clean water for their processes. For instance, the paper industry needs color-free water to make white paper and commonly treats the water it withdraws to remove color before using the water in making paper. EPA uses another example in the preamble to its water transfer rule, stating that a drinking water treatment facility that removes solids from river water before delivering the water for consumer use would require a §402 permit before returning the solids to the river. 73 Fed. Reg. 33697, 33705 (June 13, 2008). In both of these examples, the water treatment process adds chemicals to the water to remove the unwanted materials from it, i.e., adding alum to remove solids by flocculation. The materials removed from the water are thus mixed with new pollutants and cannot be returned to the water without adding the new pollutants that were not originally in the water. Even with net/gross credits, dischargers must treat the new pollutants.
39. Query whether the issue was ripe for review in the effluent guidelines, which were neutral on the issue. The issue only arose if and when EPA actually applied the effluent guidelines in permits in the manner feared by industry.
40. 40 C.F.R. §122.45(g). "Gross," of course, represents all of the pollutants in the effluent; "net" represents only those pollutants first introduced to water by the discharger.
41. 40 C.F.R. §122.45(g)(4). EPA may waive this restriction if no degradation of the receiving water will result. The restriction is inconsistent with EPA's treatment of water transfers, as will be discussed later in the chapter.

from a navigable water body and returns it after use to the same navigable receiving water, the pollutants already in the donor water prior to the industrial use (1) already have been added to the donor water by others or by nature and (2) are from the same world as the receiving water, not from the "outside world." But the rule also flows from the causal aspect of "addition" developed in this chapter. If the industrial point source owner had never withdrawn, used, and discharged the water or had never even existed, the same pollutants would still be in the same receiving water above and below the location of the same point source. Under the Eleventh Circuit's analysis in *Miccosukee*, the point source owner in this situation would not be liable because the point source is not a but-for cause of adding these returned pollutants to the navigable water. Although the Supreme Court did not discuss causation in *Miccosukee*, its holding is consistent with the Eleventh Circuit's reasoning. A counterargument might be that when the industry withdraws influent water, the water loses its navigable character, so that when the industry later discharges its effluent with the pollutants originally in the donor water to the navigable receiving water, the pollutants are newly added to navigable water.[42] This counterargument, however, is strained, while the main argument is straightforward.

EPA's net/gross regulation allows the permit writer to grant credits toward a permit's effluent limitations for the types and amounts of pollutants in an industry's intake water as long as it discharges those pollutants back to the same navigable water from which it withdrew them. Under the suggested definition of "addition," the industry did not add the pollutants to navigable water; they were already in the same navigable water and would have been in it if the point source had not existed.[43] The point source owner was simply not the but-for cause of the pollutants' presence in navigable water. Requir-

42. *See, e.g.*, National Wildlife Fed'n v. Consumers Power Co., 862 F.2d 580, 589, 19 ELR 20235 (6th Cir. 1988). A similar metaphysical argument has been made in §404 cases: dredging removes soil and vegetation from wetlands, that material becomes dredged spoil when removed from the wetland, and dredged spoil is a pollutant. Therefore, when some of the removed material is replaced in the same wetlands, the defendant added a pollutant for the first time and a §404 permit is required, even though the same material was in the wetland before and after the defendant's actions. *See* United States v. Deaton, 209 F.3d 331, 30 ELR 20508 (4th Cir. 2000).

43. That is not entirely true on a molecule-by-molecule basis. The industry cannot isolate the molecules of a pollutant in its intake water and simply transfer them to the discharge water without treatment. Those molecules mix with the molecules of the same pollutant added by the point source's operation and both are subject to whatever pollution-control technology the industry uses. The net/gross credit works by allowing the point source to subtract from total number of molecules of the pollutant its treatment must remove the number of the molecules already in its intake water. Its treatment will remove some intake molecules and some operation molecules and it will discharge some intake molecules and some operation molecules. Since the molecules are fungible, as long as both sets of molecules are treated effectively, the receiving water will not know the difference.

ing an industry to treat such pollutants would be beyond EPA's statutory authority because the industry did not add the pollutants to the navigable water.

b. Dams and Recirculation

A variant of the net/gross fact pattern arose when environmental advocates challenged EPA's failure to include dams in the §402 permit program and later challenged a pump-storage hydroelectric project for discharging pollutants without a permit.[44] As in the net/gross effluent guidelines challenges, dams and the pump-storage project withdrew donor water from navigable water and discharged it to the same navigable receiving water after use, thus recirculating the water. And as in the net/gross challenges, the substances discharged were in the donor water in some form to begin with and hence the point sources were arguably not the but-for causes of the pollutants being in the receiving water. But in these cases, the point source owners were not truly passive as to the pollutants: the owners' use of the water changed the character of the pollutants in environmentally detrimental manners. The changes the owners made in the pollutants were not merely metaphorical, but were observable, physical changes.

In *Gorsuch*,[45] environmental advocates challenged EPA's failure to regulate discharges from dams in the §402 permit program. They contended that dams were point sources that added dam-induced pollutants into downstream waters. The issue of whether dams are point sources was not argued; EPA admitted that under some circumstances, dams are point sources.[46] The pollutants identified by plaintiffs included low dissolved oxygen (DO) caused by impoundments,[47] high DO caused by water cascading over dam spillways,[48] temperature changes caused by impoundments,[49] suspended solids caused by algae growth in impoundments,[50] and similar material. The

44. National Wildlife Fed'n v. Gorsuch, 693 F.2d 156, 13 ELR 20015 (D.C. Cir. 1982) (the dams case); *Consumers Power Co.*, 862 F.2d 580 (the pump-storage project case).
45. 693 F.2d 156.
46. 693 F.2d at 165.
47. Fish need high dissolved oxygen (DO). DO diminishes in quiescent water in an impoundment, particularly in deeper water. *See Gorsuch*, 693 F.2d at 161-62.
48. Too much DO, i.e., super saturation, can cause trauma in fish associated with oxygen bubbles in the blood stream. *Gorsuch*, 693 F.2d at 164.
49. Shallow impoundments with high surface-to-volume ratios tend to raise water temperatures, particularly at the surface. Deep impoundments with low surface-to-volume ratios tend to lower temperatures, particularly in deeper levels. Different aquatic species thrive at different temperature levels: trout, for instance, like lower temperatures and bass like higher temperatures. Higher temperatures also promote algae growth. *See Gorsuch*, 693 F.2d at 163.
50. *Gorsuch*, 693 F.2d at 164.

court held that what the petitioner characterized as pollutants were not pollutants, but merely changes in water quality.[51] This will be discussed further in Chapter 2.[52] In the alternative, the court held that if the materials were pollutants, the dams did not add pollutants to the downstream water.

The National Wildlife Federation (NWF) argued that dams created the impoundments, that pollutants were formed in and because of the impoundments, and that the dam then added the pollutants from the impoundments to the waters below. This amounts to arguing that dams are the but-for causes of the pollutant in their ultimate form. EPA argued for the first time that "addition from a point source occurs only if the point source itself physically introduces a pollutant into the water from the outside world."[53] Because most dam-caused pollutants were already in the reservoir water in some form before they flowed over or through the dam, EPA argued those pollutants were not added to downstream waters by the dam, but merely flowed over or through the dam.

The U.S. Court of Appeals for the District of Columbia (D.C.) Circuit found that nothing in the statute's wording, structure, or legislative history directly addressed the "addition" issue. It concluded that because both NWF's and EPA's interpretations were reasonable, it was bound to defer to EPA's interpretation of the statute to a degree that would later become known as *Chevron* deference.[54] EPA's interpretation, however, was enunciated only in its litigation position in that and earlier cases and therefore was not entitled to what would come to be a *Chevron* level of deference.[55] Courts later rejected the D.C. Circuit's excessive deference, appropriate only

51. *Id.* at 171-74.
52. In an earlier, little-cited district court decision to the contrary, *South Carolina Wildlife Fed'n v. Alexander*, 457 F. Supp. 118, 8 ELR 20757 (D.S.C. 1978), environmental groups sued to halt the construction of one dam and the continued operations of two other dams because they discharged pollutants without §402 permits. Plaintiffs argued that water entering the impoundments behind the dams was or would be high in DO and low in metallic substances, whereas water released by the dams was or would be low in DO and high in metallic substances, both changes were caused by the impoundments behind the dams and both were detrimental to fish. The court agreed, "[h]igh quality water . . . will enter the facility . . . and low quality water . . . will be discharged. Thus the release of the water changed because of the impoundment constitutes the 'addition' of pollutants into a navigable water." 457 F. Supp. at 126. At least as to the change from high to low DO, the court was wrong; DO is depleted by, not added in, the reservoir. The court responded that low DO is pollution as defined in §502(19) and that "no reasonable purpose would be served by admitting pollution while denying the existence of a pollutant." While the argument has intuitive appeal, it ignores the fact that §301(a) makes it illegal to add unpermitted pollutants, not to create pollution, and that Congress decoupled "pollutant" and "pollution" in the CWA. At least under these sections, pollutants need not cause pollution. *See infra* Chapter 2, Section 2.d. EPA was not a party to this case, and it does not appear that its "outside world" gloss of "addition" was before the court.
53. *Gorsuch*, 693 F.2d at 175.
54. Chevron U.S.A., Inc. v. Natural Res. Def. Council, 467 U.S. 857, 14 ELR 20507 (1984).
55. United States v. Mead Corp., 533 U.S. 218 (2001).

if EPA's interpretation had been "adopted in a rulemaking or other formal proceeding,"[56] which it was not.

In *National Wildlife Federation v. Consumers Power Co.*,[57] the environmental group sued a pump-storage project for withdrawing water containing live fish from Lake Michigan; pumping it to a hilltop reservoir at night; releasing it during the day to run by gravity through a turbine to generate both electricity and coincidentally fish puree; and ultimately discharging it with fish puree, fish parts, and some surviving fish back into Lake Michigan. Because electricity is cheaper at night, when the company used it to pump water uphill, and more expensive during the day, when the company generated it by letting the water flow downhill through the turbine by force of gravity, the operation was profitable, producing energy when it was most needed and therefore most expensive.[58]

Acknowledging that the dead fish parts were pollutants, the outfall pipe was a point source, and Lake Michigan was navigable water, the U.S. Court of Appeals for the Sixth Circuit held that the defendant did not violate the CWA because it did not add pollutants to Lake Michigan: The fish parts were not from the "outside world," but instead originated as whole fish in Lake Michigan waters. The key to the court's reasoning was its proposition that the "water passing through the . . . facility never loses its status as water of the United States"[59] or, in the words of the district court, "the Lake water does not lose its status as navigable water simply because it is removed from the Lake, and since the fish never leave the Lake, they cannot be added to it from the outside world."[60] Although there was Supreme Court precedent (in another context) for the first part of this proposition, the court cited no authority for it.[61] It took pains to explain that in normal industrial use, including industrial cooling, water loses its status asnavigable somewhere between

56. Catskill Mountains Chapter of Trout Unlimited, Inc. v. City of New York, 273 F.2d 481, 490 (2d Cir. 2001).

57. 862 F.2d 580, 19 ELR 20235 (6th Cir. 1988).

58. This explains why the operation was profitable, even though it requires more energy to pump water uphill than it generates by flowing downhill through the turbine.

59. *Consumers Power*, 862 F.2d at 589.

60. *Id.* at 589, citing National Wildlife Fed'n v. Consumers Power Co., 657 F. Supp. 989, 1008, 17 ELR 20801 (W.D. Mich. 1987).

61. "[T]hat the running water in a great navigable stream is capable of private ownership is unconceivable." United States v. Chandler-Dunbar Water Power Co., 229 U.S. 53, 69 (1913); *cited with approval in* S.D. Warren Co. v. Maine Bd. of Envtl. Prot., 547 U.S. 370, 379, n.5 (2006) (stating "nor can we agree that one can denationalize waters by exerting private control over them"). Both decisions involved water diverted to generate electric power. Neither, however, considered the particular type of diversion or the issue discussed here. The observation is also consistent with the doctrine that if a water was once navigable for Commerce Clause jurisdiction, it is always navigable for that purpose. Economy Power & Light Co. v. U.S., 256 U.S. 113, 123 (1920). But, again, the context of that decision was entirely different.

the industry's withdrawal of influent from the donor water, the industry's use of the water in the industrial process, and the industry's discharge of the effluent back into the same receiving water.[62] If not, of course, the CWA would prevent little or no pollution. Thus, under the court's analysis, when a power company withdraws water from Lake Michigan to turn a turbine, the water remains navigable while it is used and returned to the lake, but if the same or a different power company withdraws the same water from the same lake to cool a turbine, even the same turbine, the water is no longer navigable when it is returned to the same lake. EPA reiterated this distinction in the preamble to its water transfer rule.[63] Neither the court nor EPA cited authority or explained the rationale for this difference. Despite revisiting the analysis annually for decades, this author is unable to explain it either, except as a results-oriented distinction or administrative and judicial metaphysics.[64]

The probable reason the court did not cite Supreme Court precedent for its proposition that the water the defendant diverted for power generation did not lose its status as water of the United States was that the precedent did not support the court's distinction between the status of water diverted for power generation and water diverted for other industrial uses. The court did not need this unexplained and probably unexplainable distinction to support its ruling, however. It could, for instance, have held the defendant's pump storage operation did not require a §402 permit because the change from live to dead fish occasioned by the defendant's operation was a change in water condition, not the addition of a new pollutant, relying on *Gorsuch.*

As in the net/gross situation, these point source owners and operators appear not to have added pollutants from their influents to the navigable receiving waters, because the donor water and the receiving water were the same body of navigable water and the pollutants were already in the donor water when the industry withdrew it. At the same time, they differed from the passive dischargers in the net/gross situation, for these point sources were not truly passive; they manipulated material or characteristics already in

62. *Consumers Power*, 862 F.2d at 589.
63. 73 Fed. Reg. 33697, 33704 (June 13, 2008).
64. There are some differences. In the cooling-water situation, the power plant adds heat to the water in the pipes, while in the electricity-generation situation the plant does not add fish. But this difference relates to the "addition" element, not the "navigable waters" element. It does not explain why the water in the plant is non-navigable in the first case but navigable in the second. In the cooling-water situation, §316(b) requires that the location, design, construction, and capacity of cooling water intake structures minimize adverse effects on the environment, while it requires no such location, design, construction, or capacity requirement for non-cooling water intake structures. But again, this has nothing to do with whether these waters are navigable. Another possible distinction is that in *Consumers Power*, the water is used to generate electricity, while in the other cases, it is used for industrial purposes. But generation of electricity is an industrial process and cooling a generator is part of the generation of electricity.

the influent water to make them different and more harmful to the aquatic biota when discharged to the receiving waters. But for their actions, the point sources would not have created, discharged, or added to navigable water fish parts, low DO, or higher concentrations of suspended solids; those substances would not otherwise be in the navigable receiving water. Whether there is an addition under either the suggested definition of "addition" or EPA's "outside world" theory, therefore, depends on whether the pollutants already in the donor water are the same pollutants that the point sources added to the receiving waters and whether they are pollutants at all, rather than water conditions. That is determined by the meaning of "pollutant" rather than by the meaning of "addition."

c. EPA's "Outside World" Theory of Addition

EPA's theory that addition "occurs only if the point source itself physically introduces a pollutant into the water from the outside world" first surfaced in the early 1980s as EPA's litigation position in *Gorsuch.*[65] Although EPA cited no statutory or other basis for it, the theory sounds logical. The "outside world" is such a nicely turned phrase that it lodges in our minds, a good example of a legal meme.[66] It is no surprise that the theory is frequently cited and seldom questioned. Like many nicely turned phrases, however, it is more pleasing than precise.

Examined closely, the theory incorporates two ideas. First, the point source in question must itself introduce the pollutant into water. Second, the pollutant must originate from the "outside world." The first idea, that the point source itself must introduce the pollutant into "the water," conflates the "addition" and "point source" elements of the CWA offense. Moreover, the Supreme Court rejected it in *Miccosukee* as contrary to the statute's definition of "point source" as a conveyance: "[A] point source need not be the original source of the pollutant: it need only convey the pollutant to navigable water."[67] Significantly, the United States agreed with the Court.[68] The mining waste decisions, discussed below, also implicitly reject the first concept.

---

65. 693 F.2d 156, 175, 13 ELR 20015 (D.C. Cir. 1982).

66. Memes are contagious phrases, tunes, and other ideas that leap into our minds and propagate by leaping from them into the minds of others. The author explored "navigable water" as a meme in Jeffrey G. Miller, *Evolutionary Statutory Interpretation: Mr. Justice Scalia Meets Darwin*, 20 Pace L. Rev. 409, 419-20 (2000).

67. South Florida Water Management District v. Miccosukee Tribe of Indians, 541 U.S. 95, 105, 34 ELR 20021 (2004).

68. *Id.* at 105-06.

The second idea is ambiguous. What does EPA's phrase "into the water from the outside world" mean? It is not clear what "the water" in EPA's phrase refers to. Is it the donor water, the receiving water, either, any navigable water or any water? The world outside "the water" will be different depending on which "water" is considered. In any event, could it make sense for the addition to be from the "inside world?" On the other hand, if the origin of the pollutant in an "addition" is inherently from the "outside world," the phrase adds nothing to the definition of "addition." No doubt EPA could develop this concept to be a meaningful part of a full definition of "addition," but EPA has not attempted to define the word.

Moreover, the "outside world" idea plays havoc with the CWA's second permit program, §404, which protects the loss of wetlands from unregulated filling. Most of the reported §404 decisions involve landclearing activities to prepare wetlands for agricultural or other uses, activities in which soil and organic material are moved from one location to another in a wetland. It is not apparent how this material is from the world outside the wetland. EPA sidesteps the issue by observing that Congress defined "pollutant" to include "dredged spoil," knowing that it was removed from water, thereby sanctioning the application of §404 to wetlands landclearing cases. While this establishes one of the four elements of the offense, "pollutant," it does not establish the other three, including "addition." If "addition" includes the "outside world" concept, it is difficult to square it with most §404 cases, unless "outside world" applies to §402 cases, but not to §404 cases, which EPA has not argued and which would be contrary to the canon of statutory construction that words be interpreted the same throughout a statute, unless the statute explicitly indicates otherwise. The incompatibility of EPA's "outside world" theory of "addition" and §404 is explored in greater detail below.

EPA's "outside world" theory of "addition" is ambiguous, unhelpful, inconsistent with §404, and contrary to the Supreme Court's ruling in *Miccosukee.* Yet, as discussed below, EPA has continued to use it, as if *Miccosukee* had never been decided. It is time that EPA either abandons the theory or builds on it to promulgate a complete definition of "addition." If EPA does not act in one of these manners, courts should recognize that the theory is without merit.

2. Additions From Stormwater Runoff

Although there was considerable skirmishing at the outset of the CWA's implementation over whether surface stormwater runoff was subject to the

§402 permit program, Congress ended much of the contention by amending the CWA in 1987 to include §402(p). In that subsection, Congress directed EPA to issue or deny permits "for discharges composed entirely of stormwater" that are either "associated with industrial activity" or discharged from "municipal separate storm sewer system[s]."[69] EPA's implementing regulations included an amendment to its definition of "discharge of a pollutant," specifying that the "definition includes additions of pollutants into waters of the United States from: surface runoff which is collected or channeled by man."[70] And in its stormwater regulations, it defined "discharge associated with industrial activity" to mean "the discharge from any conveyance that is used for collecting and conveying storm water and that is directly related to manufacturing, processing or raw materials storage areas at an industrial plant."[71]

The "composed entirely of stormwater" wording of §402(p) is curious. If there were no pollutants in the stormwater, there would be no reason to regulate the discharges. Indeed, if there were no pollutants in the stormwater, adding it to navigable water from point sources would not violate the prohibition of §301(a). If the point source owner could prove the water from these sources carried no pollutants, it would require no permit. EPA's reference to "additions of pollutants . . . from stormwater runoff" in its definition of "discharge of a pollutant" acknowledges this.

The most commonly litigated fact pattern in decisions on "addition" of pollutants from stormwater contamination involves mining operations, often inactive mining operations. For example, in *American Mining Congress v. U.S. EPA*, the plaintiff broadly challenged EPA's interpretation of the statutory phrase "discharge associated with industrial activity" in §402(p) as it applied to inactive mining operations.[72] The plaintiff argued that because there is no "activity" at an inactive mine, EPA's definition exceeded its statutory authority. The U.S. Court of Appeals for the Ninth Circuit found that EPA's reliance on "associated with" in the statutory phrase "associated with industrial activity" was reasonable, however, since nothing in that statutory phrase required temporal concurrence between the mining activity and the polluted stormwater discharge.[73] Finding no legislative history and nothing in the statute to the contrary, the court deferred to EPA's interpretation.

69. Municipal separate storm sewers carry only stormwater, not sewage, while municipal combined sewers carry both stormwater and sewage.
70. 40 C.F.R. §122.2.
71. 40 C.F.R. §122.26(b)(14).
72. 965 F.2d 759, 22 ELR 21135 (9th Cir. 1992).
73. *Id.* at 764. A similar attempt to require temporal concurrence between "addition" and the polluting activity was rejected in *National Cotton Council of America v. EPA*, 553 F.3d 927, 39 ELR 20006 (6th

Courts have universally rejected the passive owner defense when raised by owners and operators of point sources discharging pollutants in stormwater runoff from past, present, or neighboring mining operations. In these situations, courts have held that "addition" does not require the owner or operator of the point source to generate,[74] cause,[75] create,[76] or be the source[77] or origin[78] of the pollutant, but requires only that the point source convey the pollutant to navigable water. These decisions are entirely consistent with the Supreme Court's holding in *Miccosukee* that the passive owner defense is "untenable."[79]

The most thorough analysis of the issue regarding mining wastes is in *Sierra Club v. El Paso Gold Mines, Inc.*[80] There, the U.S. Court of Appeals for the Tenth Circuit noted that in the definition of "discharge of a pollutant" in §502(12), Congress modified "addition" with "any," suggesting an expansive reading of "addition," and that "viewed as a whole . . . the liability and permitting sections of the Act focus on the point of discharge, not the underlying conduct that led to the discharge."[81] It also notes that the CWA "refers to the obligations of the 'owners and operators' of a point source, suggesting that [they] . . . are responsible for a functional point source"[82] and that EPA's regulations "focus . . . on ownership of the point source, not the discharge-causing conduct."[83] EPA defines "owner or operator" as the regulated "facility or activity" and defines "facility or activity" as a "'point

Cir. 2009).

74. Natural Res. Def. Council, Inc. v. County of Los Angeles, 673 F.3d 880, 900, 41 ELR 20109 (9th Cir. 2011), *aff'd*, Los Angeles Cnty. Flood Control Dist. v. Natural Res. Def. Council, 133 S. Ct. 710, 43 ELR 20004 (2013).
75. West Virginia Highlands Conservancy, Inc. v. Huffman, 624 F.3d 159, 167-68, 40 ELR 20014 (4th Cir. 2010).
76. Catskill Mountain Chapter of Trout Unlimited v. City of New York, 273 F.3d 481, 493 (2d Cir. 2001).
77. Miccosukee Tribe of Indians of Fla. v. South Fla. Water Mgmt. Dist., 280 F.3d 1364, 1368 n.6, 32 ELR 20475 (11th Cir. 2002).
78. United States v. Law, 979 F.2d 977, 979, 23 ELR 20466 (4th Cir. 1992).
79. South Fla. Water Mgmt. Dist. v. Miccosukee Tribe of Indians, 541 U.S. 95, 34 ELR 20021 (2004).
80. 421 F.3d 1133 (10th Cir. 2005). *See also* the district court opinion, 2002 WL 33932715 (D. Colo. Nov. 15, 2002), which covered much of the same ground.
81. Sierra Club v. El Paso Gold Mines, Inc., 421 F.3d at 1143, citing §§301(e), 402(a)(1), and 101(a)(3), 33 U.S.C. §§1311(e), 1342(a)(1), and 1251(a)(3). This is perhaps somewhat simplistic, as illustrated by industrial discharges into municipal sewerage systems. The municipalities must secure §402 permits for their discharges of treated effluent into navigable water, and the industries are not required to secure §402 permits for their discharges into the municipal sewerage system; the industries are indirect discharges to the navigable water, 40 C.F.R. §122.3(b). But the industries are required to treat their wastes before discharging them to the municipal sewerage system to meet pretreatment requirements established under §307(b) and are civilly and criminally liable for not doing so under §309(a) & (c).
82. 421 F.3d at 1143-44, citing §§301(g)(2) and 308(a), 33 U.S.C. §§1311(a)(2) and 1318(a)(2).
83. 421 F.3d at 1144, citing 40 C.F.R. §122.2 (definition of "addition of any pollutant") and §122.26(b)(14)(iii).

source' or any other [regulated] facility or activity."[84] The decision also cited relevant precedent from the Ninth Circuit,[85] as well as EPA's stormwater runoff regulations,[86] to which it gave deference. Other mining decisions have rejected arguments that addition required intentional conduct[87] or a greater presence of pollutants after installation of the point source than before,[88] and that addition could not be of pollutants already on the streambed or in the stream banks.[89]

A few passive point source decisions have involved the collection and channeling of non-mining-related stormwater runoff.[90] They reached the same conclusions as the mining-related stormwater runoff cases, for the same reasons, especially the explicit coverage of stormwater runoff by Congress[91] and EPA's corresponding regulations.[92]

The mining waste decisions implicitly reject the first concept in EPA's "outside world" gloss on "addition." They held, as did the Supreme Court in *Miccosukee*, that point sources added pollutants to navigable water, even though the point sources did not introduce the pollutants into water in the first instance. Are the decisions consistent with the more-developed concept of "adding a pollutant to navigable waters from a point source when that pollutant would not otherwise be in those navigable waters?" Did the point sources cause the pollutants to be present in the receiving water? We can not be sure, because the courts did not undertake the factual analyses necessary to address those questions. It is probable that without the channeling systems and their associated point sources, some of the same pollutants in the same runoff from the same mining operations would have made their way by force of gravity to the same receiving waters. The point sources would not be but-

84. 40 C.F.R. §122.2.
85. Committee to Save Mokelumne River v. East Bay Mun. Util. Dist., 13 F.3d 305, 24 ELR 20225 (9th Cir. 1993); American Mining Cong. v. EPA, 965 F.2d 759, 22 ELR 21135 (9th Cir. 1992); Beartooth Alliance v. Crown Butte Mines, 904 F. Supp. 1168, 26 ELR 20639 (D. Mont. 1995).
86. 40 C.F.R. §122.26(b)(14)(iii) (active and inactive mining operations require stormwater runoff permits under §402(p)).
87. United States v. Earth Sci., Inc., 599 F.2d 368, 9 ELR 20542 (10th Cir. 1979). Section §309(c) requires negligent conduct for misdemeanors and knowing conduct for felonies, and knowing conduct means knowledge of facts, not law.
88. *Committee to Save Mokelumne River*, 13 F.3d 305.
89. Rybachek v. U.S. Envtl. Prot. Agency, 904 F.2d 1276, 20 ELR 20973 (9th Cir. 1990).
90. Natural Res. Def. Council v. County of Los Angeles, 673 F.3d 880, 41 ELR 20109 (9th Cir. 2011) (city liable for discharging pollutants from its storm drains into navigable water, even though others had placed pollutants in the stormwater), *aff'd*, Los Angeles Cnty. Flood Control Dist. v. Natural Res. Def. Council, 133 S. Ct. 710, 43 ELR 20004 (2013); Environmental Prot. Info. Ctr. v. Pacific Lumber Co., 469 F. Supp. 2d 803 (N.D. Cal. 2007) (stormwater from silvacultural activity requires a §402 permit, even though point source causes no net increase in pollutants reaching stream).
91. Section 402(p), 33 U.S.C. §1342(p).
92. Natural Res. Def. Council v. County of Los Angeles, 636 F.3d 1235 (9th Cir. 2011). *aff'd, Los Angeles Cnty. Flood Control Dist.*, 133 S. Ct. 710; *Envtl. Prot. Info. Ctr.*, 469 F. Supp. 2d 803.

for causes of adding these pollutants to these navigable waters; the pollutants would have made their way into these navigable receiving waters without the point sources. Of course, absent channeling systems and associated point sources, it is also probable that some of the same pollutants in the same runoff from the same mining operations would have percolated into the ground, evaporated into the atmosphere, or reached another surface water. The point sources would be but-for causes of adding these pollutants to these navigable receiving waters; these pollutants would not have made their way into these navigable receiving waters without the point sources.

In many of these cases, the owners of the point sources existed and were available as defendants, while the persons responsible for the mining waste in the runoff passing through defendants' point sources no longer existed. Thus, if anyone was responsible for violating the CWA, it had to be the point source owners. Perhaps, in their zeal to hold someone liable for violating the CWA, these decisions focused on the "point source" element of the offence and ignored the "addition" element, with its causation component. Thus, *El Paso Gold Mines* specifically avoided discussion of causation. Such caution may be unnecessary in most cases, however, because there can be two but-for causes of a violation.

Under the second part of EPA's "outside world" gloss on "addition," these decisions would be easy. The mining and other materials were from the world outside the navigable water and were therefore added by the point sources. But EPA's "outside world" theory begins with the concept that the point source must originally place the mining wastes into water.[93] This is not the case in the mining waste decisions, however, because the mining wastes were in stormwater before it entered the point source; indeed, they may have been in the stormwater before it entered the defendant's property. The results are more complicated but accurate if we use our sophisticated definition that "addition means the act of a person adding something (A) to something else (B) from a point source when (A) would not otherwise be in (B)." The mining wastes (A) are certainly added to navigable water (B) from a point source, but would they otherwise be in navigable water (B)?

In the absence of fact-finding in the decisions on this question, it is impossible to say with certainty whether the pollutants in the cases would have reached navigable water in the absence of the defendant's channeling structures. However, it is most probable that some of the mining wastes would have found their way by gravity and natural channels to navigable water without the point sources, and some would not because they would have

93. That concept, however, has been rejected by the Supreme Court in *Miccosukee*, 541 U.S. at 105.

percolated into groundwater or evaporated. As long as some of those wastes would not have found their way to navigable waters on their own, the point sources add at least those mining wastes to navigable waters. That is enough to violate the statute and require a permit.

3. Additions From Water Transfers

The final variant of the "passive point source" fact pattern are transfers of water from one water body to another, specifically from more-polluted navigable donor water to less-polluted navigable receiving water. Water transfers move water from one watershed to another for a variety of uses, of which agricultural irrigation and municipal water supply are the most common. The owners of point sources introducing the transferred water to the receiving water may be "passive," in that they do not add pollutants to the donor water being transferred. But they are active in that they add more-polluted donor water to less-polluted receiving water. EPA misleadingly claims it has consistently taken the position that when donor navigable water is transferred to other navigable receiving water, nothing is added to the receiving water from the outside world because all navigable waters are one, the "unitary navigable waters" theory.[94] EPA's theory conflates meanings of "addition" and "navigable waters," forcing this chapter to consider "navigable waters" as well as "addition" to understand and analyze EPA's theory. Until quite recently, every appeals court considering the issue has ruled against EPA's position.[95]

After these decisions, EPA significantly changed the legal landscape by promulgating its water transfer rule, for the first time[96] cloaking its interpretation of "addition" with *Chevron* deference.[97] The one court of appeals decision considering the issue after promulgation of the rule cited *Chevron* deference to uphold EPA's position.[98] Petitions for judicial review of the rule in the circuit courts were consolidated in the Eleventh Circuit, which dis-

94. South Fla. Water Mgmt. Dist. v. Miccosukee Tribe of Indians, 541 U.S. 95, 107, 34 ELR 20021 (2004). The Court points out that there is at least one EPA General Counsel Opinion to the contrary. In *Catskill Mountains v. U.S. EPA*, 2014 WL 1284544 (S.D.N.Y. Mar. 28, 2014), EPA explained that the outlier 1975 Opinion was overcome by a 1977 amendment to the CWA. *See also* Agency Interpretation, *infra* note 130, at 2-3, n.5.

95. Catskill Mountains Chapter of Trout Unlimited, Inc. v. City of New York, 451 F.3d 77 (2d Cir. 2006); Miccosukee Tribe of Indians of Fla. v. South Fla. Water Mgmt. Dist., 280 F.3d 1364, 32 ELR 20475 (11th Cir. 2002); Catskill Mountains Chapter of Trout Unlimited, Inc. v. City of New York, 273 F.3d 481 (2d Cir. 2001); and Dubois v. U.S. Dep't of Agric., 102 F.2d 1273, 27 ELR 20622 (1st Cir. 1996).

96. 40 C.F.R. §122.3(i), 73 Fed. Reg. 33708 (June 13, 2008).

97. *See* United States v. Mead Corp., 533 U.S. 218 (2001).

98. Friends of the Everglades v. S. Fla. Water Mgmt. Dist., 570 F.3d 1210 (11th Cir. 2009).

missed them for lack of jurisdiction.[99] The Southern District of New York subsequently heard a consolidated challenge to the rule and in March 2014, vacated the rule in *Catskill Mountains Chapter of Trout Unlimited, Inc. v. U.S. EPA (Catskill Mountains).*[100] This opinion is not only the most recent decision on the issue, it is also by far the most comprehensive in its analysis. Appeals, of course, are sure to continue.

a. Early Decisions

Prior to EPA's promulgation of the water transfer rule, the U.S. Court of Appeals for the First Circuit, the U.S. Court of Appeals for the Second Circuit, and the Eleventh Circuit all rejected EPA's interpretation of "addition" to exclude water transfers and EPA's unitary navigable waters theory. The First Circuit initially considered the issue in *Dubois v. U.S. Department of Agriculture,*[101] in which environmental plaintiffs challenged the U.S. Forest Service's approval of plans to expand a ski facility in a national forest. The plans included withdrawing polluted water from the East Branch of the Pemigewasset River to make snow and ultimately discharging the used water to the pristine Loon Pond, without a §402 permit. The district court held there would be no addition of pollutants to Loon Pond because all navigable waters were a "singular entity."[102] The district court also reasoned from the analogy of redistributing water from the bottom of a pond to the top of the pond, requiring no permit even if accomplished by a point source. In reversing, the First Circuit rejected this analogy as ill-conceived, because redistribution of water within the pond was redistribution within one water body, not between two water bodies.[103]

The Second Circuit analyzed the issue of whether a water transfer was an "addition" in greater depth in *Catskill Mountains Chapter of Trout Unlimited,*

99. Friends of the Everglades v. U.S. EPA, 699 F.3d 1280, 42 ELR 20222 (11th Cir. 2012). Federal appellate jurisdiction for judicial review of EPA actions under the CWA is conferred in §509(b), 33 U.S.C. §1396(b), and the water transfer rulemaking did not fall within any of the categories of actions for which §509(b) granted jurisdiction.
100. Catskill Mountains Chapter of Trout Unlimited, Inc. v. U.S. EPA, 2014 WL 1284544 (S.D.N.Y. Mar. 28, 2014) (*Catskill Mountains*).
101. 102 F.2d 1273, 27 ELR 20622 (1st Cir. 1996).
102. *Id.* at 1296.
103. The First Circuit also cited earlier decisions, but they did not deal directly with water transfers. *See* Committee to Save the Mokalumne River v. East Bay Mun. Util. Dist., 13 F.3d 305, 24 ELR 20225 (9th Cir. 1993) (overflow to stream from dam collecting polluted runoff from abandoned mining area, where the runoff would have reached the stream without the dam); Dague v. City of Burlington, 935 F.2d 1343 (2d Cir. 1991) (culvert was a point source that added polluted waters from a landfill to nearby river); United States v. M.C.C. of Fla., Inc., 772 F.2d 1501, 15 ELR 21091 (11th Cir. 1985) (propeller of tugboat added bottom sediment to nearby sea grass beds by stirring up the sediment).

Inc. v. City of New York (*Catskill I*).[104] Environmental plaintiffs challenged New York City's transfer without a §402 permit of sediment-laden water from a reservoir in the Schoharie River through the Shandaken Tunnel to Esopus Creek, a well-known and clear trout stream (or at least once clear below and still clear above the Shandaken Tunnel). The transfer was part of the city's water supply system, moving water it collected in the protected Catskill Mountains Watershed for use in the city. The Schoharie River and Esopus Creek were connected in that they were both part of the larger Hudson River Watershed, but until they both entered the Hudson River, they were separate watersheds, having no natural influence on each other. The city argued, and the district court found, that there was no addition of sediment from the Shandaken Tunnel to Esopus Creek, relying on *Gorsuch* and *Consumers Power.* The Second Circuit rejected those precedents as having "accorded unjustified deference to the EPA's interpretation of 'addition,'" because its interpretation was not developed in a rulemaking or other formal administrative process.[105] The court also found that the decisions were distinguishable on their facts: In both of the earlier decisions, navigable donor water was withdrawn from, used, and returned to the same navigable water, essentially a "recirculation of water" within the same water body, whereas *Catskill I* involved transferring water between different navigable watersheds.[106]

To explain *Gorsuch* and *Consumers Power*, the Second Circuit used the analogy of taking "a ladle of soup from a pot, lift[ing] it above the pot, and pour[ing] it back into the pot," adding nothing to the pot. If this was held to be an "addition," "EPA might as easily require a permit for Niagara Falls."[107] But the court explained that the analogy was not apt for the transfer between the Schoharie River and the Esopus: "No one can reasonably argue that the water in the Reservoir and the Esopus are in any sense the 'same,' such that the 'addition' of one to the other is a logical impossibility."[108]

The city then argued there was no "addition" in the case under EPA's "outside world" theory of addition because the city's point source did not initially introduce the pollutants into water. The Second Circuit rejected that argument as well: "The tunnel itself need not have created the pollution; it is enough that it conveys the pollutants from their original source to the navigable water," for point sources are conveyers, not creators or originators,

---

104. 273 F.3d 481 (2d Cir. 2001).
105. *Id.* at 489-91.
106. 273 F.3d at 491-92.
107. *Id.* at 491. Of course, Niagara Falls would not violate §301(a) because the Falls is neither a person nor a point source.
108. *Id.* at 492.

anticipating the Supreme Court's analysis and conclusion in *Miccosukee*. The Second Circuit agreed with EPA's interpretation that for an "addition" to occur, "a point source must *introduce* the pollutant into navigable water from the outside world,"[109] but only if "that 'outside world' is construed as any place outside the particular water body to which pollutants are introduced,"[110] or, in our nomenclature, from outside the receiving water. Using our understanding of "addition," the city added sediment to the Esopus from the tunnel: but-for the tunnel, the sediments would not be in the Esopus.

The court specifically rejected the "singular entity" or "unitary navigable waters" theory that the addition of a pollutant to one navigable water is an addition of that pollutant to all navigable waters, both as a matter of Second Circuit precedent[111] and as leading to an absurd result. Concluding that the "transfer of water from a water body contaminated with myriad pollutants to a pristine water body containing few or no pollutants" is not an addition would be "inconsistent with the ordinary meaning of the word 'addition.'"[112]

Finally, the city argued that the legislative history and the structure of the statute supported its interpretation of "addition." The court, however, found the legislative history to be "silent on the meaning of 'addition.'"[113] And the court found that the city's reliance on §101(g) (expressing congressional policy that the authority of states to allocate water use was not abrogated by the CWA) was offset by §101(a) (expressing congressional intent to restore and maintain the integrity of the nation's waters). The court found transfers of polluted water to pristine water would be inconsistent with §101(a). In the end, the Second Circuit found no reason to depart from the plain meaning of "addition."

The Eleventh Circuit considered the issue in even greater depth in *Miccosukee Tribe of Indians of Florida v. South Florida Water Management District*,[114] in which the plaintiffs challenged the pumping of canal water contaminated with agricultural nutrients into less-polluted water flowing to the Everglades, without a §402 permit. The Water Management District, like New York City in *Catskill I*, relied on *Gorsuch* and *Consumers Power*, but the Eleventh Circuit rejected that reliance for the same reasons the Second Circuit had rejected it.[115] The Eleventh Circuit determined that the appropriate water

109. *Id.* at 491, quoting *Gorsuch*, 693 F.2d at 165. *Contra Miccosukee*, 541 U.S. at 105.
110. *Id.* at 491. This is in accord with the Supreme Court's later ruling in *Miccosukee*.
111. Dague v. City of Burlington, 935 F.2d 1343, 21 ELR 21133 (2d Cir. 1991). The facts in this decision may have amounted to a water transfer, but the court did not analyze it as such.
112. *Catskill I*, 273 F.3d at 493.
113. *Id.*
114. 280 F.3d 1364, 32 ELR 20475 (11th Cir. 2002).
115. *Id.* at 1367-68.

body for determining whether a point source adds pollutants is the receiving water, not the donor water.[116] It then rejected the passive point source argument, concluding that a point source need not be the origin of pollutants to add them to navigable water. In part, it used a dictionary definition of "from" to include "by" stating that "no dispute exists on whether pollutants, in fact, are added to navigable waters . . . *by* a point source . . . ."[117] Finally, the court concluded that a point source adds pollutants to a navigable water if the "point source is the cause-in-fact of the release of pollutants" into that receiving water.[118] The point source, a pump, added pollutants from the donor water to the Everglades-bound receiving water "because the pollutants would not have entered the second body of water *but for* the change in flow caused by the point source, an addition of pollutants from a point source."[119]

In *South Florida Water Management District v. Miccosukee Tribe of Indians of Florida*,[120] the Supreme Court upheld the Eleventh Circuit's conclusion that "addition" did not require that point sources generate the pollutants they add to navigable waters. The Court underlined its rejection of the passive point source argument, describing the argument as "untenable."[121] The Court, however, based its ruling on the definition of "point source," not on the definition of "addition." The definition of "point source" as a "conveyance," "makes plain that a point source need not be the original source of the pollutant: it need only convey the pollutant to 'navigable waters.'"[122] But it remanded the case for further factual findings on whether the donor and receiving waters were "meaningfully distinct."

The government argued as amicus that the inclusion of "any" before three of the first four elements of the offense of water pollution, and its absence before the fourth, navigable water, "signals Congress' understanding that NPDES [the national pollutant discharge elimination system permit pro-

116. *Id.* at 1368.

117. *Id.* The court's conversion of "from" to "by" in the offense is unnecessary; the court's "but-for" interpretation of "addition" is justification enough for its conclusion. And it is also ill-advised, as there is already a "by" prepositional phrase in the offense "by any person" in §301(a). A sentence with two "by" prepositional phrases is awkward. Indeed, §502(12) uses "of," "from," and "to" in its prepositional phrases, while only §301(a) uses "by" in its prepositional phrases, suggesting Congress used these prepositions advisedly.

118. *Id.* at 1368-69.

119. *Id.* (emphasis in original).

120. 541 U.S. 95, 34 ELR 20021 (2004).

121. *Id.* at 105.

122. *Id.* A secondary support for the Court's conclusion was the intent of Congress to impose §402 permitting requirements on municipal sewage treatment plants, an intent that would be frustrated if POTW discharges were exempt from permitting because the pollutants they discharged all originated from industrial and domestic discharges into the sewer system. Municipal sewage treatment plants are referred to as POTWs.

gram created by §402] permits would not be required for pollution caused by the engineered transfer of one 'navigable water' into another."[123] The Court did not respond to this argument. But it did express considerable skepticism of the "singular entity" theory, relabeled as the "unitary navigable water" theory.[124] The government argued that §304(f)(F), directing EPA to develop information on how state and local programs could address "flow diversion facilities" as nonpoint sources, meant that water transfers are nonpoint source pollution. The Court noted, however, that §304(f)(F) did not "exempt nonpoint pollution sources from the NPDES program if they *also* fall within the 'point source' definition."[125]

When the government argued that its long-standing unitary navigable water interpretation should be accorded deference, the Court noted that EPA's interpretation of "addition" had not been consistent and was inconsistent with some of EPA's own regulations, citing the net/gross regulation.[126] The government further argued that it would be impractical for it to issue §402 permits for "every diversion of one navigable water into another," because "thousands of new permits might have to be issued," imposing special problems in "western States, whose water supply networks often rely on engineered transfers among various natural water bodies."[127] The Court commented, however, that such permitting might be necessary to protect water quality; the administrative burden could be lessened by issuing general rather than individual permits; and at least one state had interpreted the CWA as requiring permits for interbasin transfers.[128] The Court might also have noted that diversions of water do not require CWA permits and that the discharge of irrigation return flow to navigable water after diversion does not require CWA permits either, considerably limiting the number of permits required.

The Court commented that treating all navigable water as one would be inconsistent with the water quality standards program.[129] The Court was correct in this regard, but did not begin to understand how inconsistent

123. *Miccosukee*, 541 U.S. at 106.
124. *Id.* at 106-12.
125. *Id.* at 107.
126. *Id.* In the net/gross regulation, EPA established a process for granting dischargers credit for pollutants in their intake water, but only if they discharged to the same water body, an intrabasin transfer. As discussed above, EPA's restriction of net/gross credits to situations where the donor and receiving waters are the same is inconsistent with allowing interbasin water transfers where the transferred water is polluted.
127. *Miccosukee*, 541 U.S. at 108. The applicability of general permits for water transfers is also explored, in somewhat more detail, in Chris Reagen, *The Water Transfer Rule: How an EPA Rule Threatens to Undermine the Clean Water Act*, 83 Colo. L. Rev. 307, 35-37 (2011).
128. 541 U.S. at 108-09.
129. 541 U.S. at 107.

the unitary navigable water theory is with the water quality standards program, an inconsistency this chapter addresses in detail below. Although the Court decided the passive point source issue, it left open for the parties to argue the unitary navigable waters theory on remand. Because the Court neither sustained nor rejected the unitary navigable waters theory, it has been reargued in cases before the Second and Eleventh Circuits and district courts. In the meantime, EPA altered the legal landscape, first by issuing a detailed water-transfer interpretive statement and later by promulgating a rule exempting water transfers from the requirement of obtaining §402 permits.

b. EPA's Interpretive Memorandum and Water Transfer Rule

EPA's interpretation of the applicability of §402 to water transfers (Agency Interpretation)[130] and its subsequent proposed and final water transfer rule[131] are considered here together because they are virtually identical in content, wording, and rationale, differing only in the more formal structure and process of rulemaking and the greater deference due to a rule.[132]

After the Supreme Court's *Miccosukee* decision, EPA developed the Agency Interpretation, concluding that Congress intended water transfers that "merely convey or connect navigable waters," uninterrupted by industrial, municipal, or commercial use, be overseen by authorities other than §402. Shortly thereafter, EPA published a proposed rule to add "[d]ischarges from a water transfer" to the 40 C.F.R. §122.3 list of exclusions from the §402 permit program. Its proposed definition of "a water transfer" was a conveyance "of water of the United States to another water of the United States without subjecting the water to intervening industrial, municipal, or commercial use."[133] The preamble to the proposed rule explained its background and rationale by repeating the body of its Agency Interpretation almost verbatim.[134] Two years and 18,000 public comments later,[135] EPA

130. Memorandum From Gen'l Counsel Ann R. Klee, *Agency Interpretation on Applicability of Section 402 of the Clean Water Act to Water Transfers* (Aug. 5, 2005) [hereinafter Agency Interpretation], *available at* http://www.epa.gov/ogc/documents/water_transfers.pdf.
131. 40 C.F.R. §122.3(i).
132. *Catskill Mountains*, 2014 WL 1284544, at **34-35. While the three documents are virtually identical in their substantive content, the proposed and final rules include sections required in promulgating regulations, but not in establishing policy. For instance, the preamble to the final rule has a section responding to comments made during the comment period after publication of the proposed rule.
133. NPDES Water Transfers Proposed Rule, 71 Fed. Reg. 32887, 32895 (June 7, 2006).
134. *Compare* Agency Interpretation, *with* proposed rule, 71 Fed. Reg. 32887, 32888-92 (June 7, 2006).
135. *Catskill Mountains*, 2014 WL 1284544, at *10.

promulgated the water transfer rule, with an accompanying preamble, virtually unchanged from its earlier proposal.[136]

The first difficulty with the water transfer rule is to determine exactly what legal issue it addresses, the starting point for judicial review. Several issues are possible. First, on its face, the rule is a regulatory exemption for water transfers from the requirement that all additions of pollutants by point sources to navigable waters be in compliance with a permit. Both the wording of the regulation and its placement in a section listing such exemptions leave little doubt that it is a regulatory exception. That raises the legal issue of whether EPA has the authority to promulgate such a regulatory exemption. On judicial review, however, EPA insisted that the rule did not create a regulatory exemption.[137] At the outsets of the Agency Interpretation, the preambles to the proposed and final rule, and its argument on judicial review, EPA suggested a second legal issue by stating that the precise legal question at issue was whether water transfers are "additions" under the CWA.[138] EPA's legal analysis in that context posits that a "holistic" interpretation of the CWA leads to the conclusion that "addition" in the CWA does not include water transfers.[139] This is not a definition of "addition," but a statement of what is not an "addition," regardless of the meaning of the word.[140]

A third possible legal issue is interpreting "navigable waters" to mean that all navigable waters are the same, the unitary navigable waters theory. This flows from EPA's use of the theory to support its interpretation that transferring pollutants from one navigable water to another navigable water is not an "addition" of pollutants from the first navigable water to the second navigable water, because both waters are the same.[141] But the rule does not purport to define either "addition" or "navigable water," the rule is not codified in the definitional section of the CWA regulations, and

136. NPDES Water Transfers Final Rule, 73 Fed. Reg. 33697 (June 13, 2008), codified at 40 C.F.R. §122.3(i).

137. *Catskill Mountains*, 2014 WL 1284544, at **34-35.

138. *See* Proposed Rule, 71 Fed. Reg. 32887, 32889; Final Rule, 73 Fed. Reg. 33697, 33700; *Catskill Mountains*, 2014 WL 1284544, at **14-15.

139. *See* Proposed Rule, 71 Fed. Reg. 32887, 32889; Final Rule, 73 Fed. Reg. 33697, 33701; *Catskill Mountains*, 2014 WL 1284544, at **20-26.

140. In discussing the scope of its "interpretation" of "addition," EPA wrote that it "address[ed]" "addition." Agency Interpretation, *supra* note 130, at 18. (Addressing a term is much less than defining the term.) At the same time, it expressly declared it was not "address[ing] the meaning of . . . "navigable waters."

141. The initial conflation of "addition" and "navigable waters" was in *Consumers Power*: If live fish were in Lake Michigan waters when those waters were diverted to generate electricity, the fish in those diverted waters were not added to Lake Michigan when the diverted water was returned to the lake. EPA asserted that position in subsequent litigation, the Agency Interpretation, and the preamble to the final rule.

the definitions in that section do not incorporate or suggest any of EPA's concepts.[142]

What words in the statute, if any, does EPA's rule interpret? Why is it so difficult to identify the legal question at issue on judicial review? Is it because EPA is playing a shell game, shuttling our attention from one statutory word to another? The most straightforward approach to EPA's objective would be for it to define "navigable waters" or "waters of the United States"[143] to incorporate the unitary navigable waters theory. EPA studiously avoided doing so in the Agency Interpretation and the rulemaking, although it could not avoid discussing the theory in those documents.[144] Indeed, it has recently promulgated comprehensive amendments to its definition of "waters of the United States" and neither the proposed or final rule or the preambles accompanying them mention the theory.[145] Why is EPA so disinclined to interpret "navigable waters" to reach the result it wants? Probably because the unitary navigable waters concept is counterintuitive and the Supreme Court has already seriously questioned the viability of the theory under the CWA.[146]

142. The definitional section of the regulations is 40 C.F.R. §122.2, while the exemption section in which the water transfer rule is located is 40 C.F.R. §122.3. In its Agency Interpretation, EPA maintains it does not address the meaning of "navigable waters," and although it "addresses the meaning" of "addition," it does not define "addition." How useful would it be for the dictionary to define "addition" as "an act that does not include water transfers"?

143. Section 502(7) defines "navigable waters" as the "waters of the United States." EPA does not define "navigable waters," but instead defines "waters of the United States" in 40 C.F.R. §122.2 and uses "waters of the United States" throughout its §402 permit regulations instead of "navigable waters." The Agency must have thought that parties to CWA disputes would come to think only of "waters of the United States" when addressing CWA jurisdiction, thus evading Commerce Clause restrictions on the scope of "navigable waters." If so, that sleight of hand did not work. *See Catskill Mountains*, 2014 WL 1284544 at *5. *See infra* Chapter 3, for a more detailed discussion of this.

144. EPA began each of the documents by describing the legal issue addressed as "whether the movement of pollutants *from one* navigable water *to another* navigable water by a water transfer is [an] addition" (emphasis added) or a variant of that phrase. This alludes to multiple bodies of navigable water, not a unitary navigable water. *See* Agency Interpretation, *supra* note 130, at 2; 71 Fed. Reg. 32887, 32889; and 73 Fed. Reg. 33697, 33699. But EPA could not avoid describing and using its unitary navigable water theory. For instance, in the Agency Interpretation, it quoted from the amicus brief of the United States for the Supreme Court in *Miccosukee*, stating that the issue in that case was "whether the pumping of water increased the sum of pollutants in the navigable waters as a whole, as opposed to the particular receiving water." Agency Interpretation, *supra* note 130, at 13. In the preamble to its final rule, EPA quoted at length from the U.S. brief in *Friends of the Everglades v. South Fla. Water Mgmt. Dist.*: "Nothing is being added 'to' 'the waters of the United States' . . . by virtue of the water transfer, because the pollutant at issue is already part of the 'waters of the United States' to begin with." *See* 73 Fed. Reg. 33697, 33701 (June 13, 2008).

145. Definition of "Waters of the United States" Under the Clean Water Act, 79 Fed. Reg. 22188 (proposed Apr. 21, 2014); 80 Fed. Reg. 37054 (final publication June 29, 2015), EPA does note in the preamble to the proposed rule that it proposes "no change to the regulatory status of water transfers." *Id.* at 22189. *See also id.* at 22203, stating that jurisdictional waters may include "ditches that connect two or more 'waters of the United States.'"

146. *Miccosukee*, 541 U.S. 95, at 106-12. EPA stated in its Agency Interpretation that it developed its "holistic" theory because of the Court's "concerns" with the unitary navigable waters theory.

Instead, EPA seems to have promulgated a rule exempting water transfers from the §402 permit program. But courts have long held that EPA does not have authority to exempt additions of pollutants to navigable water from point sources from being in compliance with permits[147] and EPA admits that,[148] leaving EPA no alternative but to attempt an interpretation of "addition" to exclude water transfers. However, the plain meaning of "addition" simply does not suggest such an exemption. That ultimately leaves EPA with its "holistic" interpretation of the statute to establish that Congress did not mean §§301(a) and 502(12) to prohibit water transfers in the absence of a permit, suggesting in turn that "addition" in §§301(a) and 502(12) cannot include water transfers. EPA is playing a shell game, but at least it's understandable why EPA is playing it. Of course, if EPA has to play shell games to keep water transfers from requiring permits, the whole enterprise is dubious.

The Agency Interpretation begins by outlining the importance and pervasiveness of water transfers. It states, for instance, that the U.S. Department of the Interior's Bureau of Reclamation alone furnishes water to 140,000 western farmers through water transfers,[149] implying that as many permits might be required and farmers would be adversely affected if transfers were considered additions. This largely is a bogus issue. Water diversions from navigable water for agricultural use do not require §402 permits, because diversions take away from rather than add to navigable water.[150] As for the return to navigable water of irrigation water after use, the statutory definition of "point source" excludes "agricultural stormwater discharges and return flows from irrigated agriculture."[151] This exclusion is reinforced by §402(*l*), prohibiting EPA from requiring permits for "return flows from irrigated agriculture." The

147. These decisions go back to *Natural Res. Def. Council v. Costle*, 568 F.2d 1369, 8 ELR 20028 (D.C. Cir. 1977), rejecting EPA's attempt to create regulatory exceptions to the CWA permit program in its first set of permit issuance regulations. *See also* Northwest Envtl. Advocates v. U.S. EPA, 537 F.2d 1006, 1021-22 (9th Cir. 2008); Northern Plains Res. Council v. Fidelity Exploration & Dev. Co., 325 F.3d 1155, 1164 (9th Cir. 2003); *Catskill Mountains*, 2014 WL 1284544, at *14. *See also* dicta in Milwaukee v. Illinois, 451 U.S. 304, 318, 11 ELR 20406 (1981) ("*Every* point source discharge is prohibited unless covered by a permit.").

148. *Catskill Mountains*, 2014 WL 1284544, at *14.

149. Agency Interpretation, *supra* note 130, at 3. EPA used the same information in the preambles to its proposed and final rule. *See* proposed rule, 71 Fed. Reg. 32887, 32889 (June 7, 2006); and final rule, 73 Fed. Reg. 33697, 33699 (June 13, 2008). The U.S. Department of the Interior's Bureau of Reclamation operates only in the West.

150. Thus, for purposes of §401 state certifications, a diversion of water from a river is not a discharge requiring a certification, because diversions are the very opposite of discharges. *See* PUD No. 1 of Jefferson Cnty. v. Washington Dep't of Energy, 511 U.S. 700, 776, 24 ELR 20945 (1994) (Thomas, J., dissenting). The majority did not disagree in that case, but held that §401 certifications were required for federal activities, not federal discharges.

151. Section 502(14). *See also* §402(*l*)(1). The exclusion appears to be more appropriate from "pollutant" than from "point source," since stormwater, like any water, is conveyed; it is not a conveyance.

statutory exclusion and prohibition remove most transfers of western irrigation water from the requirement to obtain a §402 permit. Although EPA demonstrates the importance of water transfers to agricultural and municipal water supply, particularly in the West, it assumes but does not demonstrate that §402 thwarts water transfers. Indeed, EPA's three documents do not cite a single case in which §402 has blocked a western water transfer. Ironically, if §402 does thwart water transfers, EPA's water transfer rule, reaching only interbasin transfers, is less protective of transfers than the statutory exclusion that reaches both interbasin and intrabasin transfers of irrigation return flow. In any event, EPA's implicit suggestion that states are completely autonomous in the development of water resources is largely rhetorical in that most sizable water resources projects are federally funded; those 140,000 western farmers get their water from Bureau of Reclamation water transfers, i.e., from federally funded public works.[152]

Water transfers are primarily, although not entirely, western phenomena associated with irrigated agriculture and municipal water supply. Indeed, western water law is composed of highly developed state systems for allocating water use as a property right and approving transfers of water between water basins.[153] Although there was some concern at the outset of the §402 program that its implementation would conflict with state water use allocation,[154] such conflict has not proven to be of great magnitude. None of the decisions interpreting "addition" or "navigable water" in water transfers arose in a western state or a state adopting a western use allocation scheme. None of the decisions involved irrigated agriculture, probably because the exclusion of irrigation return flow resolves most potential conflict between the water quality and water quantity regimes. No reported federal decision involves an actual conflict between federal water quality requirements and state water use allocation. Neither the Agency Interpretation nor the pream-

152. *See* Reagen, *Water Transfer Rule*, *supra* note 127 at 320-35 (describing how dependent western agriculture and municipal water supplies are on water transfers). Like EPA, Reagen assumes that §402 permits would frustrate them, but does not demonstrate that §402 permits have frustrated them. Unlike EPA, he concludes that water transfers should be subject to §402 permits and that general permits may alleviate much of the burden on water transfers of regulation under §402.

153. *See generally* A. Dan Tarlock et al., Water Resource Management (5th ed. 2002).

154. *See* H.R. Rep. No. 94-418 (1976), *available at* https://bulk.resource.org/gao.gov/92-500/00006DA6.pdf. The Report to Congress of the National Commission on Water Quality, submitted March 22, 1976, made no mention of conflicts between the CWA and state water resource management use allocation. Congress established the Commission in §315, 33 U.S.C. §1325, to study the implementation of the CWA and to suggest changes or "mid-course corrections." The more than 800-page-long Staff Draft Report accompanying the Report to Congress devoted its last five pages to water resource management, noting that water quality and water quantity programs were not as yet "completely compatible," but citing no conflicts between the CWA and a water transfer and making no recommendations to make the programs more compatible.

bles to the proposed or final rule identify any such conflicts. EPA's elaborate water transfer policy construct is an incomplete cure for an undocumented and perhaps theoretical problem.

Although EPA stated in the Agency Interpretation that the precise legal issue was whether water transfers were "additions,"[155] neither that document nor the preambles to the proposed or final rule defined "addition." Indeed, EPA never discussed the meaning of "addition" in those documents, other than to reiterate its earlier description of "addition" being "from the outside world."[156] Although EPA mentioned §§301(a) and 502, it did not analyze their applicability to the issue. It relied on the precedent of *Gorsuch* and *Consumers Power*, although neither supports the Agency Interpretation or the rule and both have been discredited as persuasive precedent.[157] Nor did EPA note the Supreme Court's rejection in *Miccosukee* of the first concept of EPA's "outside world" theory of "addition."

Instead, EPA engaged in what it termed a "holistic" analysis of the CWA to determine that Congress envisioned water transfers to be regulated by an unidentified federal or state authority other than §402.[158] Although EPA did not interpret the meaning of "navigable waters,"[159] the only way that pollutants in navigable donor water could not be added to other navigable receiving water would be if all navigable waters are one, the unitary navigable water theory. The Agency Interpretation and preambles mention this, but EPA did not explicitly adopt it or incorporate it into the definition of either "addition" or "navigable waters." Indeed, EPA did not discuss the meaning of navigable water in the Agency Interpretation, the rule, or the preambles. Moreover, EPA recently promulgated a comprehensive amendment to its regulatory definition of "waters of the United States," without a hint that the definition

155. Agency Interpretation, *supra* note 130, at 2.

156. *Id.* at 10. Moreover, EPA did not acknowledge that the Supreme Court had rejected at least the first of the two components of EPA's "outside world" gloss on "addition" in *Miccosukee*.

157. *Gorsuch* and *Consumers Power* dealt with *intra*-basin transfers, while the Agency Interpretation and rule deal with *inter*-basin transfers and specifically exclude *intra*-basin transfers. Neither dams nor pump storage projects come within the rule for the same reason. *Consumers Power* is also outside the scope of the water transfer rule because the hydropower generation in that case was an industrial use of water that occurred between the diversion of the water from the lake and the return of the water to the lake, an intervening industrial use the rule excludes from the exemption. The Second and Eleventh Circuits both rejected *Gorsuch* as precedent because it gave *Chevron* strength deference to EPA's "outside world" interpretation of "addition" when no such deference was due under *United States v. Mead Corp.*, 533 U.S. 218 (2001), insofar as it did not take place in a rulemaking or other formal regulatory activity. *See Catskill I*, 273 F.3d at 491; *Miccosukee*, 280 F.3d at 1367-68 (11th Cir. 2002).

158. Agency Interpretation, *supra* note 130, at 4-8. *See also* Proposed Rule, 71 Fed. Reg. 32887, 32889 (June 7, 2006); Final Rule, 73 Fed. Reg. 33697, 33701 (June 13, 2008).

159. Agency Interpretation, *supra* note 130, at 18 n.19.

adopted the unitary waters theory.[160] No wonder that the Southern District of New York in *Catskill Mountains* had great difficulty determining exactly which CWA terms EPA claimed *Chevron* deference for interpreting.[161]

EPA's self-described "holistic" approach is to interpret particular parts of the CWA's text, i.e., "addition," in the context of the entire statutory structure to avoid absurd results.[162] EPA concludes that the CWA, interpreted as a whole, strikes a grand balance between federal water pollution control and state water use allocation. Looking at the entire statute, EPA observes that the CWA has several programs to control pollution other than the §402 permit program, most notably the state-administered nonpoint source program.[163] It admits that the CWA contains no provision specifically stating the §402 program covers or does not cover water transfers. But it contends that §§101(b), 101(g), 304(f), and 510(2), taken together, establish a grand balance between federally directed water pollution control and state-controlled water use allocation, including state supervision of water transfers.[164] EPA concludes that interpreting "addition" not to incorporate water transfers is consistent with this understanding.[165]

Interpreting a long and complex statute such as the CWA as a seamless whole rather than a disjointed jumble of sections is a positive goal[166] and one for which EPA is uniquely qualified. Despite EPA's claim that its interpretation of "addition" to exclude water transfers is "holistic," EPA fails to establish it. First, neither the sections EPA cited, the remainder of the statute, nor its legislative history explicitly establish any grand congressional design to strike a balance between water quality and water quantity. Second, the four statutory provisions on which EPA bases its interpretation provide, at best, ambiguous support for the meanings EPA thrusts upon them. Third, while EPA focuses on the three subsections and one paragraph of the CWA,[167] it

160. Definition of "Waters of the United States," *supra* note 145, 79 Fed. Reg. 22199 (proposed Apr. 12, 2014); 80 Fed. Reg. 37054 (promulgated June 29, 2015).

161. *Catskill Mountains*, 2014 WL 1284544, at **14-15.

162. Although the Agency Interpretation and the preambles to both the proposed final rules cited the interpretive canon of avoiding absurd results, none of them gave examples of absurd results from requiring water transfers to obtain §402 permits. The Agency Interpretation quotes *Natural Res. Def. Council v. Muszynski*, 268 F.3d 91, 98, 32 ELR 20203 (2d Cir. 2001), to the effect that the most helpful canons in interpreting the CWA are the whole-statute canon and the avoid-absurd-results canon. The preambles to the proposed and final Water Transfer Rule repeat this. *See* 71 Fed. Reg. 32887, 32889 (June 7, 2006); 73 Fed. Reg. 33697, 33701 (June 13, 2008).

163. Agency Interpretation, *supra* note 130, at 5.

164. *Id.* at 5-7.

165. *Id.* at 9.

166. *See* Eskridge, *supra* note 18, at 324 (interpret a statutory section "by reference to the whole act"); Scalia & Garner, *supra* note 18, at 167-70.

167. Provisions denoted by a lower-case letter are subsections and those denoted by an Arabic number are paragraphs. Thus, §§101(b) & (g) are subsections, while §510(2) is a paragraph (although paragraphs

ignores the remainder of the 200-page statute, consisting of over 500 subsections and over 800 paragraphs, most of which are unambiguously focused on promoting pollution control. Section 301, the locus of the basic prohibition against the "discharge of any pollutant" without a permit, alone has 16 subsections and 40 paragraphs. Claiming to interpret the statute as a whole to favor state regulation of water transfers without regard to federal concerns for water quality, based on weighing three short and ambiguous subsections against over 100 sections, 500 subsections and one short and ambiguous paragraph against over 800 paragraphs,[168] does not establish a "holistic" view of the statute. Finally, not only does EPA's "holistic" interpretation ignore virtually all of the statute, it is in derogation of one-half of the CWA's regulatory strategies, water quality standards, and one-half of the CWA's permit programs, §404.

The initial sentence in the first of EPA's four relied-upon statutory provisions, §101(b), recites congressional policy "to recognize, preserve, and protect the primary responsibilities and rights of States to prevent, reduce, and eliminate pollution, to plan the development and use (including restoration, preservation, and enhancement) of land and water resources . . . ." This subsection does not mention water transfers or the state allocation of rights to use water. It certainly does not state that water quantity trumps water quality. EPA may argue that it refers to western water rights when it speaks of "the primary responsibilities and rights of States . . . to plan the . . . *use* of . . . water resources," (emphasis added) for western water rights are allocated by states for public and private *use*. But "to plan" is more consistent with reference to the states' roles in designating *uses* as the initial step in establishing water quality standards under §303(c). The "to plan" in §101(b) also corresponds with the §303(3) requirement that states establish a "continuing planning process" to assure that water quality standards are attained and maintained. Allocation of water rights requires authority well beyond planning, especially when most water rights are based on a first-come first-use basis,[169] the very antithesis of planning.

The conclusion that the first sentence of the subsection refers to the state role in establishing water quality standards is supported by the remainder of the subsection, which deals with federal and state roles in implementing other portions of the CWA, its grant program for construction of publically owned treatment works, the §402 and §404 permit programs, and federal

are normally indented and that was not done in §510).

168. Since subsections incorporate the paragraphs, it may be simpler to refer to four out of 525 subsections.

169. *See* Tarlock, *supra* note 153, at 154-58.

research, technical assistance, and financial aid programs for state, local, and interstate agencies. This strongly suggests that the first sentence also speaks to federal and state roles in a CWA program, the water quality standards program, rather than to ordering CWA goals and goals external to the CWA, because none of the remainder of the subsection deals with ordering CWA goals and external goals.

The second of the CWA subsections on which EPA relies for its interpretation, §101(g), is the only one that unambiguously deals with the intersection of water quality and water quantity. It states congressional policy "that the authority of each State to allocate quantities of water within its jurisdiction shall not be superseded, abrogated or otherwise impaired by this chapter." The subsection was added to the CWA in a 1977 Amendment, cosponsored by Sen. Malcolm Wallop (R-Wyo.) and Sen. Gary Hart (D-Colo.), both representing western water-rights states.[170] The amendment, however, does not mention water transfers or hint that water quality regulation is subordinate to water transfers, nor does its legislative history. Although water transfers may be inherent in water use allocation,[171] the initial diversion of water for an allocated use is not regulated by §402. Water diversions add nothing to navigable water; they only subtract from it. Once allocated and diverted, water is used, most commonly in irrigated agriculture, after which use it is eventually returned to a navigable water, often to a different water body than it was diverted from. When the diverted water is returned, now polluted, to the same navigable water body or to another navigable water body, the question arises whether a §402 permit is required. Because §502(14) excludes "return flows from irrigated agriculture" from the definition of "point source,"[172] no §402 permit is required in most cases for the return of water from agricultural diversions. This is reinforced by §402(*l*), which forbids EPA from requiring a permit for irrigation return flow. Significantly, the CWA 1977 Amendments added all three of these provisions, §§101(g), 402(*l*), and 502(14). There is no evidence that Congress intended §101(g) to limit the reach of the §402

170. Although §101(g) was enacted five years after the 1972 CWA, amendments are to be read harmoniously with the rest of the statute, as if they were part of the original statute. *See* NORMAN J. SINGER & J.D. SHAMBIE SINGER, SUTHERLAND STATUTES & STATUTORY INTERPRETATION [hereinafter SUTHERLAND] §22:34 & 35; ESKRIDGE, *supra* note 18, at 325, citing Brown v. Yuckert, 482 U.S. 137, 149-51 (1987); Sullivan v. Finkelstein, 496 U.S. 617, 631-32 (1990) (Scalia, J., concurring in part).

171. Under western water law, diversion of water from a natural water body is a hallmark of appropriation. Once diverted and used, it may be returned to the same or a different water body. *See generally* TARLOCK, *supra* note 153, ch. 5.

172. It may have been more straightforward for Congress to have placed this exclusion in the definition of "pollutant" rather than in the definition of "point source." All of the examples of point sources given in the definition of point source are of conveyances of water rather than of types of water. Nevertheless, the intent of Congress not to regulate these agricultural discharges is clear.

program, but if Congress did so intend, it defined that limit in §§402(*l*) and 502(14), removing only irrigation return flows from the permit program.

The Conference Report for the 1977 Amendments emphasizes that §101(g) "is not intended to change existing law,"[173] an intention repeated by Senator Wallop during U.S. Senate debates.[174] Indeed, Senator Wallop acknowledged in his floor statements that water quality, §402 and §404 permits, and other measures under the CWA may legitimately and necessarily "have some effect on the method of water usage . . . and incidentally affect individual water rights."[175] He stated that the purpose of the amendment was to assure that effects on western water rights "if any, are prompted by legitimate and necessary water quality considerations."[176] Reflecting all of this, the Supreme Court held in *PUD No. 1 of Jefferson County v. Washington Dep't of Ecology* that §§101(g) and 510(2) "do not limit the scope of water pollution controls that may be imposed on users who have obtained, pursuant to state law, a water allocation.".[177]

The third of EPA's relied-upon subsections, §304(f), directs the Agency to issue guidelines to identify pollution from and pollution control techniques for a number of sources, such as "changes in the movement, flow, or circulation of any navigable waters or ground waters, including changes caused by the construction of dams, levees, channels, causeways, or *flow diversion facilities*."[178] EPA argues that the listing of these structures in a subsection regarding "non-point sources" implies that dams, levees, channels, causeways, and flow diversions are nonpoint sources rather than point sources subject to the §402 permit program. Moreover, the inclusion of channels, ditches, and dams in §304(f) does not remove them from the §402 program because the definition of "point source" specifically includes channels and ditches,[179] and EPA admits that dams may be point sources.[180] While "flow

173. Joint Explanatory Statement of the Conference Comm., *reprinted in* 3 LEGISLATIVE HISTORY OF THE CLEAN WATER ACT OF 1977 [hereinafter LEGIS. HIST.], Comm. Print of the S. Comm. on Env't and Pub. Works, at 186, 234 & 236.

174. "It is not intended to change present law," which is already established by §510(b). 3 LEGIS. HIST. 531.

175. *Id.* at 532.

176. *Id.* Senator Wallop was concerned that western water rights might be interfered with by persons using water pollution control measures to achieve other purposes. He indicated his concern was raised by the then-recent Option Papers for the Water Resource Policy study being conducted by the Water Resource Policy Council, raising options for using federal water pollution control law to achieve "Federal purposes that were not strictly related to water quality." *Id.* at 531.

177. 511 U.S. 700, 720, 24 ELR 20945 (1994).

178. §304(f)(F) (emphasis added).

179. Section 502(14), 33 U.S.C. §1362(14). Of course, flow diversions are neither point nor nonpoint sources of pollution because they divert from rather than add to navigable waters.

180. EPA admitted in *Gorsuch* that under some circumstances dams are point sources, even though they are not on the statutory list of examples. *See* 693 F.2d at 165.

diversions" are not on the list of examples of point sources, once waters are diverted for use in irrigation, they are transferred for that use in ditches, and "ditch" is listed in §502(14) as a point source. Thus, when channels and ditches discharge agricultural wastes, they are point sources, not nonpoint sources, although they may be exempt from permitting under §§402(*l*) and 502(14) if they contain only irrigation return flow.[181] If flow diversion facilities are not point sources, that does not mean water transfers are not subject to the §402 permit program, because diversions take flow from navigable waters rather than adding flow to them. It is the flow return facilities that add flow and pollutants to navigable waters. "Well" may also be inferred to be a nonpoint source by being listed elsewhere in §304(f), but is on the statutory list of examples of "point source," an unnecessary exclusion unless wells are otherwise point sources. The Supreme Court in *Miccosukee* commented that §304(f) does not mean that water transfer facilities are not point sources[182] and EPA admitted that in its Agency Interpretation.[183]

Note also that §304(f)(F) deals only with "changes in the movement, flow, or circulation" of navigable waters" . . . "by the construction" of the listed facilities, not by their operation. This wording suggests that for listed facilities that are point sources, the concern in §304(f) is not their ultimate discharge into receiving waters, but the effects of their location, design, and construction on donor waters. Although the flow-diversion portion of water transfer facilities are not regulated by §402, because flow diversions subtract from navigable water rather than adding to it,[184] the location, design, and construction of those facilities may raise water quality issues.

Although EPA does not emphasize it, §208(b)(2)(F) complements §§304(f)(F), 402(*l*), and 502(14) by requiring states to identify and provide plans to control pollution from agricultural nonpoint sources, including irrigation return flow. Significantly, §§101(g), 208(b)(2)(F), and 402(*l*), and the exemption for agricultural stormwater and irrigation return flows in §502(14), were all added to the CWA in the same set of amendments in 1977, and all originated in the Senate bill.[185] The Senate Committee Report stated that the effect of the amendments "is to exempt irrigation return flows

---

181. "Well" may also be inferred to be a nonpoint source by being listed elsewhere in §304(f), but is on the statutory list of examples of point sources in §502(14). *Compare* §304(f)(D), 33 U.S.C. §1314(f)(D), *with* §502(14), 33 U.S.C. §1362(14).
182. *Miccosukee*, 541 U.S. at 106.
183. Agency Interpretation, *supra* note 130, at 6.
184. Section 316(b) requires best available technology for cooling water intake structures associated with point sources, to minimize negative environmental impact.
185. H.R. Rep. No. 95-830, at §§5(a), 33(a)-(c) (1977); Joint Explanatory Statement of the Conference Comm., at 50, 52 & 69, *reprinted in* 3 Legis. Hist., at 186, 197, 234, 236 & 253.

from all permit requirements" and made no mention of water diversions or western water rights.[186] Similar comments were made in reporting the Conference Committee Report to the U.S. House of Representatives.[187] Indeed, all but §101(g) were "in recognition of a specific recommendation of the National Commission of Water Quality."[188] The Commission noted that "[w]hile there is little doubt that Congress defined irrigation return flows as point sources" in 1972, controlling them through the §402 permit program "has proved difficult."[189] The Commission recognized, however, that pollution from irrigation return flow was an important cause of degradation in the nation's waters and concluded it should not be exempted from coverage by the statute. Instead, the Commission recommended that irrigation return flow be addressed by more-flexible regulatory measures at the state level.[190] At the same time, the Commission did not identify a conflict between the §402 permit program and water transfers or western water rights as the root of the irrigation return flow problem. Indeed, the Commission Report did not mention water transfers or western water rights.[191] If Congress perceived that §402 interfered with water transfers, it did so in the context of irrigated agriculture, and it dealt with that issue completely in 1977 by adding language in §§208(b)(2)(F), 304(f)(F), 402(*l*), and 502(14).

The final provision supporting EPA's "holistic" analysis, §510(2), provides that nothing in the CWA shall "be construed as impairing or in any manner affecting any right or jurisdiction of the States with respect to the waters . . . of such states." This preemption preclusion provision should be read in

186. S. Rep. No. 95-370, at 35 (1977) (Clean Water Act of 1977), *reprinted in* 4 Legis. Hist. 633, 668.

187. The "conference report exempts return flows from irrigated agriculture from all permit requirements . . . and recognizes that this activity is not a point source," with no mention of water transfers or western water rights. *See* Statement on the House floor by Rep. Roberts, Chairman, H. Comm. on Pub. Works and Transp., authorizing committee for the 1977 Clean Water Act Amendments, Dec. 17, 1977, *reprinted in* 3 Legis. Hist. at 318.

188. 3 Legis. Hist. at 318. When Congress enacted the CWA in 1972, it established a National Commission to study the implementation of the statute and report its recommendations to Congress. The National Commission on Water Quality submitted its Report to Congress on March 22, 1976, with recommendations for "mid-course" corrections for the statute. *See* H.R. Rep. No. 94-418 (1976).

189. 3 Legis. Hist. at 40. The difficulty resulted from differences in geography, climate, agricultural method, and drainage, and water application methods made it impossible to develop uniform national technology-based standards.

190. National Commission on Water Quality, Report to Congress, 40-1 (Mar. 22, 1976). The Staff Draft Report issued November 1975 elaborated on this. *See* 3 Legis. Hist. at I-51 to I-54.

191. National Commission on Water Quality, *supra* note 190. Staff Draft Report at VI-54 to VI-55. The Commission staff did discuss the relationship between water quality control and water use allocation. The staff saw the two as complementary in part, because water resources are diminished when they are too polluted to be available for successive use. At the same time, the staff commented that water quality and water quantity concerns were not yet "completely compatible," but made no recommendations to make them so. The Commission itself did not note any incompatibility between water quality and water quantity and made no recommendations to reconcile any incompatibility.

conjunction with §510(1), which provides that the CWA does not preclude a state from adopting or enforcing a limitation on discharges of pollutants or a requirement respecting water pollution control, except that a state cannot adopt or enforce such a limitation or requirement less-stringent than a federal requirement under the CWA. This section does not mention water transfers and does not hint that regulation of water quality is in any way subordinate to allocation of water use. The first part of the section suggests it is limited to federal and state roles in water pollution control. The Supreme Court in *PUD No. 1* held that "[s]ections 101(g) and 510(2) preserve the authority of each State to allocate water quantity as between users; they do not limit the scope of water pollution controls . . . ."[192] Finally, as noted by the Southern District of New York in *Catskill Mountains*, the protection of state rights and jurisdiction in §510 applies "[e]xcept as expressly provided in this chapter." "Because this language does not address what other provisions of the Act 'expressly' provide[ ]," this provision has little bearing on the interpretation of those other provisions—i.e., §§301(a) and 502(12).[193] In other words, the preemption preclusion applies only when the statute is silent.

The three subsections and one paragraph analyzed above simply do not support EPA's assertions that Congress intended a grand balance in the CWA between federal water quality control and water use allocation; that Congress intended water transfers to be exempt from the §402 permit program; or that Congress intended water transfers not to be "additions" in the CWA. None of the four provisions mention water transfers. The Supreme Court has held that three of them do not so limit the CWA. The legislative sponsor of §101(g), which speaks the most directly of any of the four provisions to water resource management, agreed on the Senate floor that water allocation may be subordinate to water quality control. To interpret §301(a), the central prohibition of the statute, to exclude a major category of discharges, such as water transfers, would require that statutory provisions unambiguously establish the exclusion. These four ambiguous and minor provisions do not do so.

As the Supreme Court suggested in *Miccosukee*, EPA's interpretation of "addition" and its unitary navigable water theory are also inconsistent with one of the CWA's two grand strategies for controlling water pollution. Before the enactment of the CWA in 1972, water quality standards were the only strategy employed for that purpose.[194] With the enactment of the CWA's

192. 511 U.S. 700, 720, 24 ELR 20945 (1994).
193. *Catskill Mountains*, 2014 WL 1284544, at *23.
194. *See* Frank F. Grad, Treatise on Environmental Law §3.03.

technology-based standards requirements in 1972, water quality standards temporarily took a secondary, fallback role in pollution control, but resurfaced as the potential driver for further pollution reduction after pollution sources achieved technologically achievable pollution reduction.[195] Water quality standards and their implementation are a multistage process with interplay between EPA and states that Congress established in detail. In the first step, states designate the desired uses to be made of different navigable waters within their borders,[196] such uses as body contact sports, cold water fisheries, and drinking water without treatment. In the next step, states designate the maximum concentrations of various pollutants, known as criteria, compatible with the designated uses of the water bodies. Designating uses is a political decision, designating criteria is a scientific decision. EPA must approve a state's designated uses and supporting criteria, but EPA's role is much greater with criteria.[197] If a state does not adopt an approvable water quality standards program, EPA must do so.[198]

Once these uses and criteria are in place, effluent limitations for individual §402 permits affecting specific water bodies or segments are calculated to achieve the more stringent of state established water quality criteria or EPA-promulgated technology-based standards.[199] This inherently requires examining navigable waters as individual water bodies and separate segments of individual water bodies, rather than as a unitary whole body of navigable water. Section 302(a) restates this explicitly for more-protective water quality uses: If application of effluent limitations in a permit based on technology-based standards would "interfere" with the "attainment or maintenance of that water quality in *a specific portion of the navigable water* . . . effluent

---

195. *See generally* Oliver A. Houck, The Clean Water Act TMDL Program: Law, Policy, and Implementation (2d ed. 2002). It may seem impossible that water pollution could be treated beyond the best technology achievable to meet water quality standards. But while the CWA specifies cost criteria to be considered in promulgating technology-based standards, it does not specify them in establishing water quality standards. *Compare* §304 (a)(&)(b)(1)(B), (2)(B), and (4)(B), 33 U.S.C. §1314(a) & (b)(1)(B), (2)(B), and (4)(B). Thus, cost constraints may limit technology-based standards, while more expensive means of controlling pollution may be necessary and required to achieve water quality standards.

196. Section 303(c)(2); 40 C.F.R. §131.2; *see also* 40 C.F.R. §130.2. According to EPA's regulations, a "water quality standard defines the water quality goals of a water body, or portion thereof, by designating the use or uses to be made of the water . . . establishing the water quality goals for a specific water body."

197. Use designation is a political decision, having nothing to do with science. Establishing criteria, however, requires a scientific judgment of the concentration of particular pollutants that will or will not interfere with a chosen use. EPA has established detailed scientific guidance on criteria and requires states to follow that guidance or demonstrate that other criteria are appropriate using scientifically defensible methods. 131 C.F.R. §131.11. By contrast, EPA is wholly responsible for establishing technology-based standards. §§301(b), 304, and 306.

198. Section 303(b), (c)(4).

199. Sections 301(b), 302(b); 33 U.S.C. §§1311(b), 1312(b).

limitations . . . for such point source or sources shall be established which can reasonably be expected to contribute to the attainment or maintenance of such water quality."[200] The CWA also requires states to develop a list of "those waters within the State" that will not achieve water quality standards and "for each segment of the navigable waters"; a list of point sources preventing achievement of water quality standards for toxic pollutants; and control strategies for those point sources to achieve the standards.[201] Section 319 complements this by requiring states to identify "those navigable waters within the State, which" cannot attain water quality standards without controlling nonpoint sources and to identify "particular nonpoint sources which add significant pollution to each portion of the" identified navigable waters.[202] In developing nonpoint source control management plans under §319, states are to "develop and implement [them] . . . on a watershed-by-watershed basis."[203] EPA's implementing regulations have states submitting lists of "water quality limited segments,"[204] "the water quality of all waters of the United States" and "waters needing action,"[205] "those waters within the State" and "all navigable waters in such State" not meeting water quality standards," "each segment of navigable waters included on such list," and "a water" meeting a condition.[206]

One aspect of water quality standards merits special mention, for it overlays this entire pattern. EPA's regulations require states to establish antidegradation programs for waters that have achieved the criteria established for their designated uses.[207] This requirement helps ensure that pristine waters remain pristine. That is wholly antithetical to transferring a polluted water into a pristine water.

Water quality standards are established, and achievement of them is accomplished river by river, stream by stream, lake by lake, water segment by water segment, reflecting local conditions, local uses, and local goals. The standards result in effluent limitations for particular point sources to meet

200. Section 302(a); 33 U.S.C. §1312(a) (emphasis added).
201. Section 304(*l*)(1), 33 U.S.C. §1314(*l*). *See also* EPA's implementing regulations, 40 C.F.R. Parts 130 & 131.
202. Section 319(a)(1)(A) & (B), 33 U.S.C. §1329(a)(1)(A) & (B).
203. Section 319(b)(4), 33 U.S.C. §1329(b)(4).
204. 40 C.F.R. §131.7.
205. 40 C.F.R. §130.8.
206. 40 C.F.R. §130.10(d).
207. 40 C.F.R. §131.12. *See also* §304(d)(4)(B), 33 U.S.C. §1324(d)(4)(B). Although the 1972 version of the CWA did not mention the antidegradation policy, water quality standards existing at that time contained such provisions and §303(a) continued them in force. When Congress amended §303 in 1987, it included a reference to the antidegradation policy in §303(d)(4)(B). The Court discusses the history of the antidegradation policy in *PUD No. 1 of Jefferson County v. Washington Department of Ecology*, 511 U.S. 700, 718, 24 ELR 20945 (1994).

particular local water quality goals. Considering all navigable waters as one does not conform to or accomplish this strategy on a conceptual basis, much less on a point source-specific basis. How does a state compile a list of waters meeting specific and different conditions if all waters are one? If all navigable waters are one, they would have the same designated use and be subject to the same criteria. Perhaps, EPA considers navigable waters as one just for the purposes of defining addition, but multiple waters for other purposes. That would be an odd concept, perhaps unworkable, and contrary to the canon of construction that words are to be interpreted identically throughout a statute.[208] It is also contrary to the CWA's definition section, which provides that the definitions it provides apply throughout the statute "[e]xcept as otherwise specifically provided."[209]

The Supreme Court in *Miccosukee* commented on the inconsistency between the unitary navigable waters theory and water quality standards.[210] EPA ignored the Court's comment and did not respond to it either in the Agency Interpretation or in the preambles to its proposed and final rule. Indeed, in the preamble to the final rule, when responding to public comments that the water transfer rule was inconsistent and would interfere with water quality standards, EPA only addressed the concern as it related to water impoundments for settling sediments in mining wastes, merely replying that the rule "does not affect the permitting of such facilities"[211] and that states can establish their own water quality standards if they wish to.[212] Its assurance was wholly unresponsive to the broader question of the compatibility of the rule and the theory with water quality standards under the CWA.

EPA's "outside world" and unitary navigable water theories also create significant inconsistencies with the §404 program. In a typical §404 wetlands landclearing case, soil and vegetation are moved from one area to another area in the same wetland to convert it to a dry and level field for agricultural or other use.[213] If the unitary navigable water theory is applied to such landclearing, there is no addition, because the material redeposited in the wetland was already in the waters of the United States (the wetland) and therefore could not be added to the waters of the United States (another part of the same wetland). Responding to comments on this, EPA attempted to

208. Eskridge, *supra* note 18, at 324, citing Sullivan v. Stroop, 496 U.S. 478, 83 (1990); United Sav. Ass'n of Tex. v. Timbers of Inwood Forest Assts, Lt., 484 U.S. 335 (1998). *See also* Scalia & Garner, *supra* note 18, at 170-73.
209. Section 502, 33 U.S.C. §1362.
210. 541 U.S. at 107.
211. 73 Fed. Reg. 33697, 33703; *Catskill Mountains*, 2014 WL 1284544, at *35.
212. 73 Fed. Reg. 33697, 33705; *Catskill Mountains*, 2014 WL 1284544, at *35.
213. Avoyelles Fishermen's League, Inc. v. Marsh, 715 F.2d 897, 13 ELR 20942 (5th Cir. 1983).

avoid the issue in the preamble to the final transfer rule by pointing out that the definition of "pollutant" includes "dredged spoil," which by definition comes from water. EPA believes this makes it clear that Congress "explicitly forbade discharges of dredged material" except in compliance with a §404 permit, emphasizing that "dredged material" is a "pollutant."[214] Therefore, EPA believed the rule "would not have an effect on the 404 program."[215]

The expected origin of dredged spoil pollutants in navigable water, however, only addresses the issue of whether dredged spoil is a "pollutant." But an activity does not violate §301(a) because it satisfies one of the four elements of the offense ("pollutant"); it must meet all four elements, including "addition" and "navigable waters."[216] Under both EPA's "outside world" and "unitary navigable water" glosses on "addition" and "navigable waters," the dredged spoil does not come from a world or a water outside the wetland or the unitary navigable water. Unless "addition" and "navigable water" have different meanings for §402 and for §404, many activities currently regulated under §404 are not subject to §404 jurisdiction under EPA's interpretations because the waters from which fill material is taken and the waters to which it is redeposited are one and the same. But, again, words are to be read consistently throughout the statute under both cannons of statutory construction and the wording of §502.[217] The interpretation of "addition" in §404 cases is discussed in more detail below.

The Agency Interpretation and the preamble to the proposed rule view the CWA as constructing a grand balance between state powers to allocate water use and federal powers to protect water quality. Perhaps, perceiving the weakness of this grand balance approach, discussed above, the preamble to the final rule more modestly relies on the same three subsections and one paragraph simply to establish congressional intent that water transfers not be regulated under §402, because application of §402 to water transfers would interfere with or subvert state water use allocation, which the CWA preserves. However, the four provisions, whether individually or collectively, do

214. 73 Fed. Reg. at 33703.

215. *Id.*

216. *See, e.g.*, Waterkeeper Alliance, Inc. v. U.S. EPA, 399 F.3d 486 (2d Cir. 2005) (rejecting challenge to EPA regulation extending §402 permit requirements to concentrated animal feeding operations (CAFOs) making no additions to navigable water, even though CAFOs are statutorily defined as point sources); Community Ass'n for Restoration of Env't, 2011 WL 61882 (W.D. Wash. 2011) (unpermitted CAFO not discharging pollutants to navigable water does not violate the statute); Nelson Faria Dairy, Inc., Higbee v. Starr, 598 F. Supp. 323 (E.D. Ark. 1984), *aff'd without opinion*, 782 F.2d 1045 (8th Cir. 1985); Waterkeeper Alliance, Inc. v. Hudson, 2012 WL 6651930, 43 ELR 20010 (D. Md. Dec. 20, 2012) (unpermitted CAFO not discharging pollutants to navigable water does not violate the statute).

217. Eskridge, *supra* note 18, at 324; Scalia & Garner, *supra* note 18, at 170-73.

not support even this more modest argument, as discussed above. Moreover, the Agency Interpretation and the preambles never explain how subjecting water transfers to §402 jurisdiction has interfered or will interfere with or subvert state water use allocation. If such interference or subversion is real, surely EPA would be aware of examples of it during the 40-year history of the §402 program. But the legislative history of §101(g), added in 1977 and addressing this issue more directly than any other subsection in this statute, mentioned no examples of such interference or subversion. Nor does EPA mention any examples of such interference or subversion in its Agency Interpretation or the preambles to its proposed and final rules.

Because use of such examples could only strengthen EPA's justification for the transfer rule, EPA's failure to mention water transfers negatively impacted by §402 infers such negative impacts are few and far between, if they exist at all. The lack of relevant case law also suggests water transfers negatively impacted by §402 are not numerous or are not severely impacted. The only examples of water transfers challenged for not having §402 permits are in Florida, New Hampshire, and New York, states with no western water-law allocation schemes. Indeed, in none of those decisions is there a hint that the water diversion in question was part of a state water use allocation scheme. Moreover, despite New York City's allegations in *Catskill I* and *II* that the application of §402 would cripple the city's water supply, presumably by requiring expensive treatment of the polluted water before its transfer, the Second Circuit noted that there was considerable flexibility on how the CWA's requirement could be met under the circumstances.[218] Indeed, if the holding in *Catskill II* has crippled the city's ability to supply water to its citizens, that inability has been a well-kept secret in the dozen years following the decision. Not only did EPA fail to describe the extent of economic harm occasioned by the lack of the rule, it failed to describe the extent of environmental benefit that would be caused by the rule. What sort of a reasoned balance is that?

c. Post-Interpretive Memorandum and Water Transfer Rule Decisions

In *Catskill I*, the Second Circuit remanded the case to the district court to proceed with the remedy phase. Appealing from the district court's decision on remand, New York City asked the Second Circuit to reconsider *Catskill I*

218. *Catskill II*, 451 F.3d 77, 85. The best available technology for preventing sediments in the transferred water, for instance, might be locating intake structures at different points in the reservoir, enabling the waters withdrawn to be taken from the location the least burdened by sediments at the time.

in light of the Supreme Court's decision in *Miccosukee* and EPA's 2005 Agency Interpretation. In *Catskill Mountains Chapter of Trout Unlimited, Inc. v. City of New York* (*Catskill II*),[219] the Second Circuit held that the Supreme Court's *Miccosukee* decision confirmed rather than questioned *Catskill I*, in particular the Second Circuit's distinction between *inter*-basin (*Gorsuch* and *Consumers Power)* and *intra*-basin (*Dubois* and *Catskill I*) transfers. Indeed, as discussed below, the *Miccosukee* Court had even adopted the Second Circuit's "soup ladle" analogy designed to illustrate the differences between such transfers. The Second Circuit then considered EPA's new Agency Interpretation, looking at its "power to persuade" rather than giving it *Chevron* deference.[220] It found nothing in the document that the city had not raised, albeit in a less-detailed way, in *Catskill I.* The Second Circuit concluded:

> In the end, . . . these "holistic" arguments about the allocation of state and federal rights, said to be rooted in the structure of the statute, simply overlook its plain language . . . . It is the meaning of the word "addition" upon which the outcome of *Catskill I* turned and which has not changed, despite the City's attempts to shift attention away from the text of the CWA to its context.[221]

Although the Second Circuit did not explicitly employ the two-step *Chevron* deference test to EPA's water transfer rule, it left no doubt how it would have decided the case under *Chevron.* With regard to the first step, whether the statute is ambiguous, the court in *Catskill I* held the statute's plain meaning was clear.[222] With regard to the second step, whether the Agency's interpretation was reasonable, the court held "[n]o one can reasonably argue that the water in the Reservoir and the Esopus are in any sense the 'same'" and that such a conclusion led to absurd results.[223]

After *Catskill II*, EPA promulgated its water transfer rule and multiple parties filed petitions for judicial review in several federal courts of appeal and in the Southern District of New York. The appeals court petitions were consolidated in the Eleventh Circuit. At the time, the Eleventh Circuit already had an appeal before it, from a citizen suit in which the validity of the rule was an issue. The Eleventh Circuit stayed the consolidated petition for judicial review until it decided that appeal in *Friends of the Everglades v. South Florida*

---

219. *Catskill II*, 451 F.3d 77 (2d Cir. 2006).
220. *Catskill II*, 451 F.3d. at 82, citing United States v. Mead Corp, 533 U.S. 218, 235 (2001), and Chevron U.S.A., Inc. v. Natural Res. Def. Council, 467 U.S. 837, 14 ELR 20507 (1984). Under *Mead, Chevron* deference is due only to agency interpretations developed in rulemaking or other formal administrative procedure.
221. *Catskill II*, 451 F.3d at 82.
222. *Catskill I*, 273 F.3d at 492 & 94.
223. *Id.* at 492.

Water Management District,[224] reexamining its earlier rejection of the unitary navigable water theory.[225] It initially noted that all existing court of appeals precedents had rejected the theory,[226] but that none had considered EPA's new regulation, which was entitled to *Chevron* deference.[227] The court then undertook the familiar two-step *Chevron* analysis: (1) is the statute ambiguous; and (2) if so, is the agency's interpretation a reasonable one?

As to ambiguity, the Eleventh Circuit found that precedents either did not involve interbasin transfers, *Gorsuch* and *Consumers Power*, or did not consider whether the statute was ambiguous, *Catskill I* & *II* and its *Miccosukee* decision.[228] The court was not correct, however, in reading *Catskill I & II* to find ambiguity in the meaning of "addition."[229] After analyzing the wording of the relevant provisions and the structure of the statute as a whole, the Eleventh Circuit concluded that it was ambiguous whether "addition . . . to navigable waters" in §502(12) "referred to waters in the individual sense or as one unitary whole."[230] The court concluded that the use of the plural "waters" was not dispositive, for it could be used to denote all waters in general or specific waters in particular, for example, the waters of Mobile Bay.[231] While the absence of "any" before "navigable waters" and its presence before the other three elements of the offense was suggestive that "navigable waters" was meant generally rather than specifically, that inconsistency was not dispositive either, because Congress included or omitted "any" before "navigable water" or "navigable waters" elsewhere in the statute randomly rather than connoting either general or specific waters.[232] All of these arguments, of course, are directed at the meaning of "navigable water," not "addition."

Finally, the Eleventh Circuit concluded that the structure of the statute as a whole did not resolve the ambiguity. In this inquiry, it considered the statute's objective of restoring and maintaining the integrity of the nation's waters in §101(a), the dominant role of the §402 permit program in achieving that objective, and the absurdity of "pumping dirty canal water into a reservoir of drinking water" without a permit.[233] The court found that while "it may seem inconsistent with the lofty goals" of the CWA "to leave

224. 570 F.3d 1210 (11th Cir. 2009).
225. Miccosukee Tribe of Indians of Fla. v. South Fla. Water Mgmt. Dist., 280 F.3d 1364 (11th Cir. 2002).
226. Friends of the Everglades v. South Fla. Water Mgmt. Dist., 570 F.3d 1210, 1218 (11th Cir. 2009).
227. *Id.* at 1219.
228. *Id.* at 1220-22.
229. *Catskill I*, 273 F.3d at 492, 94; *Catskill II*, 451 F.3d at 84.
230. *Friends of the Everglades*, 570 F.3d at 1223.
231. In the absence of an indication in the statute to the contrary, the singular includes the plural and the plural includes the singular. *See* Scalia & Garner, *supra*, note 18, at 129-31. *See also* 1 U.S.C. §1.
232. *Friends of the Everglades*, 570 F.3d at 1224-25.
233. *Id.* at 1225-26.

out of the permitting process the transfer of pollutants from one navigable body of water to another, . . . it is no more so than to leave out all non-point sources, allowing agricultural run-off to create a huge 'dead zone' in the Gulf of Mexico."[234] That is a bad analogy, however, for Congress chose explicitly not to regulate nonpoint sources through the §402 permit program. It did not make an explicit choice to exempt water transfers from the §402 permit program.

In the end, the Eleventh Circuit wrongly found the statute ambiguous and, for the same reasons, found EPA's interpretation of "addition" reasonable, although the court might have reached a different interpretation on its own. The court's examination of the structure of the statute, on which it based both its *Chevron* step one and step two analyses, is significantly deficient. The court failed to examine or even to perceive the weaknesses in EPA's arguments that §§101(b), 101(g), 304(f)(F), and 510(1) established a grand balance between water quality and water quantity. And it failed entirely to consider that the rule was inconsistent with the water quality standards strategy of the statute and with the §404 permit program.

To illustrate the reasonableness of EPA's interpretation, the Eleventh Circuit posited the analogy of a hypothetical four-marble rule:

> Two buckets sit side by side, one with four marbles in it and the other with none. There is a rule prohibiting "any addition of any marbles to buckets by any person." A person comes along, picks up two marbles from the first bucket, and drops them into the second bucket. Has the marble-mover "add[ed] any marbles to buckets?" On the one hand, as the Friends of the Everglades might argue, there are now two marbles in a bucket where there were none before, so an addition of marbles has occurred. On the other hand, as the Water District might argue and as the EPA would decide, there were four marbles in buckets before, and there are still four marbles in buckets, so no addition of marbles has occurred. Whatever position we might take if we had to pick one side or the other we cannot say that either side is unreasonable.[235]

While either interpretation may be reasonable when involving solid, inert marbles, are they equally reasonable when involving four ounces of liquid or dissolved toxic pollutants in a bucket? To get equal amounts of liquid or dissolved pollutants in both buckets (two liquid ounces in each), the contents of both buckets would have to be mixed together in a larger container and then equally divided between the two buckets, considerably more complicated than playing marbles. Moreover, the end result would

234. *Id.* at 1227.
235. *Friends of the Everglades*, 570 F.3d 1228.

be two buckets with poisonous water rather than one, not consistent with the goals of the CWA, especially the antidegradation provision, which the court did not consider.

Once the Eleventh Circuit decided this citizens suit appeal, it turned to the consolidated petitions for judicial review of the water transfer rule and dismissed them for lack of jurisdiction under §509(b).[236] In the meantime, the Southern District of New York had stayed the consolidated petitions before it, pending the outcome of the petitions in the Eleventh Circuit.[237] After the Eleventh Circuit dismissed those petitions, the Southern District vacated its stay and proceeded with judicial review of the rule in *Catskill Mountains Chapter of Trout Unlimited, Inc. v. U.S. EPA*.[238] In this, the most recent decision on the issue, the district court vacated and remanded EPA's rule. Turning to the first step of the *Chevron* analysis, the court agreed with the Eleventh Circuit that the CWA was ambiguous as to whether it required water transfers to have §402 permits.[239] On the *Chevron* step-two analysis, however, the court found EPA's interpretation to be unreasonable for many of the reasons discussed above, especially because: (1) EPA's "holistic" analysis of the CWA was anything but holistic, examining only the four small parts of the statute supposedly favoring its conclusion and ignoring the overwhelming remainder of the statute not supporting it; and (2) EPA gave no reasoned justification for its decision, including an explanation of how the rule was consistent with water quality standards and the §404 permit program and an analysis of the relative harms to the economy and the environment of promulgating and of not promulgating the rule. The decision is very thorough, making it easy for the Second Circuit to affirm on appeal, especially because it is in accord with the Second Circuit's own earlier decisions. If the Second Circuit does affirm, the decision will result in a split between the Second and Eleventh Circuits on EPA's rule, setting the basis for a grant of certiorari by the Supreme Court for resolution of the issue it did not reach earlier in *Miccosukee*.

An interesting and perhaps significant aspect of the appellate decisions discussed above is that all but one develops or adopts a metaphor to describe the unitary navigable waters theory or to distinguish interbasin from intrabasin

236. Friends of the Everglades v. U.S. EPA, 699 F.3d 1280, 42 ELR 20222 (11th Cir. 2012).

237. *Catskill Mountains*, 2014 WL 1284544 at *26.

238. *Id.*

239. *Catskill Mountains*, 2014 WL 1284544, at **31-55. Although, as discussed above, it appears that the Second Circuit had already concluded that the statute was not ambiguous on that issue with regard to EPA's Agency Interpretation, it had not precisely held so because EPA had not yet promulgated its water transfer rule, so that the Second Circuit was not performing a *Chevron* analysis.

transfers.[240] The district court in *Catskill Mountains* labeled them "warring analogies."[241] These metaphors raise several questions. Why do the courts of appeal use metaphors in all but one of their water transfer decisions, but use them in only one other decision interpreting "addition"?[242] Why do courts use them so pervasively to interpret "addition," but not to interpret "pollutant," "navigable water," or "point source?"

Are the metaphors useful in determining the meaning of "addition"? We use analogies to explain an abstract or difficult idea by making it more concrete and easier to understand, suggesting that courts find the idea of the unitary navigable water theory singularly difficult to grasp. But the idea of all navigable water being one entity is not difficult to understand. What is difficult to understand is the rationale of applying that abstract concept to water pollution control. The metaphors may help understand the theory, but they do not suggest a rationale for the theory, and they certainly do not explain the meaning of "addition." The wine import ban analogy in *Friends of the Everglades*, for instance, is understandable as a means of economic protectionism, but why would we require §402 permits for transfers of water from

240. The Supreme Court in *Miccosukee* adopts the metaphor from *Catskill I* that in an intrabasin transfer, "'one takes a ladle of soup from a pot, lifts it above the pot, and pours it back into the pot . . . [not adding] soup or anything else to the pot'" 541 U.S. at 110, quoting 273 F.3d at 492 (discussed in further detail in the chapter at Section III.C.1., below). The Second Circuit developed the "soup ladle" analogy to characterize *Gorsuch* and *Consumers Power*, both involving discharges of water from one water body back to the same water body, and to distinguish them from *Catskill I*, involving the discharge from one water body to a distinctly different water body. The Second Circuit agreed that no permit was required for a transfer within the same water body, otherwise, "the EPA might as easily require a permit for Niagara Falls." 273 F.3d 292. In the case of Niagara Falls, the discharge is not "by any person" and would therefore not require a CWA permit. The district court in *Dubois*, in adopting the government's unitary navigable waters theory (in that decision called the "singular entity" theory), developed the metaphor of a "pond in which 'we place a pipe . . . and we pump the pond water from the bottom to the surface. No one would reasonably contend that internal pumping causes an "addition" of pollutants to the pond. Instead, we would consider the pumping to be a redistribution of pollutants from one part of the pond to another.'" Dubois v. U.S. Dep't of Agric., 102 F.3d 1273, 1296-97, 27 ELR 20622 (1st Cir. 1996) (quoting the unreported district court decision). The First Circuit rejected the comparison as "not at all analogous to the instant case" that featured a transfer between two distinct bodies of water that would not mingle in a state of nature. *Id.* at 1297. In *Friends of the Everglades, Inc. v. South Fla. Water Mgmt. Dist.*, 2006 WL 3635465 at *34 (S.D. Fla. 2006), a post-*Miccosukee* decision, the defendant offered, as a metaphor for the unitary navigable water theory, "a hypothetical law that bans the addition of wine to the United States. The ban would undoubtedly apply to the importation of wine from, for example, France or Italy. However, it would have no effect on the movement of wine from California to Florida, as movement between states would not result in the addition of any wine to the United States as a whole."

241. Indeed, the district court used warring analogies as one indication in its first-step *Chevron* analysis to find that the CWA was ambiguous on the meaning of "addition" in the context of water transfers.

242. The court in *Sierra Club v. El Paso Gold Mines, Inc.*, 421 F.3d 1133, 1145 (10th Cir. 2005), analogized abandoned mining operations to leaky faucets: "[I]f you own the leaky 'faucet,' you are responsible for its 'drips.'"

Manitoba[243] to Montana, but not from Idaho to Montana, when the water quality impacts are the same? The four-marble analogy raises the question of why our water pollution control statute would authorize doubling the number of polluted stream miles, as opposed to continuing pollution only where it currently exists; indeed, this is inconsistent with EPA's antidegradation policy.[244] The courts' use of metaphors to explain the unitary navigable water theory signals their unease with the theory; the metaphors they develop do not alleviate that unease.

If a point source adds a pollutant to a navigable water in the course of an interbasin transfer, that transfer should require a §402 permit. EPA has exempted such a transfer from §402, however, in its water transfer rule. EPA justifies its rule, not as an interpretation of "addition," but instead as a "holistic interpretation" of the statute, based on four brief passages that EPA sees as establishing a "grand balance" between water quality achievement and water resource management. Examined closely, none of the four passages support EPA's use of them. Even if they did, four brief passages are insufficient to support any grand balance in a 200-page-long statute. Worse, the "unitary navigable waters" theory underlying EPA's interpretation of "addition" in the rule is incompatible with water quality standards, particularly the antidegradation policy, and drastically limits the jurisdiction of the §404 permitting program. EPA's flimsy construct cannot withstand the force of the basic goals and strategies of the CWA, each rooted in long, explicit, and detailed statutory provisions. The rule and the theory are contrary to the purposes and structures of the statute, are bad policy, and should either be withdrawn by EPA or overturned on judicial review.

On the other hand, the definition of "addition" suggested in this chapter applies comfortably to water transfers, without creating interpretive difficulties. Under that definition, water transfers that add pollution to receiving waters require §402 permits. Transfers are the but-for causes of pollutants from dirty navigable donor water flowing into clean navigable receiving water; they add pollutants from the donor water to the receiving water, when the pollutants would not otherwise be in the receiving water. This will not sound the death knell of western water rights; EPA offers no evidence that §402 has interfered in any significant way with western water rights for the last 40 years, probably because the CWA itself exempts the return of agricultural irrigation water from the §402 permit program.

243. Manitoba intervened as a plaintiff in *Catskill Mountains*, seeking to overturn the rule. *See Catskill Mountains*, 2014 WL 1284544 (caption to decision).
244. 40 C.F.R. §131.12. *See also* 33 U.S.C. §303(d)(4)(B).

B. *Indirect Additions*

Industries that discharge wastes into POTWs for treatment are commonly called indirect sources because they do not add pollutants directly to navigable waters, but do so indirectly through the POTWs. In such cases, the POTWs are subject to the §402 permit program, but the indirect dischargers are not, although they are subject to regulation under pretreatment requirements established by §307 and 40 C.F.R. Part 403. Industrial dischargers to POTWs, however, are not the only indirect additions of pollutants to navigable waters. Indirect additions can be made to navigable waters from discharges onto or into adjacent land or by air emissions to nearby or even very distant navigable waters. The indirect additions from adjacent land are evasions of the CWA and should be regulated by it. The indirect additions from the air may be regulated already by the Federal Insecticide, Fungicide, and Rodenticide Act (FIFRA)[245] or the Clean Air Act,[246] possibly conflicting with regulation by the CWA. This chapter first addresses indirect additions though adjacent land and groundwater.

1. Indirect Additions From Land or Groundwater

Imagine a factory located adjacent to a river. To avoid the cost of water pollution control, the owner removes the pipe used to discharge waste to the river and instead pumps the waste through another pipe into a deep hole dug 50 feet from the river. Miraculously, the waste never overflows the hole, but appears to disappear into the earth. Of course, we know the waste percolates through the ground into the river or to groundwater that flows into the river. In citizen and government enforcement actions against such discharges under the CWA, courts generally treated them as discharges into tributary groundwater, holding the discharges violated the CWA in the absence of a permit.[247] Otherwise, of course, polluters could easily evade the prohibition of the CWA. Discharges into tributary groundwater were a small exception to the general conclusion that navigable waters do not include groundwater. EPA's new regulatory definition of waters of the United States, however, explicitly excludes groundwater from

245. 7 U.S.C. §§136-136y, ELR Stat. FIFRA §§2-35..
246. 42 U.S.C. §§7401 et seq.
247. See, for instance, *Quivira Mining Co. v. EPA*, 765 F. 2d 126, 15 ELR 20530 (10th Cir. 1985), and *Sierra Club v. Colorado Refining Co.*, 838 F. Supp. 1428, 24 ELR 20749 (D. Colo. 1993). Chapter 3, Section IV.C. addresses these decisions and this issue in detail.

waters of the United States, apparently ending the exception for tributary groundwater.[248]

The demise of the theory that discharges to tributary groundwater without a permit violate the CWA, however, does not free such dischargers from CWA liability. The theory of liability changes from a discharge to tributary groundwater to an indirect discharge to the adjacent river. The plurality in *Rapanos* commented that the CWA "does not forbid 'the addition of any pollutant *directly* into navigable waters from any point source,' but rather the 'addition of any pollutant *to* navigable waters.'"[249] This suggests the Court may be more amenable to the theory of indirect additions than to the theory of tributary groundwater. There are no reported decisions examining such a theory, but it is predictable that they will be files. When they are filed, their success will depend in large part on how compelling their individual facts are, especially the extent to which they suggest an intentional evasion of the requirements of the CWA.

Assuming such cases are unsuccessful, the action of discharging industrial wastes onto or into land adjacent to navigable water will violate RCRA,[250] except in violation of a permit as long as the wastes are RCRA hazardous wastes.[251] RCRA and the CWA do not conflict, for if a discharge is regulated under the CWA, is not regulated by RCRA.[252] Thus, if attempts to regulate these discharges under the CWA fail, the discharges will be regulated under RCRA, and if attempts to regulate them under the CWA are successful, the discharges will not be subject to duplicative regulation under RCRA.

2. Indirect Additions From the Air

Some point sources directly add pollutants to the air and indirectly add the same pollutants to water, when gravity or precipitation take them from the air to the water below. The most-detailed analyses of such indirect additions are in challenges to aerial spraying of pesticides into or near navigable water in an attempt to eradicate mosquitoes carrying the West Nile virus

---

248. 80 Fed. Reg. 37054, 37114 (June 29, 2015), to be codified at 40 C.F.R. §122.2 waters of the United States, (2)(v). It might be thought that groundwater is covered by waters "adjacent to" a navigable water, §122.2(1)(vi) & (3)(i), or by waters on 100-year floodplains, §122.2(1)(viii). But the exclusion of groundwater applies even for waters meeting "the terms of paragraphs (1)(iv) through (viii)."
249. 547 U.S. at 743.
250. 42 U.S.C. §§6901 et seq.
251. Under 42 U.S.C. §6925(a), the disposal of hazardous waste without a permit is illegal and 42 U.S.C. §1004 defines disposal as "the discharge, deposit, injection, dumping, spilling, leaking, or placing of any solid waste or any hazardous waste into or on any land or water."
252. RCRA's definition of solid waste excludes discharges through point sources regulated under the CWA, 42 U.S.C. §6003(27), and hazardous wastes are a subset of solid wastes, 42 U.S.C. §6903(5).

or other pathogens. In *No Spray Coalition, Inc. v. City of New York,*[253] the plaintiffs sought to enjoin New York City from spraying pesticides, on the grounds that the spraying constituted the addition of pollutants to navigable waters from point sources without §402 permits. The city applied the pesticides for a purpose and in a manner EPA had approved under the federal statute regulating the manufacturing, sale, and use of pesticides (FIFRA).[254] FIFRA authorizes applications of EPA-registered pesticides, for EPA-approved purposes, and in accordance with directions on EPA-approved labels.

The Southern District of New York decided the case primarily on the grounds that Congress intended pesticide application to be governed by FIFRA rather than the CWA, and that Congress did not intend the CWA's citizen suit provision to be used to enforce FIFRA, which lacks such a provision. But the court also held, with little analysis, that pesticides sprayed near water and drifting into water were not added to water. According to the court, the city "discharge[d] the insecticides into the atmosphere and not into the navigable waters. It would be stretching the language of the statute well beyond the intent of Congress to hold that the *de minimus* incidental drift over navigable waters is a discharge from a point source into those waters."[255] The court commented that the "fact that a pollutant might ultimately end up in the navigable waters as it courses through the environment," does not make its use a violation of CWA §301(a). If it did, every emission of "smoke, exhaust fumes and pesticides" would violate the statute.[256]

Two neighboring New York district courts came to the same conclusion, using both the same and different reasoning. In *Altman v. Town of Amherst,*[257] the Western District's main reasoning was that pesticides used for their intended purposes were not pollutants, also holding that FIFRA rather than the CWA governed. In *Peconic Baykeeper, Inc. v. Suffolk County,*[258] the Eastern District's main reasoning was cursory: "[a]tmospheric emission of aereal adulticides are not defined as a pollutant [and] at no time was the spray made directly to navigable water." The court also held that FIFRA governed the situation and that the spray bars attached to trucks and planes were not point sources. Finally, it deferred to EPA's policy that such spraying of pesticides was not subject to permitting under CWA §402.

253. 2000 WL 1401458 (S.D.N.Y. 2000), *rev'd,* 351 F.3d 602, 34 ELR 20007 (2d Cir. 2003).
254. 7 U.S.C. §§136-136y, ELR Stat. FIFRA §§2-35.
255. *No Spray Coalition,* 2000 WL 1401458 at *3.
256. *Id.*
257. 190 F. Supp. 2d 467 (W.D.N.Y. 2001).
258. 585 F. Supp. 2d 377 (E.D. N.Y. 2008).

The Second Circuit reversed all three decisions, on different grounds. In *No Spray Coalition, Inc. v. City of New York*,[259] it held that if the plaintiffs alleged a cognizable cause of action for a violation of the CWA, they could maintain a citizen suit under it. The court, however, did not address whether the plaintiffs in the case alleged a cognizable CWA cause of action, in particular whether discharges of pesticides into the atmosphere near navigable water could constitute an addition to the navigable waters into which they drifted or whether spraying pesticides in compliance with FIFRA could violate the CWA. In *Altman v. Town of Amherst*, the Second Circuit found that the district court had granted a motion for summary judgment on an insufficient record and remanded the case to proceed with discovery, allowing plaintiffs to make their case that defendant had added pollutants to navigable waters from point sources.[260] The court stated that until EPA "articulates a clear interpretation of current law—among other things, whether properly used pesticides released into or over waters of the United States can trigger the requirement for a NPDES permit—the question of whether properly used pesticides can be pollutants that violate the CWA will remain open."[261]

Before the Second Circuit decided *Peconic Baykeeper, Inc. v. Suffolk County*,[262] EPA, hoping to avoid such litigation or at least affect its outcome, issued an interpretive statement and promulgated a rule exempting from the §402 permit requirements the application of registered FIFRA pesticides for FIFRA-approved purposes and in accordance with FIFRA-approved labels. The exemption covered both (1) direct application to navigable waters for control of waterborne pests and (2) application to land adjacent to navigable waters for control of pests on adjacent land, from which pesticides drift into navigable waters.[263] The Second Circuit thereafter decided *Peconic Baykeeper*, cognizant of the rule, which the Sixth Circuit by then had overturned in *National Cotton Council of America v. U.S. EPA*,[264]discussed immediately below. Because the Sixth Circuit had stayed its mandate, the rule remained in effect and authorized many of the defendant's actions. On the facts found by the district court, however, some of the spraying appeared not to be in conformity with the FIFRA label, and the Second Circuit reversed the district court's ruling to that extent.

---

259. 351 F.3d 602 (2d Cir. 2003).
260. 47 Fed. Appx. 62 (2d Cir. 2002).
261. *Id.* at 67.
262. 600 F.3d 180 (2d Cir. 2010).
263. 40 C.F.R. §122.3(h).
264. 553 F.3d 927, 39 ELR 20006 (6th Cir. 2009).

Although judicial review of EPA's pesticide application rule in *National Cotton Council* focused on EPA's interpretation of "pollutant," it also considered the meaning of "addition." EPA argued that pesticides are not pollutants, while admitting that excess pesticides and pesticide residues are pollutants.[265] It then argued that pesticide applicators spray pesticides (nonpollutants), but do not spray excess pesticides or pesticide residue (pollutants). Thus, EPA argued, pesticide applicators do not add pollutants to the water because the pesticides they spray are not "pollutants" and they do not spray excess pesticides or pesticide residue, which are pollutants. Excess pesticides and pesticide residue "are not created [and hence are not added] until later, presumably after they are already in the water."[266] But the Sixth Circuit concluded that EPA offered no support, other than this descriptive narrative, "for its assertion that a pesticide must be 'excess' or 'residue' at the *time of discharge* [addition] if it is to be considered as discharged *from a point source.* This omission of authority is understandable, as none exists."[267] Indeed, EPA's semantic gymnastics[268] ignore the reality that when the pesticide applicator sprays pesticides, it is simultaneously spraying molecules of pesticide, a few of which hit their targets and most of which do not, but instead fall into water, although it is impossible to tell at the point of spraying which molecules of the sprayed material will be in each category.

The court rejected EPA's attempt to tie "addition" to "point source" in a temporal sense, finding it "unsupported by the Act" and contrary to the purpose of the permit program to prevent harmful discharges, which include "discharges which are innocuous at the time they are made but extremely harmful at a later point."[269] It found that the plaintiff's position conformed to EPA's "outside world" gloss on "addition," because the pesticide applicator adds pesticide residue or excess to the water from the outside world. Moreover, although not noted by the court, spraying the pesticide from a

---

265. This mirrors earlier arguments by EPA that consumer products used for their intended purposes are not waste under the RCRA, governing the management and disposal of hazardous waste. For instance, EPA determined that lead shot fired over water at ducks or skeet, missing them, and falling in the water were consumer products used for their intended products and were not waste under EPA's definition of that term. *See* Cordiano v. Metacom Gun Club, Inc., 575 F.3d 199 (2d Cir. 2009). The argument might be applicable to pollutants fitting only within one of the subcategories of "waste" in §502(6), e.g., "chemical waste."

266. *National Cotton Council*, 553 F.3d at 939.

267. *Id.*

268. EPA's semantic gymnastics in *National Cotton Council* are reminiscent of the metaphysics of *United States v. Deaton*, 209 F.3d 331, 30 ELR 20508 (4th Cir. 2000) (discussed *infra* in Section III.C.1.), in which dredged material is metamorphosed into pollutants when the dredge lifts it from the water.

269. *National Cotton Council*, 553 F.3d at 939.

point source is the but-for cause of the pesticide residue and excess entering navigable water.[270]

In *Chemical Weapons Working Group, Inc. v. U.S. Department of the Army*,[271] environmental advocates challenged the Army's incineration of chemical warfare agents, in part because CWA §301(f) prohibits the "discharge . . . [of] . . . any . . . radiological, chemical, or biological warfare agent." The state of Utah had issued a Clean Air Act (CAA)[272] permit authorizing emissions from the U.S. Army's incinerator. The plaintiffs contended that some of those emissions would be added to unspecified navigable waters by "atmospheric deposition," in violation of §301(f).[273] The Tenth Circuit rejected the claim, concluding that the "stack emissions constitute discharges into the air—not water,"[274] labeling them "indirect discharges."[275] In addition, the court rejected the plaintiffs' argument because it would create a regulatory conflict between the CWA and the CAA, enacted by Congress to regulate such emissions.[276]

In both of these situations, it is important to remember that all four elements of §301(a) must be met before the CWA is violated, and the only issue considered in this chapter is whether these types of emissions into the air constitute additions to water. The two other significant questions are whether these materials are pollutants, especially those whose purpose is introduction in, on, or into the water and whether point sources convey them to the water or merely into the air. Even if all four elements are met in a given instance, it is possible that another statute governs or that regulation by another statute could influence how EPA or courts interpret one of the elements.

None of the decisions in these cases analyze the "addition" element in any depth, although many of them allude to it. The most that can be deduced from the decisions is that spraying pesticides directly on water may be addition, while long-range land and water deposition from smokestacks are not

---

270. *See* Weinberger v. Romero-Barcelo, 456 U.S. 305, 12 ELR 20538 (1982), a citizen suit against Navy bombing practice in which many bombs missed their targets and landed in the sea. Although the case is similar on the facts, the Court did not analyze the "addition" issue.
271. 111 F.3d 1485, 27 ELR 21130 (10th Cir. 1997).
272. Clean Air Act (CAA), 42 U.S.C. §§7401-7671q, ELR Stat. CAA §§101-618.
273. *Chemical Weapons Working Grp.*, 111 F.3d at 1489-91.
274. *Id.* at 1490.
275. *Id.* at 1490 n.3. "Indirect discharges" was a confusing choice of words because "indirect dischargers" under the CWA refers to industrial dischargers to municipal sewage treatment systems. *See* 40 C.F.R. §122.2. While those indirect dischargers do not require §402 permits, they are subject to directly applicable and directly enforceable technology-based pretreatment standards for toxic pollutants and to general prohibitions against interfering with or passing untreated through a municipal sewage treatment plant. *See* §307(b), 33 U.S.C. §1317(b).
276. 111 F.3d at 1490-91. *See also* United States ex rel. McKeown v. Port Auth. of N.Y. & N.J., 162 F. Supp. 2d 173 (S.D.N.Y. 2001).

direct addition, but may be indirect addition. While spraying pesticides from an airplane and spewing air pollutants from smokestacks appear to be similar, in that both emit material into the air and some of the material predictably finds its way into water by natural means, the actions are very different in terms of their underlying facts. The underlying factual differences in turn result in differences in the feasibility of regulating them within the CWA's structure. And those differences, in turn, affect the analysis of how to interpret the CWA *in pari materia* with the other statutes.

The underlying factual differences between these two situations are apparent. The pesticide-spraying plane flies above water, points the nozzles of its spray bar at the water, the target pests it aims at are in or fly above the water, and most of the pesticide inevitably enters the water without making contact with the targets. The pesticides enter the water from the plane much as they would pour into the water from a horizontal industrial or municipal outfall pipe terminating a few feet above the receiving water. So described, it is difficult to characterize this as anything other than direct addition of the pesticides to the water. The smokestack, however, is not above the water, but is on dry land, perhaps many miles from water. It points its emissions straight up into the air. It has no targets in or above the water; its purpose is to dispose of waste into the air. Some of its emissions eventually come to rest on land, some on water, and some wander to the upper atmosphere. Some of its emissions ultimately may fall into hundreds of separate water bodies, located in many states downwind from the smokestack. In turn, any one of these waters may receive pollutants from many upwind smokestacks in different states. It is difficult to characterize the smokestacks as adding this material to any particular water, except indirectly and inadvertently.

Direct addition by spraying pesticides over water is susceptible to regulation under the CWA because there is an identifiable pollutant, an identifiable point source, an identifiable addition, and an identifiable navigable receiving water, usually located in the same state in which the spraying occurs. Issuing or denying a CWA permit for spraying pesticides on a particular water body adds no complications to the normal CWA permit issuance process, requiring only the determinations of whether there are applicable technology-based standards (there are none at present, except limitations on or prohibitions of spraying), whether best management practices are available to minimize the waste pesticide and pesticide residue that reach the water, and whether the spraying will interfere with the attainment of the designated water quality standards applicable to the receiving water. Most of the relevant environmental protection can be provided by issuing a CWA permit requiring best management practices or by denying the permit because the spraying will violate water quality standards.

On the other hand, emissions from smokestacks are far more difficult to regulate under the CWA. Because EPA has not promulgated a technology-based standard under the CWA for smokestack emissions, it would have to base the permit's effluent limitations in a CWA permit for smokestack emissions largely on preventing the emissions from violating the water quality standards of the waters into which they fell.[277] Indeed, when issuing a CWA permit, EPA or the state must ensure that the pollutants discharged will not violate water quality standards "established pursuant to *any* State law or regulations."[278] Moreover, the permit issuer must notify both EPA and downwind states whose waters are affected of the proposed permit, allowing them to comment and request a public hearing.[279] If EPA is the permit issuer, it could establish effluent limitations in the permit to meet affected water quality standards in states downwind from the permitted source.[280] If a state is the permit issuer, EPA presumably could veto a permit that failed to provide effluent limitations to meet affected water quality standards in downwind states.[281]

Applying water quality standards to smokestack emissions, however, presents two complications. First, the emissions from a smokestack, particularly a tall one, may fall into many downwind water bodies in many states. The only way to determine which water bodies smokestack emissions might enter is by developing a mathematical model, by definition only an approximation of reality.[282] Once each of the hundreds of receiving water bodies are identified, the states in which each receiving water body is located must calculate the total maximum daily load (TMDL) for each relevant pollutant to determine whether the smokestack emissions would have to be limited. If they do, the states in which the affected water bodies are located would have to develop waste-load allocations to determine how much of the required emission reductions would be allocated to the partic-

277. In the absence of promulgated effluent guidelines for technology-based standards, the permit writer could devise technology-based standards on a case-by-case basis using best engineering judgment. *See* §402(a). On the other hand, promulgating a technology-based standard for controlling emissions from smoke stacks is more appropriately done by air pollution regulators than by water pollution regulators.
278. Section 301(b)(1)(C) (emphasis added).
279. Section 402(b)(3) & (5).
280. *See* Arkansas v. Oklahoma, 503 U.S. 91, 22 ELR 20552 (1992).
281. Section 402(d).
282. *See, e.g.*, Ohio v. U.S. EPA, 784 F.2d 224, 16 ELR 20447 (6th Cir. 1986), in which the court held that EPA's reliance on a computer-generated model to calculate emissions limitations for smokestacks was arbitrary and capricious.

ular smokestack. This invites interstate disputes, because the state with the receiving water develops TMDLs for its waters and has every incentive to disproportionately burden out-of-state pollution sources. A single smokestack whose emissions eventually fall into hundreds of water bodies could be subject to hundreds of such exercises for a variety of pollutants. Worse, the exercises for each water body would have to consider each smokestack whose emissions enter that water body, and there may be hundreds of such smokestacks.

Developing TMDLs considering traditional water pollution sources took decades and protracted litigation,[283] strongly suggesting that the more difficult task of controlling thousands of indirect smokestack sources in hundreds of TMDLs would take the better part of this century and convert most litigation lawyers into TMDL warriors. Water pollution permits are simply not designed to control smokestack air pollution. Similar complications prompted the Supreme Court to reject the considerably easier application of injured downstream states' common-law torts to point source pollution from upstream states.[284]

Programs have been developed under the CAA, however, to deal with just such problems. It is far more feasible to regulate smokestack emissions under the CAA than under the CWA. While it is beguiling for friends of clean water to interpret the CWA to regulate all activities producing emissions that eventually enter water, particularly in cases where most or all of the material inevitably falls into nearby navigable water[285] (atmospheric deposition is a major source of water pollution, accounting, for example, for most of the mercury in the Great Lakes and the acidification of lakes in the Northeast and neighboring Canada),[286] nothing in the CWA speaks explicitly to regulating such indirect additions. "Addition" could be interpreted to cover them or not cover them. Each of the actions at issue, however, is routinely and typically regulated by another federal statute, the CAA or FIFRA. Those other statutes authorize EPA to take the impact on water quality into account

283. *See* Houck, *supra* note 195.

284. *See* International Paper Co. v. Ouellette, 479 U.S. 481, 495-500, 17 ELR 20327 (1987).

285. A pioneering article makes the case that atmospheric deposition of pollutants into navigable water may violate the basic prohibition of the CWA. *See* Amil Anthony, *Shotguns, Spray, and Smoke: Regulating Atmospheric Deposition of Pollutants Under the Clean Water Act*, 29 UCLA J. Envtl. L. & Pol'y 215 (2011). Amil Anthony's argument might be stronger if augmented with an element-by-element analysis. However, he underestimates the interference with the CAA and FIFRA of regulating air emissions and pesticide application under the CWA. While his argument may not lead to a revolutionary expansion of CWA jurisdiction, it may provide a template in particularly deserving cases, much as Mary Christina Wood's article, *Regulating Discharges Into Ground Water: The Critical Link in Pollution Control Under the CWA*, 12 Harv. Envtl. L. Rev. 369-426 (1988), did for CWA citizen suits against point source addition of pollutants to tributary groundwater.

286. Anthony, *supra* note 285, at 221-32.

in regulating air emissions and pesticide applications, enabling EPA to prevent water pollution without invoking the CWA.[287] The question in both situations is how best to interpret the statutes involved to implement them without conflict.

The underlying concept of *in pari materia* comes into play when two statutes may govern the same situation. *In pari materia* posits that "similar statutes should be interpreted similarly"[288] and "interpreted together, as if they were one law."[289] A corollary is that statutes should be interpreted not to conflict with each other.[290] If a conflict between statutes cannot be avoided, the more-specific statute or the latest in time governs.[291] The initial question is whether statutes are *in pari materia*. Justice Antonin Scalia writes that statutes relating to the same subject, having one object in view, or acting in one system fall within that category. Under that definition, surely the CAA and the CWA are *in pari materia*: Both protect the ambient environment, have similar goals, have similar structures, share common tactics and strategies to control discharges of pollutants to air and water, and even have very similar provisions.[292] FIFRA is further afield; it is a product-control statute with a completely different structure than the CWA, using different tactics and strategies, and having no similar provisions.[293] As a product-control statute, its first purpose is to assure that pesticides are efficacious,[294] as well as not unreasonably harmful to the environment.[295] Although FIFRA is not as closely related to the CWA as is the CAA, Justice Scalia has observed that

---

287. CAA §109(b), 42 U.S.C. §7409(b), requires EPA to promulgate primary and secondary air quality standards, primary standards protecting public health and secondary standards protecting public welfare, both of which could include effects on water, *see* CAA §302(h), 42 U.S.C. §7602(h). Air quality standards are much like water quality criteria under the CWA: they trigger many of the regulatory requirements under their respective statutes. Moreover, the CAA includes a regional cap-and-trade program to control acid rain deposition from midwestern coal-fired power plants affecting waters in the northeastern United States and eastern Canada. *See* CAA §§401-416, 42 U.S.C. §§7651-7651*o*. FIFRA §3(c)(5)(C), 7 U.S.C. §136a(c)(5)(C), provides as one criterion for EPA's registration of a pesticide for a particular use that "it will perform its intended function without unreasonable adverse effects on the environment." "Environment," of course, includes "water." *See* FIFRA §2(j), 7 U.S.C. §136(j).

288. Eskridge, *supra* note 18, at 327.

289. Scalia & Garner, *supra* note 18, at 252. *See also* Sutherland, *supra* note 170, at §51:2.

290. Sutherland, *supra* note 170, at §51:2.

291. *Id.*

292. Both have ambient standards, NAAQS, national ambient air quality standards, and WQS, water quality standards, that are to be met by controlling pollution. Both use technology-based standards. Both are implemented by cooperative federalism. Both have similar permitting, inspection, enforcement, and citizen suit provisions.

293. Instead of controlling the discharge of pollutants by individual sources, FIFRA registers pesticides.

294. FIFRA §3(c)(5)(A), 7 U.S.C. §136a(c)(5)(A) (that "[the product's] composition is such as to warrant the proposed claims for it").

295. FIFRA §3(c)(5)(C), 7 U.S.C. §136(5)(C) (that "[the product] will perform its intended function without unreasonable adverse effects on the environment").

all statutes are part of a single *corpus juris*, and courts should interpret them accordingly.[296]

Interpreting statutes *in pari materia* avoids interpreting a statute to conflict with another statute, and avoids creating requirements on the regulated public that conflict with requirements imposed on it by other statutes. The *in pari materia* analysis of whether the CWA applies to spraying pesticides regulated by FIFRA is different from whether the CWA applies to emitting pollutants from smokestacks regulated by the CAA. The CWA and FIFRA do not perform the same functions in protecting the environment. Under FIFRA, EPA registers pesticides for uses and applications that will not result in "unreasonable effects on the environment," including water.[297] By its nature, EPA's determination in the registration process is whether a pesticide will have unreasonable effects on water generally, rather than on particular water bodies, for it is not clear to what water bodies pesticides will be applied until they are manufactured, sold, and about to be used, well after FIFRA registration. Under the CWA, however, EPA's determinations of whether to issue or deny a permit and of what effluent limitations to include in a permit are based, in part, on protecting individual water bodies and the water quality standards designated for them. Moreover, FIFRA's "unreasonable effects of the environment" standard involves an "economic, social, and environmental" risk/benefit analysis,[298] while the water quality standards designated under the CWA for particular water bodies are not limited by economic impact or any other risk/benefit analysis.[299]

If EPA denies a CWA §402 permit for spraying a pesticide on a particular water body or adds conditions beyond those already imposed by FIFRA on such spraying to achieve its water quality standards, the denial or extra conditions would not interfere with FIFRA's goals of providing that approved pesticides are efficacious and do not unreasonably adversely affect the environment. There is no conflict if both statutes apply; the more-stringent requirements govern in any particular situation, providing the greatest protection of the environment. Finally, issuing or denying a CWA permit for spraying pesticides on a particular water body adds no complications to the normal CWA permit issuance process, requiring only the determinations of whether there are applicable technology-based standards (there are none at present), whether management practices are available to minimize the pes-

296. Scalia & Garner, *supra* note 18, at 252-56. *See also* Sutherland, *supra* note 170, at §51.3, "various statutes relating to environmental policy" are *in pari materia*.

297. FIFRA's definition of "environment" includes water. *See* FIFRA §2(j), 7 U.S.C. §136(j).

298. FIFRA §2(bb), 7 U.S.C. §136(bb).

299. CWA §§303 & 304(a), 33 U.S.C. §§1313 & 1314(a). *But see* CWA §302(b), 33 U.S.C. §1312(b).

ticide that reaches the water, and whether the spraying will interfere with attainment of the designated water quality standard for the particular and known water body.

On the other hand, the CWA and CAA perform many of the same or similar functions in protecting the environment. Both provide for the issuance of permits to sources of emissions or discharges of pollutants to protect the environment. Both apply technology-based standards for pollution reduction and both condition permits to meet ambient standards in the air or water. They are *in pari materia* and should be interpreted in harmony. Of course, they could be in harmony if both applied to a particular smokestack and the more-stringent of the two governed. That is how the CWA reconciles differences between federal and state water pollution requirements.[300] And the imposition of the more-stringent standard does not interfere with the environmental protection afforded by the superseded statute, since by definition it provides greater environmental protection. But because the CAA primarily regulates smokestacks, and some emissions from many or most smokestacks eventually fall into water, requiring CWA permits for smokestack emissions could result in the CWA superseding the CAA in many or most cases, a result almost certainly not intended by Congress. Moreover, the issuance of CWA permits to smokestacks imposes such great complications on CWA permit issuance process that it is doubtful Congress intended the CWA to apply in such cases.

Concluding that spraying pesticides onto navigable water is a classical addition of pollutants to navigable water from a point source under the CWA does not conflict with FIFRA. Moreover, the considerations necessary to issue or deny a CWA §402 permit for spraying pesticides into water do not differ substantially from the considerations necessary to issue or deny §402 permits for adding pollutants to navigable water from other point sources. On the other hand, concluding that emissions from a smokestack require a CWA permit because some of those emissions eventually enter navigable water would conflict with the CAA by largely superseding it. Moreover, the difficulties in issuing CWA permits for smokestack emissions are overwhelming. While it may be tempting to define the CWA's jurisdiction to include regulation of emissions from the occasional smokestack whose emissions largely fall on and significantly pollute an adjacent body of water, this author knows of no principled way to differentiate between smokestacks that sufficiently pollute a water body to be regulated under the CWA and those that pollute many water bodies insufficiently to be regulated under it.

300. CWA §510, 33 U.S.C. §1370.

While the above analysis is directed toward pesticides sprayed from airplanes and emissions from smokestacks, the same analysis can be performed to determine whether any activity discharging pollutants into the air that come to rest, in whole or in part, in water is regulated by the CWA or another statute, usually the CAA. For instance, EPA regulates paint-spraying under the CAA,[301] but would spray-painting a bridge or a shipyard when some paint falls into the water also require a CWA permit?

C. *Section 404 Decisions*

Section 301(a) declares that the addition of pollutants to navigable waters from point sources is unlawful, except in compliance with a §402 or §404 permit. The decisions reviewed earlier in this chapter primarily concerned additions not in compliance with §402 permits. The following decisions concern activities not in compliance with §404 permits. The only structural differences between the two is that §402 authorizes EPA and states with approved programs to issue permits for discharges of "pollutants," while §404 authorizes the U.S. Army Corps of Engineers (the Corps) and states with approved programs to issue permits for discharges of "dredged or fill material."[302] The §502(6) definition of "pollutant" includes "dredged spoil," which appears to be synonymous with "dredged material," even though that definition does not include "fill material." Most materials used as fill, however, fall under the broad §502(6) definition of "pollutant."[303] This poses a potential ambiguity: Under which section should a permit be issued when the fill material added to navigable water consists of or contains pollutants? Under §402 because the material contains pollutants, or under §404 because the pollutants are fill material? Section §402(a) suggests a solution to this problem by authorizing EPA to issue permits for the discharge of pollutants, "except as provided in [§404.]" Accordingly, the Corps is to issue permits under §404 for the addition of dredged or fill material and EPA is to issue permits under §402 for the addition of other pollutants.[304] EPA and the Corps define "fill material" to mean "material [that] has the effect of . . . [c]hanging the bottom elevation of water."[305]

Section 404 has come to be seen as the guardian of wetlands. Its scope, however, is too narrow to achieve that goal. Most reported §404 decisions

301. 40 C.F.R. §§11511, 11514, 11516, 11519 & 11522.
302. The approval of state programs is more limited under §404 than under §402. *Compare* §404(g), *with* §402(b).
303. For instance, biological material, rock, and sand. *See* §502(6).
304. Coeur Alaska, Inc. v. Southeastern Alaska Conservation Council, 557 U.S. 261 (2009).
305. 40 C.F.R. §232.2.

concern the issue of whether particular landclearing and related wetlands development activities require §404 permits. By definition, §404 only regulates additions of material to wetlands or other navigable water, not extractions of material or water from wetlands or other navigable water. Landclearing or drainage activities do not inherently fill or add material to wetlands. Indeed, if wetlands developers could sweep a giant vacuum device over a wetland and suck up all the vegetation, other loose material, and water, they could clear and dry the wetland, without adding anything to it or filling it in any way, as long as they emptied the vacuum cleaner elsewhere on dry land. Thus, owners of wetlands can drain them to dry land with impunity, as long as they add no pollutants while doing so, even if they intend later to fill the dried former wetlands.[306] In terms of damage to wetlands, removal of wetland soils harms wetlands as much or more than filling wetlands with wetlands soils, but §404 regulates only the latter.

Section 404 landclearing decisions run a spectrum from moving unwanted material from one part of a wetland to fill other parts of the same wetland, which almost all courts consider addition,[307] to the sort of incidental fallback that occurs when digging a hole in sand and some sand trickles back from the shovel into the hole, which almost all courts hold is not addition.[308] Most of the decisions on the spectrum in between broadly interpret "addition" to require §404 permits. After examining these decisions, the chapter will consider how they square with EPA's "from the outside world" gloss on "addition" and the Agency's theory of unitary navigable water.

306. *See, e.g.*, Save Our Cmty. v. U.S. EPA, 971 F.2d 1195 (5th Cir. 1992). *See also* Orleans Audubon Soc'y v. Lee, 742 F.2d 901, 15 ELR 20030 (5th Cir. 1984) (the deliberate drainage of a swamp is not the discharge of fill material); and Avoyelles Sportsmen's League, Inc. v. Marsh, 715 F.2d 897, 13 ELR 2094 (5th Cir. 1983). The notion that, once drained, a wetland may be filled because it is no longer navigable water, however, is at odds with the well-established doctrine that once a water is navigable, it is always navigable. United States v. Appalachian Elec. Power Co., 311 U.S. 377, 408 (1940).

307. *See, e.g.*, United States v. Cundiff, 555 F.3d 200, 39 ELR 20025 (6th Cir. 2009); Borden Ranch P'ship v. U.S. Army Corps of Eng'rs, 261 F.3d 810, 32 ELR 20011 (9th Cir. 2001); United States v. Deaton, 209 F.3d 331, 30 ELR 20508 (4th Cir. 2000); United States v. M.C.C. of Fla., Inc., 772 F.2d 1501, 15 ELR 21091 (11th Cir. 1985); *Avoyelles Sportsmen's League, Inc.*, 715 F.2d 897; United States v. Fabian, 522 F. Supp. 2d 1078 (N.D. Ind. 2007); N.C. Shellfish Growers Ass'n v. Holly Ridge Assoc., LLC, 278 F. Supp. 2d 654 (E.D.N.C. 2003); United States v. Hummel, 2003 WL 1845365 (N.D. Ill. 2003); United States v. Bay-Houston Towing Co., 33 F. Supp. 2d 1033 (N.D. Ill. 1998); United States v. Sinclair Oil Co., 767 F. Supp. 200, 21 ELR 21323 (D. Mont. 1990).

308. *See, e.g.*, National Mining Ass'n v. U.S. Army Corps of Eng'rs, 145 F.3d 1399, 28 ELR 21318 (D.C. Cir. 1998); National Ass'n of Home Builders v. U.S. Army Corps of Eng'rs, 2007 WL 259944 (D.D.C.); American Mining Cong. v. U.S. Army Corps of Eng'rs, 951 F. Supp. 267, 27 ELR 20589 (D.D.C. 1997).

1. Redeposit

Most redeposit decisions are rooted in *Avoyelles Sportsmen's League, Inc. v. Marsh*,[309] in which an environmental group challenged a soybean farmer's clearing of woodlands that exhibited wetlands characteristics, without a §404 permit. The defendants and the government had argued before the district court that landclearing by the mere removal of wetlands vegetation without a §404 permit was not an addition and therefore not a violation of §301(a). The district court rejected this argument as "untenable" because it would "frustrate the ecological purposes of the CWA."[310] The U.S. Court of Appeals for the Fifth Circuit did not reach the issue on appeal, because the district court's factual findings demonstrated that defendant's landclearing activities included filling small sloughs, partially filling large sloughs, digging holes, and burying logs in them, rendering this more than a "mere removal" case.[311] The court held that "'addition' . . . may reasonably be understood to include 'redeposit,'" finding that definition consistent with both the purpose and the legislative history of the CWA.[312] The court noted in passing that EPA's "outside world" theory might jeopardize the application of §404, but did not address the issue because no party raised it.[313]

Avoyelles involved landclearing activity in which defendants moved materials considerable distances, for redeposit in the same wetlands. Other decisions consider more modest movement of material, many of them involving "sidecasting." Sidecasting occurs when soil is removed by digging a hole, often a ditch, and the soil is then placed beside the hole, an activity that moves material but does not appear to add new material. In *United States v. Deaton*,[314] the defendant argued that sidecasting could not be addition, because the ordinary meaning of addition is the introduction of new material or an increase in the amount of material already there, while in sidecasting, there is no new material or increase in the amount of material.[315] The U.S. Court of Appeals for the Fourth Circuit concluded that the CWA does not prohibit the addition of "material," but rather prohibits the addition of a "pollutant." Earth and vegetable matter are not "dredged spoil" before they are excavated by defendants, but once they are removed from the ditch, they

309. *Avoyelles Sportsmen's League, Inc.*, 715 F.2d 897.
310. Avoyelles Sportsmen's League, Inc. v. Alexander, 473 F. Supp. 525, 536, 11 ELR 20315 (W.D. La. 1979).
311. *Avoyelles Sportsmen's League, Inc.*, 715 F.2d at 923.
312. *Id.*
313. 715 F.2d at 924, n.43.
314. 209 F.3d 331, 30 ELR 20508 (4th Cir. 2000).
315. *Deaton*, 209 F.3d at 335.

became "dredged spoil," which is a "pollutant." When defendants redeposit the material in the wetland, they add a pollutant where there had been no pollutant before.

The court's semantic sleight of hand is ingenious, almost metaphysical. But it ignores that much, if not all, of the soil and vegetative matter moved were "pollutants" under the statute's definition, even before they became dredged spoil.[316] The Supreme Court also impliedly rejected the theory in *Miccosukee* when it adopted the Second Circuit's *Catskill I* soup-ladle analogy for addition.[317] In it, the soup in the pot represents navigable water (whether one body of navigable water or unitary navigable waters). Let's say the soup is cream of asparagus soup; it consists of water and some biological material (cream, asparagus pieces, and asparagus residue), just as wetlands bottoms consists of water and biological material (wetlands vegetation and the organic portion of the soil). The ladle lifts soup from the pot and pours it back in, adding nothing, according to the Supreme Court and the Second Circuit. Why would pouring the cream and asparagus residue and pieces from the soup back into the soup pot add pollutants to the water where there were none before?[318] Because they were not pollutants before they were taken from the soup, but they were pollutants when they are replaced in the soup? That is a strained reading. Strained reading, of course, is far from plain meaning, the preferred reading of "addition."[319]

The Fourth Circuit in *Deaton* justified its reading of the statute by reference to the statute's purpose; dredged spoil is as harmful to the waters of the United States whether it was dredged from the same or different waters.[320] That, however, depends entirely on what the court means by harmful. If the harm is pollution, moving indigenous biological material, sand, or soil from one part of a natural wetland to another part of the same wetland is far less

---

316. The definition lists "biological material" as a pollutant and "biological material" includes vegetation and the organic portion of the soil. *See* §502(6). The definition states that biological material and other substances are pollutants when they are "discharged into water." Although the courts engaged in this metaphysical transformation do not cite the phrase, it does support their arguments. However, the phrase seems to be a meaningless redundancy. It certainly would have been an odd and roundabout way for Congress to have established that redepositing or sidecasting dredged spoil requires a §404 permit. When Congress enacted this definition, it considered §404 to regulate the disposal of spoil from river and harbor dredging far from the site of the dredging. In this context, the dredged material would not have been from the same place as the disposal site, and the metaphysical transformation would have been unnecessary.

317. South Fla. Water Mgmt. Dist. v. Miccosukee Tribe of Indians, 541 U.S. 95, 110, 34 ELR 20021 (2004); *Catskill I*, 273 F.3d 481, 492.

318. The D.C. Circuit rejected such a notion in *National Mining Ass'n v. U.S. Army Corps of Eng'rs*, 145 F.3d 1399, 28 ELR 21318 (D.C. Cir. 1998).

319. *See* Jarecki v. G.D. Searle & Co., 367 U.S. 303, 307-08 (1961) (where "[t]he statute admits a reasonable construction which gives effect to all of its provisions . . . we will not adopt a strained reading").

320. United States v. Deaton, 209 F.3d 331, 336, 30 ELR 20508 (4th Cir. 2000).

harmful than moving dredged spoil laden with heavy metals and toxic pollutants from a navigation channel or port bottom to a natural wetland, the initial focus of §404. Moreover, if the purpose of §404 is to prevent the loss of wetlands, the harm consists of the destruction of wetlands by digging drainage ditches in them, not regulated by §404, rather than by redepositing or sidecasting some of the wetlands dug to form the ditches, apparently regulated by §404.

Despite the strained reasoning of *Deaton,* subsequent sidecasting and similar decisions followed it. Addressing the argument that sidecasting adds nothing new to the wetland, the Sixth Circuit in *United States v. Cundiff*[321] combined *Deaton*'s metamorphosis theory with the *Avoyelles* conclusion that any other interpretation of addition would read §404 out of the statute.[322] *Cundiff* further supported its decision with a nod to *Chevron* deference,[323] but since neither EPA nor the Corps defined "addition" in regulations, it is not apparent how *Chevron* deference is relevant. A Michigan district court held in *United States v. Bay-Houston Towing Co., Inc.*[324] that sidecasting in the construction of drainage canals in a peat-harvesting operation was not incidental fallback, but instead was "purposeful relocation"—in other words, "addition."[325]

Borden Ranch Partnership v. U.S. Army Corps of Engineers[326] held that "deep ripping" of wetlands constitutes addition. Deep ripping pokes holes in a relatively impermeable layer of soil beneath a wetland, so that the wetland drains. In the process, the soil is "wrenched up, moved around, and redeposited somewhere else."[327] When the defendant argued that he added nothing new to the wetland, the Ninth Circuit cited *Deaton* and *Avoyelles* and rejected an argument that only incidental fallback took place. An Illinois district court in *United States v. Hummel*[328] even held that material excavated to create a ditch for sewer lines in a wetland was added to the wetland when the material was placed on top of the sewer lines to fill the ditch. Perhaps, the most extreme of these decisions is *United States v. M.C.C. of Florida, Inc.,*[329] where the Eleventh Circuit held that a tugboat propeller in essence dug a channel by repeatedly traversing the same area, pushing bottom sediment

321. 555 F.3d 200, 39 ELR 20025 (6th Cir. 2009).
322. *Id.* at 213-14. *See also Avoyelles,* 715 F.3d at 924, n.43.
323. *Cundiff,* 555 F.3d at 214.
324. 33 F. Supp. 2d 596, 29 ELR 21011 (E.D. Mich. 1999).
325. *Bay-Houston Towing,* 33 F. Supp. 2d at 605.
326. 261 F.3d 810, 32 ELR 20011 (9th Cir. 2001).
327. *Borden Ranch,* 261 F.3d at 815.
328. 2003 WL 1845365 (N.D. Ill 2003) (citing *Deaton* and distinguishing deliberate redeposit of material from incidental fallback).
329. 772 F.2d 1501, 15 ELR 21091 (11th Cir. 1985).

onto adjacent sea grass beds, thereby adding dredge spoil to the sea grass beds by redepositing the material onto the ocean bed, a dubious application of *Avoyelles.*[330] There are similar decisions,[331] and few contrary ones.[332]

2. Incidental Fallback

The "redeposit" decisions discussed above essentially hold that moving soil and biological material from one part of a wetland to another part of the same wetlands is an "addition" of the material to the other part of the wetland. As the points of removal and redeposit converge, however, that understanding becomes increasingly strained. When the two points are the same, the ordinary understanding of "addition" collapses, and the metaphysics of *Deaton* is all that remains. That helps explain why courts have drawn the "addition" line at incidental fallback.

Environmental plaintiffs challenged the Corps' initial exclusion from its regulatory definition of "discharge" the de minimis soil movement that occurs during normal dredging operations. The Corps settled the case and, pursuant to the settlement, replaced the exemption with explicit coverage of any redeposit of dredged materials including "incidental fallback."[333] The National Mining Association challenged the jurisdiction of the Corps to regulate "incidental fallback" as an "addition" of dredged or fill material. Incidental fallback occurs when material is excavated from water or wetlands by lifting and moving it elsewhere and some of the sediment incidentally falls back into the water or wetlands from which it is removed. The Corps argued that under *Avoyelles* this was "addition." But that decision involved removing substantial amounts of material and intentionally redepositing it some distance away from where it was removed. It did not involve removing sub-

330. *M.C.C. of Florida*, 772 F.2d at 1506. The material moved from developing a channel through grass beds was not added to navigable water for two reasons. First, it was not from the outside world. Second, the material was not a "pollutant" since it was not "discharged into water," §502 (6); it was already in the water and never left the water. The metamorphosis theory will not work either, for the sediment was never out of the water.

331. *See, e.g.*, United States v. Fabian, 522 F. Supp. 2d 1078 (N.D. Ind. 2007) (moving material, including earth, dirt, and sand laterally in a wetland constituted addition); United States v. Sinclair Oil Co., 767 F. Supp. 200, 21 ELR 21323 (D. Mont. 1990) (moving streambed material to redirect a channel constituted the redeposit or addition of fill material).

332. *See, e.g.*, United States v. Wilson, 133 F.3d 251, 28 ELR 20299 (4th Cir. 1997) (holding that sidecasting was not addition in a criminal prosecution; two of the three judges on the panel filed concurring opinions, and the decision appears superseded by the Fourth Circuit's *Deaton* decision); United States v. Hallmark Constr. Co., 30 F. Supp. 2d 1033 (N.D. Ill. 1998) (construction of a farm pond where the only discharge is incidental fallback did not constitute addition).

333. This history is recited in detail in *American Mining Cong. v. U.S. Army Corps of Eng'rs*, 951 Supp. 267, 269-71, 27 ELR 20589 (D.D.C. 1997).

stantial amounts of material and inadvertently redepositing a small amount of it at or very near the point of removal.

The D.C. district court in *American Mining Congress v. U.S. Army Corps of Engineers* held that "incidental fallback" was not "addition" for several reasons.[334] Most importantly, incidental fallback is an inevitable consequence of dredging, and §404 does not regulate dredging (removing), it only regulates filling; dredging is regulated by 33 U.S.C. §403. Second, legislative history indicates that Congress did not intend §404 to regulate "the small-volume incidental discharge that accompanies excavation and landclearing activities."[335] Third, the Corps took the position for 18 years prior to the challenged regulation, that §404 did not regulate incidental fallback,[336] a position also taken by courts.[337] Finally, Congress had rejected amendments to expand the jurisdiction of §404 to include such incidental discharge.[338]

The D.C. Court of Appeals affirmed,[339] rejecting the Corps' metamorphosis theory that wetlands sediment becomes a "pollutant" (dredged spoil) once it is removed from a wetland and thus when it falls back, it is added as a pollutant for the first time. The court stated that regardless of any "legal metamorphosis," it "fail[ed] to see how there can be an addition of *dredged material* when there is no addition of *material*," reasoning that "Congress could not have contemplated that the attempted removal of 100 tons of that substance could constitute an addition simply because only 99 tons of it were actually taken away."[340] The plaintiffs argued that the inability to regulate redeposit of dredged material would read §404 out of the statute, an observation made by the Fifth Circuit in *Avoyelles*.[341] The court responded the Corps could regulate some forms of redeposit, but not all redeposit, at least not incidental fallback redeposit.[342] Although the Corps promulgated an interim revised rule, it did not satisfy regulated industry, and skirmishes over the details of the incidental fallback exclusion continued.[343]

---

334. *Id.* at 273.
335. *Id.*
336. *Id.* at 274.
337. Salt Pond Assocs. v. U. S. Army Corps of Eng'rs, 815 F. Supp. 766, 23 ELR 21026 (D. Del. 1993); United States v. Lambert, 18 ERC 1294, 13 ELR 20045 (M.D. Fla. 1981), *aff'd*, 695 F.2d 536, 13 ELR 20436 (11th Cir. 1983).
338. *American Mining*, 951 F. Supp. at 276.
339. National Mining Ass'n v. U.S. Army Corps of Eng'rs, 145 F.3d 1399, 28 ELR 21318 (D.C. Cir. 1998).
340. *Id.* at 1404.
341. *Id.* at 1405. *See Avoyelles*, 715 F.2d at 924, n.43.
342. *National Mining*, 145 F.3d at 1405.
343. American Mining Cong. v. U.S. Army Corps of Eng'rs, 120 F. Supp. 2d 23 (D.D.C. 2000).

3. The Outside World, and the Unitary Navigable Waters Theories

Defendants in §404 cases could use EPA's outside world gloss on "addition" and its unitary navigable waters theory as the bases for arguing that §404 does not prohibit many unpermitted wetlands-clearing operations. When a land developer moves material from one part of a wetland, a navigable water, to another part of the same wetland, the same navigable water, the material arguably does not come from outside the wetland's or navigable water's world. This is underscored by the Supreme Court's adoption of the Second Circuit's analogy of taking a ladle of soup from a pot, lifting it from the pot, and pouring it back into the pot. Nothing is added to the pot. If EPA's "outside world" gloss is applied to movement of soil within the same wetlands, no §404 permit is required for most clearing and filling of wetlands and most reported decisions upholding CWA actions against such activity are ill-founded, unless one part of a wetland is "outside world" to another part of the same wetland.

This "outside world" gloss by itself, however, would not affect the original purpose of §404, which was to regulate disposal of material dredged from harbors and navigation channels in noncontiguous wetlands.[344] In such activities, dredged material is taken from one navigable water, a river or harbor, and discharged to another navigable water, an unrelated wetland, so that the material dredged is from outside the wetland, and a §404 permit would be required.

The combination of the "outside world" gloss and unitary navigable waters theory, however, undermines §404's applicability in both situations. Whether water and material discharged to a wetland originate from the same wetland, from a different wetland, or from anywhere in navigable water, if all navigable waters are one, the dredged materials do not come from the outside world. Even that, however, would not eliminate §404 entirely; it would still apply to filling wetlands with material not originating in water. But that is not what Congress intended. It enacted §404 in 1972 to apply to discharges of material from dredging river channels and harbors and discharging them into nonadjacent waters, including wetlands, and amended it in 1977 to affirm that it applied, inter alia, to filling wetlands, including by landclearing activities. The "outside world" gloss and the "unitary navigable water" theory are inconsistent with these congressional objectives.

---

344. "Congress understood 'discharge of dredged material' to mean open water disposal of material removed during the digging or deepening of navigable waterways." *See American Mining*, 951 F. Supp. at 273 (citing pertinent legislative history).

The Fifth Circuit in *Avoyelles* recognized in a footnote that the "outside world" theory could jeopardize the operation of §404, but did not address the issue because no party raised it.[345] The court surmised that no party raised the issue because "'dredged' material is by definition material that comes from the water and a requirement that all dredged spoil come from outside the water world would effectively remove the dredge-and-fill provision from the statute."[346] In other words, the Fifth Circuit suggested that the "outside world" theory could not apply to §404 because its application would render §404 without meaning.

Its logic is wrong on at least two grounds. First, the "outside world" theory would not eliminate the application of §404; the section would still apply to adding dredged spoil that originated from outside the wetland being filled, including the original target of §404, dredged spoil from harbors and navigation channels. Second, if the "outside world" theory did eviscerate §404, the most natural conclusion is that the theory is inconsistent with the CWA and the theory does not apply to any part of the statute. EPA's suggestion that the theory applies to §402, but does not apply to §404 requires interpreting "addition" differently under §402 and §404, counter to the canon requiring us to "interpret the same . . . terms in a statute in the same way."[347] It also ignores the fact that critical use of "addition" is in the definition of "discharge of a pollutant" in §301(a), which prohibits the discharge of a pollutant except in compliance with a §402 or §404 permit. EPA's argument would require interpreting the same word "addition" differently depending on whether the allegation was that it was not in compliance with a §402 or a §404 permit.

In response to comments that its proposed water transfer rule and the related unitary navigable waters theory would undermine §404, EPA adopted the Fifth Circuit's comments in the Agency's preamble to the final water transfer rule: "Because Congress explicitly forbade discharges of dredged material except as in compliance with . . . [§404], today's rule has no effect on the 404 permit program."[348] EPA's explanation misses the mark, as did the Fifth Circuit. The Fifth Circuit's observation merely establishes that dredged material is a "pollutant" and has nothing to do with whether there is an "addition" of that pollutant; all elements of the §301(a) prohibition must be met before a §404 permit is required. While EPA's "outside world" theory

---

345. Avoyelles Sportsmen's League, Inc. v. Marsh, 715 F.2d 897, 924 n.43, 13 ELR 20942 (5th Cir. 1983).
346. *Id.*
347. Eskridge, *supra* note 18, at 324, citing Sullivan v. Stroop, 496 U.S. 478, 484 (1990); United Sav. Ass'n. of Tex. v. Timbers of Inwood Forest Assts., Ltd., 484 U.S. 365 (1998). *See also* Scalia & Garner, *supra* note 18, at 170-73.
348. 73 Fed. Reg. 33697, 33703 (June 13, 2008).

of "addition" does not itself entirely eviscerate the §404 program, when combined with EPA's unitary navigable waters theory, they eviscerate most of the reach of §404. If all navigable waters are one, then moving dredged spoil from one part of a navigable water or wetland to another part of the same navigable water or to an altogether different wetland does not require a permit because the waters or wetlands are the same.

The current dominant purpose of §404 is to regulate the disposal of dredge and fill material into wetlands, most often material redeposited within a single wetland by landclearing activities. Courts have universally interpreted "addition" to include redepositing such material from one part of a wetland to another part of the same wetland. Their interpretation is compatible with our suggested definition of addition, as well as with EPA's "outside world" theory of addition, so long as the part of the wetland from which the material is taken is located at some remove from the part of the wetland into which the material is redeposited. Although courts are comfortable with this, they have balked in applying "addition" to situations where the place of extraction and the place of deposit are virtually identical, as with incidental fallback. Sidecasting is so close to incidental fallback that decisions interpreting "addition" to include sidecasting push the meaning of "addition" about as far as it can go. EPA's unitary navigable waters theory, however, would eliminate the application of §404 in cases in which spoil is moved from navigable water, including wetlands, to navigable water, including wetlands, whether the navigable water or wetlands are the same or different. EPA avoids this conclusion by stating that its unitary navigable waters theory applies to §402 but not to §404.

EPA's explanation falls before the canon that statutory terms are to be interpreted consistently throughout a statute, unless specifically provided otherwise in the statute. Moreover, the definition of "navigable waters" in §502 provides that "[e]xcept as otherwise specifically provided, when used in this chapter . . . [t]he term 'navigable waters' means . . . ." The CWA does not specifically provide that "navigable waters" has different meanings in §402 and §404. EPA's unitary navigable waters theory is incompatible with §404 and is therefore incompatible with the CWA. For that reason and others discussed in this chapter, EPA's unitary navigable waters theory must be rejected in its entirety.

IV. Conclusion

EPA has not promulgated a definition of "addition," although it has interpreted the element as addition "from the outside world" and as excluding water transfers. These interpretations neither define "addition" nor withstand scrutiny. Indeed, the Supreme Court has rejected the first concept in EPA's "outside world" theory, holding in *Miccosukee* that a point source need not be the original source of a pollutant for a point source to add it to navigable water. The Eleventh Circuit's *Miccosukee* decision suggested that "addition" incorporates the notion of but-for causation, a principled and useful distinction between actions that add or do not add pollutants. The definition suggested at the outset of this chapter incorporates that notion: "'addition' is the act of a person adding a pollutant to navigable waters from a point source, when that pollutant would not otherwise be in those navigable waters."

With few exceptions, when applied to decisions interpreting "addition" in the difficult fact patterns examined above, the definition suggested in this chapter either leads to results consistent with the decisions or may have done so if the decisions had made more-detailed findings of fact to address the causation aspect of "addition." Section 404 sidecasting redeposit decisions push "addition" to its outer limit and cross that limit when it comes to incidental fallback. EPA's "outside world" gloss on "addition," its unitary navigable waters theory, and the water transfer rule based on them are all contrary to the CWA's water quality standards program, eliminate most of the traditional jurisdiction of its §404 program, and are unsupported by the CWA. EPA should disavow the two theories and withdraw the rule. If EPA fails to do so, courts should reject the theories, overturn the rule, and give no deference to the Agency's interpretations supporting them. Finally, EPA should promulgate this chapter's suggested definition of "addition" in 40 C.F.R. §122.2.

Chapter 2. "Pollutant"

The Clean Water Act (CWA)[1] prohibits "the discharge of any pollutant by any person," unless in compliance with several listed sections in §301(a).[2] The listed sections authorize the issuance of two types of CWA permits and specify their substantive requirements.[3] In §502(12) the statute defines "discharge of a pollutant" to mean "any addition of any pollutant to navigable waters from any point source."[4] In sum, the subsection prohibits (1) any addition (2) of any pollutant (3) to navigable waters (4) from any point source (5) by any person, except in compliance with a CWA permit. U.S. Supreme Court Justice Ruth Bader Ginsburg has called this the "core command" of the CWA.[5]

The CWA defines "pollutant" to mean a list of 18 specific substances and categories of substances "discharged into water," with two specific exclusions in §502(6). The categories include biological and radioactive materials, and solid, chemical, industrial, municipal, and agricultural wastes. While many substances may not fall into one of the substance categories, when such substances are discharged into water, most become waste and fall within one of the waste categories. As a consequence, it is difficult to imagine a substance discharged into water that is not included in one of the categories. The definition of pollutant is qualitatively different in this regard from the definitions of two other elements defined in the CWA, "point source" and "navigable water."[6] Both of them have statutorily enunciated opposites. "Point sources" are confined, discrete conveyances, but the CWA recognizes their opposites: nonpoint sources such as surface runoff. "Navigable waters" may be a broad term, but the CWA recognizes their opposites: non-navigable waters such as groundwater. And, of course, water has its opposite: dry land. But what is the

1. Clean Water Act, 33 U.S.C. §§1251-1387, ELR Stat. FWPCA §§101-607.
2. 33 U.S.C. §1311(a).
3. The U.S. Environmental Protection Agency (EPA) issues permits under CWA §402, 33 U.S.C. §1242, to regulate water pollution, and the U.S. Army Corps of Engineers (the Corps) issues permits under CWA §404, 33 U.S.C. §1344, to regulate the filling of wetlands and other navigable waters.
4. 33 U.S.C. §1362(12).
5. Coeur Alaska, Inc. v. Southeast Alaska Conservation Council, 557 U.S. 261, 298 (2009). The author has elsewhere called the subsection "the basic prohibition" of the CWA. *See* Jeffrey G. Miller et al., Introduction to Environmental Law: Cases and Materials on Water Pollution Control 141 (2008).
6. CWA §502(7) & (14), 33 U.S.C. §1362(7) & (14).

opposite of a "pollutant"? A substance that does not pollute? We will see that the U.S. Congress divorced the meanings of pollutant and pollution. Or is the opposite of a pollutant a substance that is not listed or not part of a listed category? Name a substance that does not fall into a listed category when it is discharged into water.

The definition of pollutant is extremely broad. As a consequence, almost all decisions considering whether a particular substance is a pollutant should be answered positively. Because it is fairly clear that most substances are pollutants, fewer decisions should consider whether the pollutant element of the offense is met than whether some other elements of the offense are met. Finally, most courts should use the plain meaning of the definition to determine whether a substance is a pollutant. Once courts have held that a variety of substances are pollutants, precedent also should become a frequently used short-cut interpretive device. Courts should not often engage in extensive interpretations of pollutant. The interesting pollutant decisions, therefore, will be those holding substances not to be pollutants and decisions using multiple interpretive devices. These decisions should point to the fault lines between pollutants and their opposites, whatever they are. Or as Supreme Court Justice Sonia Sotomayor has commented, they are the "muddy, unclear and gray cases."[7]

The chapter begins by examining congressional actions illuminating the meaning of pollutant. It parses the statutory definitions of pollutant and related terms; identifies the contexts in which Congress used the term in the statute and how the term relates to other defined terms in the statute; and reviews relevant legislative history. Next, the chapter catalogues the substances that courts have held are, or are not, pollutants; and analyzes EPA and judicial interpretations of pollutant. Finally, the chapter identifies and discusses prominent or difficult issues that EPA and the courts have resolved in their interpretations.

7. Remarks by Supreme Court Justice Sonia Sotomayor at Pace Law School (Nov. 12, 2012) (commenting on cases that comprise the Supreme Court's docket).

I. Legislative and Administrative Definitions of Pollutant and Related Terms

A. *Statutory Definitions*

1. Pollutant

Although "pollutant" has a familiar common meaning as something that pollutes, §502(6) defines it to mean a list of 18 specific substances or categories of substances[8] "discharged into water," with two exclusions.[9] EPA's regulatory definition is virtually identical.[10] Because the definition states that pollutant means a list of specific substances or categories of substances, the list is an exclusive one and nothing else can be a pollutant.[11] The exclusive nature of this list is emphasized by its contrast with the only other statutory definition of an element containing a list, the definition of "point source," which "includes" a nonexclusive list of examples. Although the definition of pollutant is an exclusive one, meaning only the specific substances or categories of substances listed can be pollutants, the categories are broad enough to cover virtually all substances and wastes generated by human activity.[12] Because of the specificity of the listed substances and the breadth of the listed categories, courts should interpret the definition by using plain meaning more often than using other devices for statutory construction.[13] For the same reasons, courts should hold most substances

8. CWA §502(6), 33 U.S.C. §1362(6). Dredged spoil, incinerator residue, sewage sludge, munitions, wrecked or discarded equipment, rock, sand and cellar dirt (specific substances); solid waste, sewage, garbage, chemical wastes, biological materials, radioactive materials, industrial, municipal, and agriculture waste (categories of substances). Heat, which is included in the list, is not a substance or material, although it could be a waste. Some of the substances on the list could be considered either a material or a category of material, e.g., dredged spoil or sewage.
9. *Id.* The exclusions are: (1) sewage from vessels regulated by CWA §312, 33 U.S.C. §1322; and (2) water, gas, or other material pumped into wells to facilitate production of petroleum products or pumped onto wells for disposal of wastes pursuant to state permits.
10. It adds "filter backwash" to and drops "cellar dirt" from the list, qualifies "radioactive materials" as those not regulated under the Atomic Energy Act, 42 U.S.C. §§2011 et seq., and restates the two exceptions in minor ways.
11. An exclusive definition, using "means," is confined to the specifics of the definition, while an inclusive definition, using "includes," is not. *See* Sierra Club, Lone Star Chapter v. Cedar Point Oil Co., Inc., 73 F.3d 546, 565, 26 ELR 20522 (5th Cir. 1996). The difference is emphasized for the definition of pollutant because it contained an inclusive list in its U.S. Senate version, but was changed to contain an exclusive list in the Conference Committee. *See* ANTONIN SCALIA & BRYAN A. GARNER, READING LAW: THE INTERPRETATION OF LEGAL TEXTS 132-33 (2012).
12. The breadth of the categories included in the list "tends to eviscerate any restrictive effect" of the definition. *See Cedar Point*, 73 F.3d at 565; *see also* National Wildlife Fed'n v. Gorsuch, 693 F.2d 156, 172, 13 ELR 20015 (D.C. Cir. 1982).
13. Courts used plain meaning in 60 of the 70 decisions interpreting pollutant. Courts employed the next most commonly used canon, precedent, in 30 of those decisions. Count the number of times

to be pollutants, except under unusual circumstances. As discussed below, one such circumstance is when another statute regulates the substance or the activity producing it, potentially creating a conflict between application of the two statutes.

2. Toxic Pollutant

The CWA also defines "toxic pollutant" in §502(13).[14] It might be assumed that the CWA regulates only substances that have some toxic effect. This is not the case, however, because substances Congress listed as pollutants include nontoxic substances, e.g., rock and sand. Indeed, by defining and using the phrase "toxic pollutant," Congress identified a subset of pollutants subject to advanced pollution control. While §301(a) declared it illegal to discharge *any* pollutant except in compliance with a permit, and §301(b)(a) established treatment requirements for the discharge of *all* pollutants, §§301(b)(2)(A) and 307(a) established more stringent pollution control for the discharge of *toxic* pollutants. A substance may be a pollutant without being a toxic pollutant, although a substance may not be a toxic pollutant without being a pollutant.[15]

3. Pollution

The absence of any reference to pollution in the definition of "pollutant" has occasionally confused lawyers and judges because the common understanding of pollutant is a substance that causes pollution. The plain meaning of pollutant, however, is not relevant to interpreting the term as it is used in the CWA, because the statute specifically states that "[e]xcept as otherwise specifically provided, when used in this [statute] . . . [t]he term 'pollutant' means . . ." the definition discussed above. The list of materials or categories of materials that are pollutants includes substances that in the common understanding do not pollute, for example, rock, sand, and some biological material such as indigenous live fish. Few would argue that their favorite waterway is free of pollutants or pollution only when it

that interpretive devices 22 and 23 are cited *infra* in Column E of Table B for Pollutant.

14. The term "means those pollutants or combinations of pollutants . . . which . . . cause death, disease, behavioral abnormalities, cancer, genetic mutations, physiological malfunctions . . . or physical deformations, in such organisms or their offspring."

15. Dague v. City of Burlington, 732 F. Supp. 458, 469-70, 20 ELR 21001 (D. Vt. 1989), *aff'd*, 935 F.2d 1343, 21 ELR 21133 (2d Cir. 1991), *rev'd on other grounds*, 505 U.S. 557, 22 ELR 21099 (1992). In *Headwaters, Inc. v. Talent Irrigation Dist.*, 243 F.3d 526, 532-33, 31 ELR 20535 (9th Cir. 2001), the U.S. Court of Appeals for the Ninth Circuit commented regarding a pesticide that "it would be absurd to conclude that a toxic chemical directly poured into water is not a pollutant."

is without native fish swimming in it or a sand or rock bottom. Congress was aware of the concept of pollution when it drafted the CWA; indeed, Congress defined pollution and used it in the title[16] and throughout the statute.[17] The CWA defined pollution as the "man-induced alteration of the chemical, physical, biological, and radiological integrity of water,"[18] without mentioning pollutant. Thus, Congress decoupled "pollutant" and "pollution" by defining each without reference to the other. A "pollutant" is a substance falling within a statutory list of substances or categories of substances, without regard to whether it causes pollution. "Pollution" is a negative effect on the integrity of water caused by human activity, without regard to whether it is caused by a pollutant. It is worth noting that plain meaning is useful in interpreting the words used in the statutory definition of pollutant, although it is not useful in interpreting the word pollutant as it is used in the statute.

B. *Legislative History*

The legislative history of the CWA indicates that Congress intended its definition of pollutant to be expansive, leaving EPA with the authority to define what the term means under particular circumstances. Section 502(6) incorporated the definition of pollutant from the U.S. Senate bill, S. 2770, with a few changes.[19] The Senate Committee Report accompanying S. 2770 indicated that the Senate bill defined pollutant to avoid litigation over whether particular materials were subject to CWA jurisdiction, suggesting a broad interpretation of the term. That suggestion is supported by the Committee's explicit adoption of "the basic formula [from the Refuse Act, but adding]

16. The original Federal Water Pollution Control Act was enacted on June 30, 1948, c. 758, 62 Stat. 1155, and was amended several times before the Federal Water Pollution Control Amendments of 1972, Oct. 18, 1972, Pub. L. No. 92-500, cast the statute substantially in its present form. The Clean Water Act Amendments of 1977, Dec. 27, 1977, Pub. L. No. 95-217, made some mid-course corrections to the 1972 legislation. The statute has commonly been referred to thereafter as the Clean Water Act.
17. For instance, the first section used "pollution" six times, CWA §101, 33 U.S.C. §1251, and the penultimate section used it four times, CWA §60, 33 U.S.C. §1386.
18. CWA §502(19).
19. The enacted statutory definition eliminated from the Senate's version "but not limited to" after "means," and "other waste" after "agricultural waste." S. Rep. No. 1236 (Conf. Rep.), at 143-44 (1972), 1 Legis. Hist. 281, 326-27. The U.S. Court of Appeals for the Fifth Circuit commented that "while . . . the elimination of [these phrases] may be interpreted as an attempt to limit the scope of the definition . . . we think that the retention of such broad terms in the definitions suggests that the conference committee may have determined that the eliminated phrases were simply redundant," in view of the broad coverage of "'solid waste,' . . . 'chemical wastes,' 'biological materials,' . . . and 'industrial, municipal, and agricultural waste. . . .'" Sierra Club, Lone Star Chapter v. Cedar Point Oil Co., Inc., 73 F.3d 546, 566, 26 ELR 20522 (5th Cir. 1996). *See also* National Wildlife Fed'n v. Gorsuch, 693 F.2d 156, 173, 13 ELR 20015 (D.C. Cir. 1982).

. . . municipal discharges to it, so [that] before *any* material can be added to navigable waters authorization must first be granted by the Administrator."[20] The Refuse Act of 1899[21] was indeed broad, prohibiting the discharge of "*any* refuse matter of *any* kind or description *whatever*" to navigable water or its tributaries without a permit, with the exception of liquid waste from streets or sewers.[22] Adding discharges from streets and sewers to the Refuse Act's already expansive concept of regulated substances provided the CWA with an even broader reach of "pollutant."

On the Senate floor, Sen. Ted Stevens (R-Alaska) asked Sen. Edmund Muskie (D-Me.), the author and chief sponsor of the CWA in the Senate, whether "pollutant" included fish parts discharged in various marine environments. Senator Muskie replied: "I do not get into the business of defining or applying these definitions to particular kinds of pollutants. That is an administrative decision to be made by the Administrator [of EPA]."[23] Courts have generally recognized the authority of EPA to define "pollutant" and other jurisdictional terms of §301(a),[24] based on this and similar evidence.[25] In *National Wildlife Federation v. Gorsuch*, the U.S. Court of Appeals for the District of Columbia (D.C.) Circuit stated: "Strong signals in the Congressional history [indicate] that [Congress] entrusted EPA with at least some discretion which 'pollutants' and sources of pollutants were to be regulated under the NPDES program."[26]

20. S. Rep. No. 92-414, 1972 U.S.C.C.A.N. 3742 (emphasis added). *See* United States v. Hamel, 551 F.2d 107, 110-11, 7 ELR 20253 (6th Cir. 1977).
21. 33 U.S.C. §407 (emphasis added). The Refuse Act provided the model for the permit program established in CWA §402.
22. *See* United States v. Pennsylvania Indus. Chem. Corp., 411 U.S. 655, 663-66, 3 ELR 20401 (1973) (tracing the evolution of the Refuse Act and its broad coverage).
23. 117 Cong. Rec. 38838 (daily ed. Nov. 2, 1971).
24. *See Gorsuch*, 693 F.2d at 167, 173-74. Based on a similar congressional colloquy, the district court in *Natural Res. Defense Council v. Train*, 396 F. Supp. 1393, 1396, 1401, 5 ELR 20401 (D.D.C. 1975), held that the CWA gave EPA discretion to define the meaning of "point source." The D.C. Circuit agreed. *See* Natural Res. Defense Council, Inc. v. Costle, 568 F.2d 1369, 1382, 8 ELR 20028 (D.C. Cir. 1977).
25. Some courts have articulated their authority to make such determinations in the absence of an EPA determination. *See, e.g.*, Sierra Club, Lone Star Chapter v. Cedar Point Oil Co., Inc., 73 F.3d 546, 26 ELR 20522 (5th Cir. 1996); U.S. PIRG v. Atlantic Salmon of Maine, LLC, 215 F. Supp. 2d 239, 246, 32 ELR 20535 (D. Me. 2002).
26. *See Gorsuch*, 693 F.2d at 173.

II. Judicial Interpretations of Pollutant

A. *Substances Held to Be a Pollutant or Not to Be a Pollutant*

1. Held to Be a Pollutant

Courts have held the following substances to be a "pollutant": acid mine drainage[27]; blood[28]; bombs, ordnance, and spent shot[29]; cement and shotcrete[30]; changes in water conditions[31]; chemical waste, including chlorine residue and alum sludge[32]; demolition debris[33]; dredged and fill material[34]; fecal coliform[35]; live and dead fish, fish parts and feces, and shellfish feces[36];

---

27. Committee to Save Mokelumne River v. East Bay Mun. Util. Dist., 13 F.3d 305, 24 ELR 20225 (9th Cir. 1993); Beartooth Alliance v. Crown Butte Mines, 904 F. Supp. 1168, 26 ELR 20639 (D. Mont. 1995).
28. United States v. Plaza Health Laboratories, 3 F.3d 643, 23 ELR 21526 (2d Cir. 1993).
29. Weinberger v. Barcelo-Romero, 456 U.S. 305, 309, 12 ELR 20538 (1982) (bombs); Long Island Soundkeeper Fund, Inc. v. New York Athletic Club, 1996 WL 131863 (S.D.N.Y. 1996) (spent shot and target fragments).
30. United States v. Pozsgai, 999 F.2d 719, 23 ELR 21012 (3d Cir. 1993) (concrete rubble and cement blocks); United States v. Schalloom, 998 F.2d 196 (4th Cir. 1993).
31. National Wildlife Fed'n v. Gorsuch, 530 F. Supp. 1291, 12 ELR 20268 (D.D.C. 1982), *rev'd*, 693 F.2d 156, 13 ELR 20015 (D.C. Cir. 1982); South Carolina Wildlife Fed'n v. Alexander, 457 F. Supp. 118, 8 ELR 20757 (D.S.C. 1978).
32. U.S. Steel Corp. v. Train, 556 F.2d 822, 7 ELR 20419 (7th Cir. 1977); Hudson River Fishermen's Ass'n v. City of New York, 751 F. Supp. 1088, 21 ELR 20647 (S.D.N.Y. 1990).
33. United States v. Bradshaw, 541 F. Supp. 880, 12 ELR 20629 (D. Md. 1981).
34. Coeur Alaska, Inc. v. Southeast Alaska Conservation Council, 557 U.S. 261, 302-03 (2009); Rapanos v. United States, 547 U.S. 715, 723, 745, 36 ELR 20116 (2006); Leslie Salt Co. v. United States, 55 F.3d 1388, 1391, 25 ELR 21046 (9th Cir. 1995); Town of Norfolk v. U.S. Army Corps of Eng'rs, 968 F.2d 1438, 1445, 22 ELR 21337 (1st Cir. 1992); Bersani v. EPA, 850 F.2d 36, 39, 18 ELR 20874 (2d Cir. 1988); United States v. Brink, 795 F. Supp. 2d 565, 575 (S.D. Tex. 2011); Gouger v. U.S. Army Corps of Eng'rs, 779 F. Supp. 2d 588, 603 (S.D. Tex. 2011); Alabama Rivers Alliance v. U.S. Army Corps of Eng'rs, 697 F. Supp. 2d 1251, 1258 (N.D. Ala. 2009); D'Olive Bay Restoration and Pres. Comm., Inc. v. U.S. Army Corps of Eng'rs, 513 F. Supp. 2d 1261, 1268 (S.D. Ala. 2007); Florida Wildlife Fed'n v. U.S. Army Corps of Eng'rs, 401 F. Supp. 2d 1298, 1308 (S.D. Fla. 2005); City of Shoreacres v. Waterworth, 332 F. Supp. 2d 992, 1015 (S.D. Tex. 2004); National Wildlife Fed'n v. Norton, 332 F. Supp. 2d 170, 185 (D.D.C. 2004); Kentuckians for Commonwealth, Inc. v. Rivenburgh, 204 F. Supp. 2d 927, 932, n.5 (S.D.W. Va. 2002), *rev'd on other grounds*, 317 F.3d 425 (4th Cir. 2003).
35. North Carolina Shellfish Growers Ass'n v. Holly Ridge Assocs., LLC, 278 F. Supp. 2d 654 (E.D.N.C. 2003) (soil and vegetation).
36. National Wildlife Fed'n v. Consumers Power Co., 657 F. Supp. 989, 17 ELR 20801 (W.D. Mich. 1987) (live fish, dead fish, and fish parts), *rev'd on other grounds*, 862 F.2d 580, 19 ELR 20235 (6th Cir. 1988); U.S. PIRG v. Atlantic Salmon of Maine, LLC, 215 F. Supp. 2d 239, 32 ELR 20535 (D. Me. 2002) (non-native live fish, fish feces); *contra* Association to Protect Hammersley, Eld, and Totten Inlets v. Taylor Res., Inc., 299 F.3d 1007 (9th Cir. 2002) (shellfish feces).

listed toxic pollutants[37]; animal manure[38]; municipal solid waste[39]; pesticides, pesticide waste, and pesticide residue[40]; petroleum products[41]; produced water[42]; rock, rubble, bricks, and sand[43]; sediment[44]; sewage[45]; soil and vegetation[46]; and stormwater.[47]

37. Dague v. City of Burlington, 732 F. Supp. 458, 469-70, 20 ELR 21001 (D. Vt. 1989).
38. Concerned Area Residents for the Environment v. Southview Farm, 34 F.3d 114, 24 ELR 21480 (2d Cir. 1994) (cow manure); Higbee v. Starr, 598 F. Supp. 323 (W.D. Ark. 1984) (hog manure); United States v. Frezzo Bros., 461 F. Supp. 266, 9 ELR 20139 (E.D. Pa. 1978), *aff'd*, 602 F.2d 1123, 9 ELR 20556 (3d Cir. 1979) (composted horse manure).
39. Resource Invs., Inc. v. U.S. Army Corps of Eng'rs, 151 F.3d 1162, 28 ELR 21407 (9th Cir. 1998).
40. National Cotton Council of Am. v. U.S. EPA, 553 F.3d 927, 39 ELR 20006 (6th Cir. 2009) (sprayed pesticide residue and waste); Headwaters, Inc. v. Talent Irrigation Dist., 243 F.3d 526, 31 ELR 20535 (9th Cir. 2001).
41. United States v. Eidson, 108 F.3d 1336, 27 ELR 20853 (11th Cir. 1997); United States v. Hamel, 551 F.2d 107, 7 ELR 20253 (6th Cir. 1977).
42. Produced water is pumped from groundwater or accompanies other products produced by drilling (e.g., oil and gas). Northern Plains Resource Council v. Fidelity Exploration and Dev. Co. 325 F.3d 1155 (9th Cir. 2003) (unaltered groundwater); Sierra Club, Lone Star Chapter v. Cedar Point Oil Co., Inc., 73 F.3d 546, 26 ELR 20522 (5th Cir. 1996) (produced water or its constituents).
43. Coeur Alaska, Inc. v. Southeast Alaska Conservation Council, 557 U.S. 261, 268 (2009) (crushed rock); United States v. Pozsgai, 999 F.2d 719, 23 ELR 21012 (3d Cir. 1993) (concrete rubble, cement blocks); Minnehana Creek Watershed Dist. v. Hoffman, 597 F.2d 617, 9 ELR 20334 (8th Cir. 1979) (rock and sand used in constructing dam and rip rap); Kentuckians for Commonwealth, Inc. v. Rivenburgh, 204 F. Supp. 2d 927, 932, n.5 (S.D.W. Va. 2002) (mining overburden agreed by parties to be a pollutant), *rev'd on other grounds*, 317 F.3d 425 (4th Cir. 2003); Hanson v. United States, 710 F. Supp. 1105, 19 ELR 21074 (E.D. Tex. 1989) (dirt, rock, bricks); United States v. Bradshaw, 541 F. Supp. 880, 12 ELR 20629 (D. Md. 1981 (demolition debris and sand).
44. Borden Ranch P'ship v. U.S. Army Corps of Eng'rs, 261 F.3d 810, 32 ELR 20011 (9th Cir. 2001) (soil and vegetation); United States v. Deaton, 209 F.3d 331, 30 ELR 20508 (4th Cir. 2000) (soil); Driscoll v. Adams, 181 F.3d 1285, 29 ELR 21387 (11th Cir. 1999) (sand and silt were the two primary constituents of sediment); United States v. Wilson, 133 F.3d 251, 28 ELR 20299 (4th Cir. 1997) (native soil); Rybachek v. EPA, 904 F.2d 1276, 20 ELR 20973 (9th Cir. 1990) (dirt and gravel); United States v. M.C.C. of Florida, Inc., 772 F.2d 1501, 15 ELR 21091 (11th Cir. 1985) (vegetation and sediment); Hughey v. JMS Dev. Corp, 78 F.3d 1523, 26 ELR 20924 (11th Cir. 1996) (sediment in rainwater flowing from construction site); *Minnehaha Creek Watershed Dist.*, 597 F.2d 617 (rock and sand); Tungett v. Papierski, 2006 WL 51148 (E.D. Tenn. 2006) (sediment, soil, dirt); North Carolina Shellfish Growers Ass'n v. Holly Ridge Assocs., LLC, 278 F. Supp. 2d 654 (E.D.N.C. 2003) (soil, sand, and dirt); Pronsolino v. Marcus, 91 F. Supp. 2d 1337, 30 ELR 20460 (N.D. Cal. 2000); *Bradshaw*, 541 F. Supp. 880 (demolition debris and sand).
45. United States v. Gulf Park Water Co. 972 F. Supp. 1056 (S.D. Miss. 1997).
46. *Borden Ranch*, 261 F.3d 810 (vegetation); *M.C.C. of Florida*, 772 F.2d 1501 (sea grass); *Tungett*, 2006 WL 51148 (trees, organic debris).
47. *Driscoll*, 181 F.3d 1285; *Hughey*, 78 F.3d 1523 (construction site runoff); United States v. Frezzo Bros., 461 F. Supp. 266, 9 ELR 20139 (E.D. Pa. 1978) (mushroom growing waste runoff), *aff'd*, 602 F.2d 1123, 9 ELR 20556 (3d Cir. 1979).

2. Held Not to Be a Pollutant

Courts have held the following substances not to be a "pollutant": air pollutants[48]; changes in water condition[49]; clear water[50]; pesticides, pesticide wastes and pesticide residue[51]; radioactive materials[52]; rock and sediment in stormwater[53]; and shellfish parts and feces.[54] For four of these seven substances or types of substances, there are contrary decisions. Ten of the 13 narrow interpretations were rendered in citizen suits[55] and nine of the 13 were decided during or after 1996.[56] Most of the negative decisions were grounded on avoiding interference with pervasive regulation by another statute of the substance at issue or the activity producing it.

B. *Listed Substances and Listed Categories of Substances*

Some decisions held a substance to be a pollutant because it is specifically listed in the definition of pollutant.[57] But most decisions held a substance to be a pollutant because it fell within a category of substance listed in the

48. U.S. EPA ex rel. McKeown v. Port Auth. of New York and New Jersey, 162 F. Supp. 2d 173 (S.D.N.Y. 2001). *See also* Chemical Weapons Working Grp. v. Dep't of Army, 111 F.3d 1485, 27 ELR 21130 (10th Cir. 1997).
49. National Wildlife Fed'n v. Gorsuch, 693 F.2d 156, 13 ELR 20015 (D.C. Cir. 1982); *contra* South Carolina Wildlife Fed'n v. Alexander, 457 F. Supp. 118, 8 ELR 20757 (D.S.C. 1978).
50. Orleans Audubon Soc'y v. Lee, 742 F.2d 901, 15 ELR 20030 (5th Cir. 1984).
51. Fairhurst v. Hagener, 422 F.3d 1146 (9th Cir. 2005); Peconic Baykeeper, Inc. v. Suffolk Cnty., 585 F. Supp. 2d 377 (E.D.N.Y. 2008), *rev'd in part*, 600 F.3d 180 (2d Cir. 2010); Altman v. Town of Amherst, New York, 190 F. Supp. 2d 467 (W.D.N.Y. 2001), *rev'd*, 47 Fed. Appx. 62 (2d Cir. 2002).
52. Train v. Colorado PIRG, 426 U.S. 1, 6 ELR 20549 (1976); Waste Action Project v. Dawn Mining Corp., 137 F.3d 1426, 28 ELR 21035 (9th Cir. 1998).
53. Kentuckians for Commonwealth, Inc. v. Rivenburgh, 317 F.3d 425 (4th Cir. 2003); Town of Norfolk v. U.S. Army Corps of Eng'rs, 968 F.2d 1438, 22 ELR 21337 (1st Cir. 1992); United States v. United Homes, Inc., 1999 WL 117701 (N.D. Ind. 1999). The court did not address whether contaminated river water was a pollutant in *Hudson River Fishermen's Ass'n v. City of New York*, 751 F. Supp. 1088, 1102, 21 ELR 20647 (S.D.N.Y. 1990), *aff'd without opinion*, 940 F.2d 649, 21 ELR 21226 (2d Cir. 1991).
54. Association to Protect Hammersley, Eld, and Totten Inlets v. Taylor Res., Inc., 299 F.3d 1007 (9th Cir. 2002); *contra* National Wildlife Fed'n v. Consumers Power Co., 862 F.2d 580, 19 ELR 20235 (W.D. Mich. 1987); U.S. PIRG v. Atlantic Salmon of Maine, LLC, 215 F. Supp. 2d 239, 32 ELR 20535 (D. Me. 2002).
55. See *supra* notes 48-54, for decisions with narrow interpretations, and *infra* Column D of Table B, for Pollutant for decisions in citizen suits.
56. By contrast, the author's research has found that 37 of the 68 decisions were during or after 1996, *see supra* decisions identified in notes 48-54.
57. *See, e.g.*, Weinberger v. Barcelo-Romero, 456 U.S. 305, 309, 12 ELR 20538 (1982).

definition,[58] frequently biological material,[59] or chemical,[60] agricultural,[61] solid[62] or industrial waste.[63] "Biological material" covers all living or formerly living carbonaceous matter, including substances in fossilized form such as a fossil fuel. "Solid waste," "chemical waste," "industrial, municipal, and agricultural waste," however, describe the byproducts or leftover results of human activity. Indeed, they cover most conceivable byproducts and residuals of human activity. Moreover, when most non-waste materials are discharged to water, they become waste. For instance, commercially valuable petroleum products become waste when they are spilled from a vessel at sea.[64] Although not noted in the decisions, a substance could fall into more than one listed category. Leftover tar from road paving, for instance, could be biological material, chemical waste, solid waste, municipal waste, and industrial waste. In the few decisions holding a substance not to be a pollutant, the substance at issue might have fit within a listed category, but the court nevertheless held it not to be a pollutant for other reasons.[65]

---

58. Congress' use of "broad generic terms" indicated legislative intent to capture as pollutants more than the substances specifically listed. *See* United States v. Hamel, 551 F.2d 107, 7 ELR 20253 (6th Cir. 1977); *Hudson River Fishermen's Ass'n*, 751 F. Supp. at 1101.
59. National Cotton Council of Am. v. U.S. EPA, 553 F.3d 927, 39 ELR 20006 (6th Cir. 2009) (biologically based pesticides); *Consumers Power*, 862 F.2d 580 (live fish, dead fish, and fish parts); North Carolina Shellfish Growers Ass'n v. Holly Ridge Assocs., LLC, 278 F. Supp. 2d 654 (E.D.N.C. 2003) (fecal coliform); *Atlantic Salmon of Maine*, 215 F Supp. 2d 239 (fish feces); National Wildlife Fed'n v. Consumers Power Co., 657 F. Supp. 989, 17 ELR 20801 (W.D. Mich. 1987) (live fish, dead fish, and fish parts); United States v. Frezzo Bros., 461 F. Supp. 266, 9 ELR 20139 (E.D. Pa. 1978) (composted horse manure). In the district court decision upheld in *Hamel*, 551 F.2d at 110, "the government contended successfully . . . that gasoline could be subsumed under 'biological materials,'" although the appeals court upheld the decision on a different basis.
60. *National Cotton Council*, 553 F.3d 927 (chemical pesticide residue); Headwaters, Inc. v. Talent Irrigation Dist., 243 F.3d 526, 31 ELR 20535 (9th Cir. 2001) (chemical pesticide residue); Sierra Club, Lone Star Chapter v. Cedar Point Oil Co., Inc., 73 F.3d 546, 26 ELR 20522 (5th Cir. 1996) (produced water); U.S. Steel Corp. v. Train, 556 F.2d 822, 7 ELR 20419 (7th Cir. 1977) (steel-making waste); *North Carolina Shellfish Growers*, 278 F. Supp. 2d 654; Dague v. City of Burlington, 732 F. Supp. 458, 20 ELR 21001 (D. Vt. 1989), *aff'd*, 935 F.2d 1343, 21 ELR 21133 (2d Cir. 1991), *rev'd on other grounds*, 505 U.S. 557, 22 ELR 21099 (1992) (leachate from landfill).
61. Concerned Area Residents for the Environment v. Southview Farm, 34 F.3d 114, 24 ELR 21480 (2d Cir. 1994) (cow manure); *Atlantic Salmon of Maine*, 215 F. Supp. 2d at 246 (waste from fish farm); Higbee v. Starr, 598 F. Supp. 323 (W.D. Ark. 1984).
62. United States v. Schallom, 998 F.2d 196 (4th Cir. 1993) (sand and cement); *Higbee*, 598 F. Supp. 223.
63. Northern Plains Resource Council v. Fidelity Exploration and Dev. Co. 325 F.3d 1155 (9th Cir. 2003) (produced water from gas extraction); United States v. Eidson, 108 F.3d 1336, 27 ELR 20853 (11th Cir. 1997) (tank bottoms pumped from gas station); *Cedar Point*, 73 F.3d 546 (produced water from oil drilling).
64. *See infra* Sections II.B.1.b.
65. For instance, in *Association to Protect Hammersley, Eld and Totten Inlets v. Taylor Res., Inc.*, 299 F.3d 1007 (9th Cir. 2002), the Ninth Circuit held that feces and shells from mussel-growing and harvesting facilities were not pollutants, although they were biological material. The court was largely motivated by the incongruity of holding the byproducts of shellfish propagation to be pollutants, when the goal of the CWA is to protect and promote the propagation of fish and shellfish. The Supreme Court in

I. Waste Categories and Material Categories

Most of the broad categories of pollutants are designated "waste(s)" or "materials."[66] Although the CWA does not define either term, material is a broader category than waste.[67] Material describes the substance before, during, and after human use and even a substance that is not associated with human use.[68] Waste, on the other hand, describes material that remains after human use or after abandonment without use, such as oil spilled from a vessel at sea.[69] One court, however, has held that both waste and material must be associated with human or industrial activity to be a pollutant, using the *ejusdem generis*[70] canon of construction and noting that the substances listed in the definition of pollutant were associated with human or industrial activity.[71] That analysis, however, is flawed, as discussed below.[72]

Why does the definition of pollutant include all material of a biological nature, but only waste material of a solid, chemical, industrial, municipal, or agricultural nature? When do the latter materials become waste? Although

---

Train v. Colorado PIRG, 426 U.S. 1, 6 ELR 20549 (1976), and the Ninth Circuit in *Waste Action Project v. Dawn Mining Corp.*, 137 F.3d 1426, 28 ELR 21035 (9th Cir. 1998), held radioactive source, byproduct and special nuclear materials not to be pollutants, despite the specific listing of "radioactive materials" in the CWA definition of "pollutant," because those radioactive materials were already regulated by the Atomic Energy Act, which Congress intended to be the sole regulator of such materials.

66. "Materials" are listed in the plural. "Chemical wastes" is in the plural, but the other categories of waste are in the singular. There is no apparent reason why some of the categories are singular and others plural. None of the courts interpreting pollutant found this difference to be significant nor do any of them note that a party to the cases argued the difference was meaningful. This may not be surprising because in the absence of a contrary indication in the statute, the singular includes the plural and the plural includes the singular. *See* SCALIA & GARNER, *supra* note 11, at 129-31, where the authors trace this canon back to Blackstone and Bentham and a 1278 English case. *See also* 1 U.S.C. §1, which provides that unless otherwise provided, singular nouns in the *United States Code* include the plural and vice versa.
67. For instance, hazardous materials regulated by the Hazardous Materials Transportation Act, 49 U.S.C. §§5101 et seq., are more extensive, see §5102(2) and §5103(a), than hazardous wastes regulated by RCRA, 42 U.S.C. §6903(5).
68. "Material" means "(1) the elements, constituents, or substances of which something is composed or can be made (2) matter that has qualities which give it individuality and by which it may be categorized." MERRIAM-WEBSTER'S COLLEGIATE DICTIONARY 715 (10th ed.).
69. "Waste" means "damaged, defective, or superfluous material produced by a manufacturing process: as . . . (3) an unwanted by-product of a manufacturing process, chemical laboratory or nuclear reactor . . . refuse from places of human or animal habitation . . . Garbage: Rubbish (2) Excrement . . . Sewage." *Id.* at 1329. *See Northern Plains Research Council*, 325 F.3d at 1161.
70. The *ejusdem generis* canon has been defined in a treatise as "[i]nterpret[ing] a general term to reflect the class of objects reflected in the more specific terms accompanying it." *See* WILLIAM N. ESKRIDGE JR., DYNAMIC STATUTORY INTERPRETATION 323 (1987).
71. Association to Protect Hammersley, Eld, and Totten Inlets v. Taylor Res., Inc., 299 F.3d 1007 (9th Cir. 2002).
72. For instance, while some sand and rock, listed as pollutants, may be associated with human activity, most are not.

the case law does not directly address these questions under the CWA, a few decisions examine whether particular substances are solid waste under the Resource Conservation and Recovery Act (RCRA),[73] the statute primarily designed to regulate the management and disposal of solid and hazardous waste. Whether a substance is a solid waste under RCRA has proven to be one of the most difficult legal questions under that statute,[74] suggesting the question could be a difficult one under the CWA.

a. Biological Materials

The term "biological material" raises dilemmas that courts have addressed directly or indirectly. The U.S. Court of Appeals for the Sixth Circuit used plain meaning to hold that fish parts and whole fish, dead or alive, were biological material and therefore pollutants in *National Wildlife Federation v. Consumers Power Co.*[75] There, an electric-generation facility withdrew water with live whole fish from a lake, ran the water through a turbine to produce electricity, and discharged the water back into the lake with a puree of dead fish parts, dead fish, and some live fish. A district court followed the Sixth Circuit's lead, holding that a salmon farm discharged pollutants, including live fish.[76] Finally, in *Association to Protect Hammersley, Eld, and Totten Inlets v. Taylor Resources, Inc.*,[77] the U.S. Court of Appeals for the Ninth Circuit found "biological material" to be ambiguous, possibly meaning all biological material or just biological waste resulting from human or industrial activity. It held that mussel shells and feces from a mussel-harvesting operation were not pollutants because one purpose of the CWA was the propagation of shellfish.[78] Although the reasoning of *Hammersley* is flawed, as discussed immediately below, the three decisions can perhaps be reconciled if the lake fish in *Consumers Power* were indigenous, the salmon in *Atlantic Salmon* were

73. Resource Conservation and Recovery Act (RCRA), 42 U.S.C. §§6901-6992k, ELR Stat. RCRA §§1001-11011.
74. RCRA defines "solid waste" in 42 U.S.C. §6904; EPA elaborates the definition in 40 C.F.R. §261.2. Under RCRA, hazardous waste is a subset of solid waste. Determining whether material is a solid waste is often the most difficult legal issue under RCRA, primarily because one person's waste is often another person's raw material. *See* Jeffrey G. Miller & Craig N. Johnston, The Law of Hazardous Waste Disposal and Remediation 149-71 (2d ed.). RCRA also regulates the disposal of nonhazardous solid waste, but not as comprehensively as hazardous waste.
75. National Wildlife Fed'n v. Consumers Power Co., 862 F.2d 580, 583, 19 ELR 20235 (6th Cir. 1988). The Sixth Circuit had earlier held oil to be a pollutant, although by a more circuitous route because oil is a biological material, perhaps hoping to avoid the question of whether fish and people are pollutants under the CWA. United States v. Hamel, 551 F.2d 107, 7 ELR 20253 (6th Cir. 1977).
76. U.S. PIRG v. Atlantic Salmon of Maine, LLC, 215 F. Supp. 2d 239, 246-48, 32 ELR 20535 (D. Me. 2002).
77. 299 F.3d 1007 (9th Cir. 2002).
78. *Id.* at 1015-17.

not indigenous, and the origin of the mussels in *Hammersley* was ambiguous. Such a reconciliation of the decisions would serve the statutory purpose of providing a "balanced, *indigenous* population of fish, shellfish and wildlife,"[79] although none of the courts made this distinction or suggested any other reconciliation. Giving meaning to the distinction between indigenous and non-indigenous species, however, would raise the subsidiary question of how long species must be in place to be indigenous.

In *Hammersley* the court held mussel shells and feces from a mussel-harvesting operation were not pollutants, even though they were biological material, because to be a pollutant, a substance must be a waste from a human or industrial activity. It came to that conclusion using two interpretive devices: *ejusdem generis* and avoiding absurd results.[80] *Ejusdem generis* posits that if a statute lists examples of regulated substances, other substances must be of the same nature as the listed substances to be regulated by the statute. The court observed that the examples listed in the definition of "pollutant" were all wastes from human or industrial activity and that the mussel shells and feces were not, hence the shells and feces could not be "pollutants."

The court was wrong on several counts. First, the statutory definition listed "biological material," not "biological waste." As discussed above, "material," of course, is different from and extends beyond "waste." Congress' use of both material and waste in its list of substance categories that are pollutants suggests that it knew the difference between the words and intended different results by using them. Second, all of the other substances listed are not waste from human activity. Rock and sand are listed, and while they may be waste from human activity, most often they are not. In any event, the relation to human activity need not be reflected in the nature of a pollutant, because it is already captured in the fifth element of the water pollution offense, the clause "by any person" in §301(a). Third, although mussel shells and feces are not man-made, their presence in large quantities at the harvesting operation was man-induced. Indeed, the term used by the court for the defendant's activity, "harvesting operation," is significant, for the mussels did not grow of their own accord on the operation's equipment, instead the defendant planted their seeds there; the operation included both production and harvesting.[81] Finally, Congress made it clear in §318[82] that discharges

79. CWA §316(a), 33 U.S.C. §1326(a) (emphasis added).
80. *Hammersley*, 299 F.3d at 1016.
81. The operation was conducted from a raft and consisted of planting mussel seeds on ropes hung from the bottom of the raft and anchored on the bottom. Mussels grew on the ropes, which were hoisted on board the raft at intervals for harvesting. *Hammersley*, 299 F.3d at 1016.
82. 33 U.S.C. §1328. *See* 40 C.F.R. §122.24(a). EPA may designate aquaculture operations as point sources requiring permits.

from aquacultural activities are subject to control by §402 permits, just as are discharges from concentrated animal feeding operations (CAFOs).[83] Both of those operations are very similar to growing and harvesting mussels in a confined area. Because the CWA does not define "aquaculture," a defendant might argue that it means only raising fish, not shellfish. But the dictionary definition of the word is the "culture of sea, lake and river foodstuffs, as fish, oysters, seaweed, etc."[84] Moreover, the CWA considers protection of fish and shellfish in the same phrases.[85]

The Ninth Circuit's "absurd results" reasoning in *Hammersley* may appear to be more difficult to dismiss. The court noted that the purpose of the CWA is to "provide for the . . . propagation of fish, shellfish, and wildlife."[86] The defendant's activity included propagation of shellfish, a goal that the CWA specifically espouses rather than an activity that the CWA specifically prohibits or regulates. But the CWA translates the goal elsewhere as "the propagation of a balanced, *indigenous* population of shellfish, fish and wildlife."[87] Under the facts of the case, the mussels probably are not indigenous and the defendant's activities, therefore, may interfere with a balanced population of shellfish and fish.[88] The distinction is not necessary, however, because Congress made it clear in §318 that aquacultural activities are subject to regulation by §402 permits, overcoming any contrary inferences from the goals of the statute. The applicable canon of interpretation is that the specific (§318's specification that aquacultural activities must have §402 permits) governs over the general (§101(a)'s policy of providing for the propagation of fish and shellfish).

Under the Sixth Circuit's reasoning in *Consumers Power*, a state fish hatchery would need a §402 permit to stock the state's streams with native trout. Surely, Congress did not intend that result. Perhaps, this is a perfect dilemma for EPA to resolve by amending its definition of pollutant to exclude indigenous live fish or by issuing a general permit authorizing federal and state fish and game authorities to add hatchery-raised indigenous fish stock to

83. Section 502(6) includes CAFOs within the list of example point sources.
84. *See* Webster's New Int'l Dictionary of the English Language 135 (1958). "The science, art and business of cultivating marine or freshwater foodfish or shellfish, such as oysters, clams, salmon and trout, under controlled conditions." *See* American Heritage Dictionary of the English Language 89 (4th ed.).
85. CWA §§101(a) & 316(a), 33 U.S.C. §§1251(a) & 1326(a).
86. CWA §101(a)(2), 33 U.S.C. §1251(a)(2).
87. CWA §316(a), 33 U.S.C. §1326(a) (emphasis added). Although this language is specified for the control of thermal discharges, it is an apt amplification of the statutory goal and there is no reason to believe Congress' concern for indigenous fish was confined to their survival from changes in heat.
88. Association to Protect Hammersley, Eld, and Totten Inlets v. Taylor Res., Inc., 299 F.3d 1007, 1010 n.1 (9th Cir. 2002).

navigable water. While the Agency is rulemaking on fish and shellfish, it might as well include people so that water theme parks do not need §402 permits for waterslides into lakes and rivers. EPA already has promulgated a permit by rule for burial at sea under the Ocean Dumping Act, demonstrating that it is possible to acknowledge that human bodies are pollutants, but allowing a desirable body-disposal activity to continue under conditions that will ensure that it is not harmful to human health or the environment.[89] If EPA can do so for human bodies under the Ocean Dumping Act, surely it can do so for live fish and human swimmers under the CWA. EPA might be tempted instead to promulgate an exclusion from requiring a permit in 40 C.F.R. §122.3 for the same activity. This is not as wise a course as the other options, however, because, as discussed below, courts have repeatedly held that EPA does not have the power to exempt from the permit program any discharges that the statute requires to have a permit.

b. Consumer Products

One of the most frequently litigated issues in the interpretation of pollutant is whether a consumer product becomes a waste when it is used for its intended purpose.[90] The question arises because a product's intended use may bring it to rest in navigable water without further human intervention. Most products become waste when we throw them away after use (for example, when we throw paper into the waste basket or leave our garbage cans on the street for municipal collection), or abandon them after use or even before use (for example, when we spill unused oil from a vessel at sea). But some products, such as bullets or pesticides, come to rest in navigable water because we use them for their intended purposes. Does firing a bullet at a duck flying over water constitute the addition of a solid or chemical waste to that water when we do not hit the bird? How can firing a bullet at a bird, the intended use of the bullet, constitute disposal of a waste or addition of a waste to water? Courts have held that substances can be pollutants no matter how useful they are,[91] and that the purpose of a substance is irrelevant to whether it is a pollutant.[92]

89. Marine Protection, Research, and Sanctuaries Act, 33 U.S.C. §§1401 et seq. For Ocean Dumping Act, see §1412. EPA's regulations, 40 C.F.R. §229.1, authorize burial at sea of bodies, cremated ashes, and accompanying flowers and decomposable wreathes, subject to conditions on location.

90. The same issue arises under RCRA.

91. *See, e.g.*, Hudson River Fishermen's Ass'n v. City of New York, 751 F. Supp. 1088, 1101, 21 ELR 20467 (S.D.N.Y. 1990).

92. *See, e.g.*, Minnehaha Creek Watershed Dist. v. Hoffman, 597 F.2d 617, 626-27, 9 ELR 20334 (8th Cir. 1979).

The Supreme Court addressed a similar question under the Refuse Act, holding in *United States v. Standard Oil Co.* that commercially valuable aviation fuel became refuse,[93] indeed became a pollutant,[94] once it was spilled into water. Oil and hazardous substance spill cases under the CWA generally involve this same principle because commercially valuable oil or chemicals are abandoned when they are spilled and are no longer useful products.[95] Once a bullet misses the waterfowl it was aimed at and falls into the water, it too becomes refuse. Neither the owner of the spilled oil nor of the errant bullet intended to throw it away or intended it to enter water. But by the very act of spilling oil in water or firing shot over water and missing the intended target, the substance lost its value, and the owner abandoned it to fall into water.[96]

i. Munitions

The Supreme Court acknowledged that a bomb dropped into ocean water when it missed its target on a practice range was a pollutant because the definition of pollutant includes "munitions," and bombs are "munitions."[97] The first cases in the civilian arena in which this issue surfaced were suits by environmental groups under the CWA and RCRA against gun ranges for discharging spent lead shot and skeet target fragments into Long Island Sound and other water bodies.[98] The plaintiffs challenged these operations, in part, on the ground that the defendants were point sources adding pollutants to navigable water without a CWA permit. The amounts of lead some gun ranges added to the environment were enormous.[99]

Courts had no trouble finding that the spent shot "which lands in navigable waters constitutes a pollutant within the meaning of the CWA,"[100]

93. "There is nothing more deserving of the label 'refuse' than oil spilled into a river." United States v. Standard Oil Co., 384 U.S. 224, 229-30 (1966).
94. "The word 'refuse' includes all . . . pollutants apart from those flowing from streets and sewers and passing there-from in a liquid state." *Id.* at 230.
95. *See, e.g.*, United States v. Hamel, 551 F.2d 107, 7 ELR 20253 (6th Cir. 1977).
96. The analogy is not exact, however, for the spilled oil was never used for its intended purpose, while the bullet may have been used for its intended purpose, if the intended purpose was to be fired at a bird and to hit a bird over or near water.
97. Weinberger v. Barcelo-Romero, 456 U.S. 305, 308-09, 12 ELR 20538 (1982). The Court describes the district court's holding that bombs are pollutants because bombs are munitions, and munitions are included on the exclusive list of pollutants in CWA §502(6). The Court did not question that holding.
98. Connecticut Coastal Fishermen's Ass'n v. Remington Arms Co., 989 F.2d 1305, 1313, 23 ELR 20699 (2d Cir. 1993); Long Island Soundkeeper Fund, Inc. v. New York Athletic Club, 1996 WL 131863 (S.D.N.Y. 1996).
99. For instance, at one gun range, 2,400 tons (nearly five million pounds) of lead shot had been discharged on surrounding land and water. *Connecticut Coastal Fishermen's Ass'n*, 989 F.2d at 1308.
100. *Long Island Soundkeeper Fund*, 1996 WL 131863 at *15.

effectively meaning that it fell within the chemical or solid waste categories in the CWA's definition of pollutant. Courts had considerably more trouble determining whether the spent shot was a solid waste under RCRA,[101] an elaborately defined term under that statute. Neither RCRA's statutory definition nor EPA's regulatory definition of solid waste under RCRA specifically addressed whether lead shot fired at a rifle range was or became a solid waste. But EPA consistently interpreted its RCRA regulatory definition of solid waste to exclude consumer products when used for their intended purposes and particularly to exclude spent ammunition.[102] Indeed, it promulgated a RCRA rule to define when spent military munitions are and are not solid waste, specifying that they are not solid waste when they are "used for their intended purpose."[103] It could be argued that because the RCRA regulation addresses only military munitions and not civilian munitions, spent civilian munitions are still pollutants. The logic of the regulation, however, applies equally to military and civilian spent munitions.[104] Should solid waste be defined identically under RCRA and the CWA? Because their goals are both to protect public health and the environment, arguably they should be, but that would require EPA rulemaking under the CWA.

ii. Pesticides

The issue of whether consumer products used for their intended purposes are wastes when discharged into water arises in other contexts, particularly with

101. RCRA defines "hazardous waste" as a subset of "solid waste," 42 U.S.C. §6903(5). It then defines "solid waste" as "discarded material," 42 U.S.C. §6903(27), but it also charges EPA with the responsibility to define hazardous waste, which carries with it the responsibility to define solid waste, 42 U.S.C. §6921. EPA's regulatory definitions of solid and hazardous waste are narrower than the statutory definitions. In *Connecticut Coastal Fishermen's Ass'n*, 989 F.2d 1305, the U.S. Court of Appeals for the Second Circuit held that lead shot fired at a rifle range met RCRA's congressional definition of solid waste, leaving open the question of whether it met EPA's regulatory definition. The significance between the different statutory and regulatory definitions is that material must meet the statutory rather than the narrower regulatory definition of solid waste to provide courts with jurisdiction for citizen suits seeking to remediate imminent and substantial endangerment of health or the environment from solid or hazardous waste under 42 U.S.C. §6972(a)(1)(B), but must meet the narrower regulatory definition of solid waste to provide courts with jurisdiction for citizen suits against disposal of hazardous waste without or in violation of a RCRA permit under 42 U.S.C.§6972(a)(1)(A). The court later decided in *Cordiano v. Metacom Gun Club, Inc.*, 575 F.3d 199 (2d Cir. 2009), that such lead shot was not a RCRA solid waste within EPA's regulatory definition. The end result is that plaintiffs may bring citizen suits against gun clubs for causing an imminent and substantial endangerment of health or the environment, regardless of whether they are violating RCRA, but not against the gun clubs for operating without or in violation of a RCRA permit.
102. *Cordiano*, 575 F.3d at 207-09.
103. 40 C.F.R. §266.202.
104. Why did EPA not cover both military and civilian munition in the same rule? Could it be that the U.S. Department of Defense's lobbying carries greater weight than the NRA's lobbying?

regard to the application of pesticides in or near an aquatic environment. Decisions involving such releases are complicated by the relevance of another statute administered by EPA, the Federal Insecticide, Fungicide, and Rodenticide Act (FIFRA),[105] governing the registration and use of pesticides (insecticides, fungicides, or rodenticides). To register a pesticide, EPA must approve uses that can be made of the pesticide and directions for applying the pesticide; thereafter, both are required to be printed on the label.[106] In deciding whether to register a pesticide, what uses to allow for it, and what methods of application to require for it, EPA is to "prevent unreasonable adverse effects on the environment," including water.[107] FIFRA prohibits the distribution, sale, or use of an unregistered pesticide or the use of a registered pesticide unless in compliance with EPA-approved uses and directions for its application.

When public health officials in and around New York City discovered in 2000 that mosquitoes carried the newly arrived and deadly West Nile virus, they ordered the spraying of pesticides to eradicate mosquitoes. Because mosquitoes breed in water, the resulting campaign sprayed pesticides over and near open water and wetlands. In *No Spray Coalition v. City of New York*,[108] environmental groups brought a CWA citizen suit to enjoin the spraying, arguing that it would damage the environmental and public health, but would be ineffective in suppressing West Nile virus. The district court, apparently irked by the plaintiff's attempt to interfere with the implementation of public health protection measures,[109] refused to issue an injunction, in part because Congress did not intend plaintiffs to use the CWA citizen suit provision to enforce prohibitions of FIFRA, which had no such provision. Although the court did not address whether pesticides, when properly applied to water according to the approved uses and labels, were pollutants, it held that because the pesticides were sprayed into the air rather than into the water, they were not added to navigable water.[110]

Shortly thereafter, citizens filed similar suits against nearby spraying. In *Altman v. City of Amherst*,[111] the district court dismissed the complaint, hold-

105. Federal Insecticide, Fungicide, and Rodenticide Act (FIFRA), 7 U.S.C. §§136-136y, ELR STAT. FIFRA §§2-35.

106. FIFRA §3(c)(1)(C), 7 U.S.C. §136a(c)(C).

107. FIFRA §§2(j) & 3(a), 7 U.S.C. §§136(j) & 136a(a).

108. No Spray Coalition v. City of New York, 2000 WL 1401458 (S.D.N.Y. 2000), *rev'd*, 252 F.3d 148, 31 ELR 20707 (2d Cir. 2001).

109. The district court noted that the citizens filed suit "despite the unusual unanimity of governmental agency opinion that this spraying is in the best interests of preserving public health" and that "[f] ortunately for the community, that question is decided by public health and environmental officials," not by the courts. 2000 WL 1401458 at *1.

110. *Id.* at *3.

111. Altman v. City of Amherst, 190 F. Supp. 2d 467 (S.D.N.Y. 2001), *rev'd*, 47 Fed. Appx. 52 (2d Cir. 2002).

ing that "pesticides, when used for their intended purpose, do not constitute a 'pollutant' . . . and are more appropriately regulated under FIFRA."[112] The court was persuaded that because FIFRA had no citizen suit provision, Congress did not intend that CWA citizen suits be brought against the spraying of pesticides in accordance with label instructions approved by EPA under FIFRA. In *Peconic Baykeepers, Inc. v. Suffolk County,*[113] another New York federal district court conflated "addition" and "pollutant," stating that: "Atmospheric emission of aerial adulticides are not defined as pollutants[;] at no time was the spray made directly to navigable water." It held that FIFRA rather than the CWA governed and deferred to EPA's Interpretive Statement (discussed below) that no CWA permit was required.

The U.S. Court of Appeals for the Second Circuit reversed *No Spray,* holding that CWA §505 authorized plaintiffs to maintain a CWA citizen suit if they alleged a violation of the CWA.[114] The Second Circuit also reversed *Altman,* because the record was incomplete and because the plaintiff had not been able to conduct discovery to make its case that properly applied pesticides were pollutants. The Second Circuit stated that the question of whether properly applied pesticides could be pollutants under the CWA would remain open unless and until EPA articulated an interpretation of the CWA on the issue.[115] Finally, the Second Circuit in part reversed *Peconic Baykeepers,*[116] noting that the Sixth Circuit had subsequently overturned EPA's CWA pesticide exemption rule in *National Cotton Council,*[117] discussed below, but that the Sixth Circuit had stayed its mandate pending petitions for rehearing and rehearing en banc, which were denied. Because, on the facts found by the district court, some of the defendant's spraying was not in conformity with the pesticide's FIFRA-approved label, the Second Circuit reversed the lower court's decision to that extent.

In the meantime, two decisions of the Ninth Circuit addressed the relationship between the CWA and FIFRA in regulating the application of pesticides to navigable water in accordance with their EPA-approved labels. In *Headwaters,*[118] decided in 2001, an irrigation district added a chemically derived herbicide to irrigation canals (considered navigable waters for the purposes of the case) to kill aquatic weeds choking the canals. EPA had registered the herbicide under FIFRA for that purpose, and the defen-

112. *Altman,* 190 F. Supp. 2d at 471.
113. 585 F. Supp. 2d 377 (E.D.N.Y. 2008), *rev'd in part,* 600 F.3d 180 (2d Cir. 2010).
114. *No Spray Coalition,* 252 F.3d 148.
115. Altman v. City of Amherst, 47 Fed. Appx. 52 (2d Cir. 2002).
116. Peconic Baykeepers, Inc. v. Suffolk Cnty., 600 F.3d 180 (2d Cir. 2010).
117. 553 F.3d 927, 39 ELR 20006 (6th Cir. 2009).
118. 243 F.3d 526, 31 ELR 20535 (9th Cir. 2001).

dant applied the herbicide in accordance with the instructions on its EPA-approved label. When the defendant added the herbicide to the stream, the active ingredient in the herbicide remained in the water after it had served its intended purpose. The court held that the remaining pesticide was either excess pesticide or pesticide residue, in either case a "chemical waste," and therefore a "pollutant," the discharge of which required a §402 permit.

In a later Ninth Circuit decision, *Fairhurst v. Hagener*,[119] the Montana Department of Fish, Wildlife, and Parks had added a pesticide to streams to kill invasive non-indigenous fish for the purpose of restoring the population of indigenous fish. EPA had registered the chemically derived pesticide under FIFRA for that purpose, and Montana applied the pesticide in accordance with the EPA-approved label. As unlikely as it may seem, the uncontested facts established that no excess chemical pesticide was discharged to the water and that once the pesticide performed its intended purpose, there was no residue chemical pesticide and hence no chemical waste. At the time that the Ninth Circuit considered the issue, EPA had issued an Interpretive Statement and Notice of Proposed Rulemaking on the Application of Pesticides to Waters of the United States in Compliance With FIFRA.[120] Under EPA's interpretation, pesticides applied for their intended purposes, in accordance with their EPA-approved labels, and leaving no excess or residue pesticide in the water, were not pollutants under the CWA. This was consistent with the *Headwaters* ruling, and the Ninth Circuit deferred to EPA's interpretation. EPA's interpretation conceded that excess pesticides or residual pesticides are pollutants, but concluded they are not added to navigable waters by aerial spraying point sources because the point sources spray them into the air, not into the water. The Ninth Circuit did not address this spraying scenario, however, because it was beyond the facts of the case.

EPA ultimately promulgated a CWA rule incorporating the substance and reasoning of its Interpretive Statement.[121] The rule exempted from the CWA §402 permit program pesticides applied directly to, over, or near water in full compliance with FIFRA.[122] The Sixth Circuit resolved chal-

119. 422 F.3d 1146 (9th Cir. 2005).

120. *Id.* at 1149-50. Indeed, before the district court issued its opinion, EPA had issued an Interim Interpretive Statement. The Interim Interpretive Statement and the Interpretive Statement were similar in reasoning and conclusion. Both are addressed below.

121. 71 Fed. Reg. 68483 (Nov. 27, 2006), codified at 40 C.F.R. §122.3(h).

122. *Id.* at 68486-87. The rule was consistent with *Fairhurst*, in which the Ninth Circuit had held that a chemical pesticide applied to water was not a pollutant when no excess or residual pesticide remained in the water. The rule is not convincingly consistent with *Headwaters*, however, because that decision commented that chemical pesticide excess or residue in water would be a pollutant without considering EPA's argument that the defendant did not add the pollutant from a point source.

lenges to the rule in the most detailed decision considering the legality of discharging pesticides into water without a CWA permit in *National Cotton*.[123] EPA began its argument before the court, as it had in the preamble to its rule, by stating that pesticides are either biological or chemical in composition and can be pollutants only if they are "biological material" or "chemical wastes." It then argued that when chemically derived pesticides are applied for an EPA-approved use, they are not chemical wastes. EPA conceded that excess pesticides and pesticide residue are pollutants because they are biological material or chemical wastes. Finally, it argued that biologically derived pesticides applied for their EPA-approved use cannot be biological material, because it would be absurd for biological pesticides applied for their approved use to be treated differently than chemical pesticides applied for their approved use.[124]

The Sixth Circuit considered three different dictionary definitions of waste, and concluded that under any of the definitions, chemical pesticides are not chemical waste, but that excess pesticides and pesticide residue "are wastes of the pesticide application."[125] Thus far, it agreed with EPA. But then EPA argued that the point sources only sprayed pesticides into the air for their intended uses; the point source did not spray excess pesticides or pesticide residues. At the "time of discharge, the pesticide is a non-pollutant and the excess pesticide or pesticide residues are not created until later, presumably after they are already in the water."[126] Therefore, EPA argued that by spraying pesticides, the point source did not spray (add) excess pesticides and pesticide residues into water, and thus did not require a CWA §402 permit. The court rejected EPA's assertion that "a pesticide must be 'excess' or 'residue' at the *time of discharge* if it is to be considered as discharged *from a point source*."[127] The court found that requiring a temporal connection in "the discharge of a pollutant" was "unsupported by the Act . . . [and] contrary to the purpose of the permitting program, which is 'to prevent harmful discharges.'"[128] The court noted that the purpose of the statute was to prevent the discharge of pollutants that would harm water quality, whether the harm occurred immediately upon discharge or later.[129]

123. National Cotton Council of America v. U.S. EPA, 553 F.3d 927, 39 ELR 20006 (6th Cir. 2009).
124. *Id.* at 934-35.
125. *Id.* at 936, quoting from the preamble of EPA's rule, 71 Fed. Reg. at 68487.
126. *Id.* at 939.
127. *Id.*
128. *National Cotton Council*, 553 F.3d at 939, quoting National Ass'n of Home Builders v. Defenders of Wildlife, 127 S. Ct. 2518, 2525 (2007).
129. A similar attempt to require a temporal connection between "addition" and polluting activities was rejected in American Mining Cong. v. U.S. EPA, 965 F.2d 759, 764, 22 ELR 21135 (9th Cir. 1992).

As to biologically based pesticides, the court could find no rationale for holding them not to be "biological materials" rather than falling into a more limited hypothetical category such as "biological waste." Congress specifically used the broader "material" category for biological substances. Treating biological pesticides as pollutants is also consistent with the §502(19) definition of pollution, which includes the "biological . . . integrity of water," for "[a]dding biological pesticides to water undeniably alters its biological integrity."[130] Since Congress decoupled "pollutant" from "pollution," however, this is not overwhelmingly persuasive support for the court. Alternatively, it could have argued the broader interpretation of biological material promotes the objective of the CWA as stated in §101(a) "to restore and maintain the . . . biological integrity of the Nation's waters." The court also noted that other courts had construed "biological material" broadly to include live fish, dead fish, fish parts, and fish feces and urine.[131] In the end, the court held that pesticide residue or excess was unambiguously chemical molicide and adulticide waste or biological material.[132]

iii. Water Supply Treatment Residue

Lead shot and pesticides are not the only consumer products that may become pollutants after use for their intended purposes. In *Hudson River Fishermen's Association v. City of New York*,[133] the city augmented its water supply in times of water shortage by pumping water from the Hudson River into an aqueduct, which ultimately discharged the water into a reservoir that was part of the city's water supply system. The city conceded that the reservoir was navigable water. Because the river water was not clean enough to serve as drinking water without treatment, the city added alum and chlorine at the pumping station to precipitate out solids and to kill pathogens as the water made its way to the reservoir through the aqueduct. By the time the aqueduct discharged the river water to the reservoir, the chemicals had performed their intended purposes and the aqueduct discharged alum floc (the solids precipitated by the alum) and chlorine residual (the chlorine remaining

130. *National Cotton Council*, 553 F.3d at 938.
131. Citing National Wildlife Fed'n v. Consumers Power Co., 862 F.2d 580, 17 ELR 20801 (W.D. Mich. 1987); U.S. PIRG v. Atlantic Salmon of Maine, LLC, 215 F. Supp. 2d 239, 246, 32 ELR 20535 (D. Me. 2002). The Sixth Circuit distinguished *Hammersley*, holding mussel feces and shells not to be biological material because that decision dealt with the "result of natural biological processes, not the result of a transforming human process. *See National Cotton Council*, 553 F.3d at 938 n.6.
132. *National Cotton Council*, 553 F.3d at 940.
133. 751 F. Supp. 1088, 21 ELR 20647 (S.D.N.Y. 1990), *aff'd without opinion*, 940 F.2d 649, 21 ELR 21226 (2d Cir. 1991).

after killing the pathogens) to the reservoir, along with the treated water. The environmental group sued the city for discharging alum floc and chlorine residue to the reservoir without a permit.

The city argued it was adding useful chemicals at the pumping station to perform a public health benefit, as the defendants had argued in the pesticide application cases. The district court found that the city's actions were no different than those of other water treatment plants, which commonly added alum to precipitate out solids, which they then "filtered out, backwashed from the filter, and disposed of as waste."[134] Indeed, water treatment plants routinely have CWA §402 permits limiting their discharges of solids and alum.[135] Chlorine is used to treat publicly owned treatment works (POTWs) effluent for pathogens, and chlorine residual is routinely regulated in CWA §402 permits for POTWs.[136]

The city's argument focused on the wrong substances and the wrong receiving water. The city added alum and chlorine to the water in the aqueduct at the pumping station, at which point they were useful products rather than chemical wastes or residues. Moreover, no one contended that the water in the aqueduct was navigable. But that was not what the plaintiff environmental group challenged; instead, it challenged the subsequent discharge of the alum sludge and chlorine residual into the reservoir. At that point, the reservoir was concededly navigable water and the chemicals had served their useful purposes and were either excess or residue—chemical wastes in either case. The court commented that it "is indisputable that a pollutant is a pollutant no matter how useful it may earlier have been."[137]

iv. Conclusion

Some may wonder why EPA would abandon the environmental high road by not regulating the discharges of spent lead shot and spent or excess pesticides into water. Cynics might conclude that EPA did not want to confront the National Rifle Association and the agricultural lobby. There is undoubtedly some truth in that. On the other hand, it is not clear that Congress intended the CWA to regulate additions to water of consumer products whose intended

134. *Id.* at 1102.
135. *Id.* at 1097-98.
136. Interestingly, residual chlorine is limited in terms of both maximum levels and minimum levels in POTW effluent. *See Russian River Watershed Prot. Comm. v. City of Santa Rosa*, 142 F.3d 1136, 1139, 28 ELR 21265 (9th Cir. 1998). The maximum level, of course, protects life in the receiving water. The minimum level ensures that the POTW is adding enough chlorine to do its job.
137. Hudson River Fishermen's Ass'n v. City of New York, 751 F. Supp. 1088, 21 ELR 20647 (S.D.N.Y. 1990).

use involves addition to water. This ambiguity also troubles the analyses of the discharges under other elements of the CWA offense.[138]

The meaning of waste is inherently uncertain for two primary reasons. First, what may be waste to a person who is disposing of a substance may be a useful substance for another person,[139] an issue that has not yet arisen in the decisions defining pollutant in the CWA, but underlies the issue of defining solid waste in RCRA. Second, the intended use of a consumer product may result in the product coming to rest in water, for example, spent bullets fired at birds over water but missing their targets. These and similar cases can be expected to recur in other contexts. Under these circumstances, it is curious why EPA has not promulgated a rule defining pollutant or waste to exclude consumer products, particularly regulated consumer products that are used for their intended purposes, or fashioned a general permit or permit by rule to authorize such discharges, as it has done for military munitions under RCRA[140] or burial at sea of human bodies under another statute, as discussed *supra* in Section II.B.1.a. Of course, the Agency attempted to promulgate a rule exempting pesticide applications from requiring §402 permits. But early in the CWA's implementation, courts held that EPA could not exempt any category of discharges of pollutants from point sources to navigable waters from the requirement of securing a CWA permit.[141] It is no surprise that EPA lost its bid to promulgate a similar rule exempting discharges of pesticides from the requirement of securing CWA permits.[142] EPA might be more successful using a different regulatory strategy, as suggested *supra* in Section II.B.1.b.ii.

138. For instance, what are the point sources: the rifle ranges or the rifle barrels? If the latter, is EPA to issue permits to all hunters or rifle owners for them to fire over water? That would be a thankless and probably impossible task. The analytical problem may be analogous to issues that arise under the "addition" element of the offense, where, with respect to aerial spraying of pesticides, the addition seems to be to the air rather than to the water. *See supra* Chapter 1, Section III.B. In *U.S. EPA ex rel. McKeown v. Port Authority of New York and New Jersey*, 162 F. Supp. 2d 173 (S.D.N.Y. 2001), the plaintiff sued to abate pollution from tunnel toll booths. The court commented that the pollutants at issue emanated from countless vehicle exhausts rather than from the toll booths.

139. When you leave an unwanted chair at the curbside in front of your house and it is taken by someone else for use, it was waste for a moment, but not for long. One of RCRA's objectives is to minimize the disposal of waste by encouraging "recycling and reuse." *See* 42 U.S.C. §6902(a)(6). Material destined for recycling is waste to the disposer, but a useful material to the recycler.

140. 40 C.F.R. §266.202, a rule that has survived judicial review. *See Military Toxics Project v. EPA*, 146 F.3d 948, 28 ELR 21350 (D.C. Cir. 1998).

141. *See* National Cotton Council of Am. v. U.S. EPA, 553 F.3d 927, 39 ELR 20006 (6th Cir. 2009); Northwest Envtl. Advocates v. U.S. EPA, 537 F.3d 1006, 1021-22 (9th Cir. 2008); Northern Plains Resource Council v. Fidelity Exploration and Dev. Co. 325 F.3d 1155, 1164 (9th Cir. 2003); NRDC v. Costle, 568 F.2d 1369, 8 ELR 20028 (D.C. Cir. 1977); Catskill Mountains Chapter of Trout Unlimited v. U.S. EPA, Nos. 08-CV-5606, -8430, 2014 WL 1284544 (S.D.N.Y. Mar. 28, 2014).

142. *National Cotton Council*, 553 F.3d 927.

C. *Discharged Into Water*

Under §502(6), the listed substances and listed categories of substances are pollutants only if they are "discharged into water." But the §301(a) prohibition, elaborated by §502(12), already specifies that only the addition of a pollutant from a point source into navigable *water* is a violation. The "discharged into water" phrase in §502(6) appears to be redundant with the "into . . . water" already in §502(12), making the offense: the addition of a pollutant *into water* from a point source *into* navigable *water*. This makes no sense. "Water" in "discharged into water" is not modified by "navigable," making "water" a more expansive term, perhaps including groundwater. Accordingly, the offense would read as follows: the addition of a pollutant into water, including groundwater, from a point source into navigable water. That makes no sense either. Section 502(16) defines the freestanding "discharge" to include "discharge of a pollutant," which is already defined as addition of a pollutant into navigable water from a point source in §502(12). That interpretation would make the offense even more circular and nonsensical. Statutes are to be interpreted to give every word a meaning[143] and to avoid redundancies.[144] Avoiding the redundancy of "discharge" and "water" in §502(6), however, results in circularity, absurdity, and nonsense, which are to be avoided even more than redundancy.

The analysis is complicated by the fact that Congress included a similar redundancy in its definition of "point source" to mean exclusively a "discernible, confined and discrete conveyance . . . from which pollutants are or may be discharged," with an inclusive list of examples, including a pipe, §502(14). This, together with the "discharged into water" phrase in §502(6) makes the offense read: the addition of any listed substance into water from a pipe from which listed substances are added into navigable waters. This suggests the offense is a two-staged activity: the addition of a substance into water, and the subsequent addition of their mixture by a conveyance into navigable water. While that describes many industrial and municipal effluents, nothing else in the statute suggests the prohibition excludes addition of pollutants directly into navigable water without first being mixed with water.

143. Senne v. Village of Palentine, Ill., 695 F.3d 617, 621 (7th Cir. 2012). *See* Scalia & Garner, *supra* note 11, at 174-79.

144. Mackey v. Lanier Collection Agency & Serv., Inc., 486 U.S. 825, 837 (1988). Eskridge, *supra* note 70, at 324, citing Kungys v. United States, 485 U.S. 759, 778 (1988); South Carolina v. Catawba Indian Tribe, Inc., 476 U.S. 498, 510 n.22 (1986).

Perhaps, the drafters of these definitions did not want to slander the substances listed as pollutants unless they were actually added to navigable water. We do not want children to recoil from eating broccoli because it is a "pollutant" (a "biological material"), regardless of whether it is discharged into water. Similarly, perhaps the drafters did not want to imply that all point sources were illegal unless they actually added pollutants to navigable water. Such caution was unnecessary, however, because all four elements of the offense must be met before a discharge violates the CWA unless it is in compliance with a permit. Although when interpreting a statute we are supposed to avoid redundancy and to give meaning to every word, a few redundancies, meaningless words, and scrivener's errors[145] may be inevitable in a 200-page statute. These redundancies are two of them.

Only two decisions have identified this issue. In the first, *United States v. Pozsgai*,[146] the defendants in an enforcement action for filling a wetland without a §404 permit argued that they had not discharged substances into water and, therefore, the substances were not pollutants. They argued that the "into navigable waters" phrase in §502(12) defined the geographic jurisdictional reach of the statute, and the "into water" phrase in §502(6) limited the definition of "pollutant." Though it seemed to be a clever argument, the U.S. Court of Appeals for the Third Circuit rejected it in favor of reading "navigable waters" in §502(12) to modify "water" in §502(6), an interpretation it found more consistent with the legislative history.[147] The court's interpretation, however, is circular and ignores the recurrence of similar redundancies in these definitional phrases.[148] The second decision, *Pronsolino*,[149] was a dispute over whether §§303(d) and 319 require total maximum daily loads (TMDLs) for water polluted only by nonpoint sources. While the California federal district court recognized the issue posed by "into water" in the definition of pollutant in §502(6), it did not need to address that issue to resolve the TMDL dispute before it.

As much as courts abhor redundancy in statutes, there is no apparent way around the redundancy between "into navigable water" in §502(12), the primary definition of the offense, and "into water" in §502(62), defining an element in §502(12).

145. *See* Scalia & Garner, *supra* note 11, at 234-39.
146. 999 F.2d 719, 23 ELR 21012 (3d Cir. 1993).
147. *Id.* at 726-27.
148. Section 502(12) similarly defines the "discharge of a pollutant" to mean addition of a pollutant from a "point source," which it defines in §502(14) as a conveyance from which a pollutant may be "discharged."
149. 91 F. Supp. 2d 1337, 30 ELR 20460 (N.D. Cal. 2000).

D. *Must a Pollutant Cause Pollution?*

As discussed above, Congress decoupled "pollutant" and "pollution" by defining each without reference or regard to the other. The §502(6) definition of pollutant does not hint that a pollutant is something causing pollution; the §502(19) definition of pollution does not hint that pollution is a condition caused by a pollutant. Most courts agree.[150] Indeed, in an important early decision, *National Wildlife Federation v. Gorsuch*,[151] the D.C. Circuit deferred to EPA's argument that low dissolved oxygen, cold, and oxygen supersaturation resulting from water flowing over dams were not pollutants, but instead were "water conditions," e.g., "pollution," and therefore the dams from which these conditions followed did not require §402 permits.

Despite the decoupling of pollutant and pollution, some courts have determined particular substances to be pollutants or not to be pollutants in part on the basis of their negative impact or lack of negative impact on water quality.[152] Because the overall purpose of the CWA is "to restore and maintain the chemical, physical, and biological integrity of the Nation's waters"[153] and the CWA's definition of pollution is the "man-made or man-induced alteration of the chemical, physical, biological, and radiological integrity of water," the very purpose of the CWA, not surprisingly, is to reduce pollution. The frequently used canon of construction to interpret a remedial statute to effectuate its purpose,[154] therefore, suggests that "pollutant" be interpreted to accomplish "pollution" control, even though the two words are decoupled by their statutory definitions. On the other hand, the definition of pollutant is so broad that use of this additional canon should not be necessary to hold that a substance is a pollutant. Moreover, while the fact that a substance

---

150. Demonstration of a negative impact on water quality is not necessary for a substance to be a pollutant. *See, e.g.*, Northern Plains Res. Council v. Fidelity Exploration and Dev. Co., 35 F.3d 1155, 1161 (9th Cir. 2003); Committee to Save Mokelumne River v. East Bay Mun. Util. Dist., 13 F.3d 305, 309, 24 ELR 20225 (9th Cir. 1993); Minnehaha Creek Watershed Dist. v. Hoffman, 597 F.2d 617, 626-27, 9 ELR 20334 (8th Cir. 1979); U.S. PIRG v. Atlantic Salmon of Maine, LLC, 215 F. Supp. 2d 239, 247 n.3, 32 ELR 20535 (D. Me. 2002); Long Island Soundkeeper Fund, Inc. v. New York Athletic Club, 1996 WL 131863 (S.D.N.Y. 1996).

151. 693 F.2d 156, 13 ELR 20015 (D.C. Cir. 1982).

152. For decisions where a substance was deemed a pollutant because of negative impact on water quality, see *Borden Ranch Partnership v. U.S. Army Corps of Engineers*, 261 F.3d 810, 814-15, 32 ELR 20011 (9th Cir. 2001); and *United States v. Deaton*, 209 F.3d 331, 336, 30 ELR 20508 (4th Cir. 2000). In *Hammersley*, the Ninth Circuit held that a substance was not a pollutant because of its neutral or positive impact on water quality. *See* Association to Protect Hammersley, Eld, and Totten Inlets v. Taylor Resources, Inc., 299 F.3d 1007, 1016 (9th Cir. 2002).

153. CWA §101(a).

154. Eskridge, *supra* note 70, at 327, citing Reves v. Ernst & Young, 494 U.S. 56 (1990).

causes pollution might suggest that the substance is a statutory pollutant,[155] the reverse is not true. After all, Congress defined pollutant to include substances not causing pollution, such as rock and sand.

One problem with using the concept of pollution as a driver for interpreting pollutant is that the meaning of pollution is not altogether clear. While the statutory definition of pollution as the "chemical, physical, biological and radiological integrity of water" appears meaningful, what does it in fact mean? What is water integrity?[156] Surely, it cannot mean pure water, which scarcely if ever occurs in a state of nature and would not nourish fish and shellfish if it did. Is water with integrity water as it existed at a particular time and place? Before the industrial revolution? Before the arrival of European colonists in North America? Before the arrival of the original colonists in the western hemisphere?[157] Or does it relate to "the protection and propagation of fish, shellfish, and wildlife and . . . for recreation in and on the water," a goal of the CWA?[158] If so, which fish and shellfish?[159] Indigenous or introduced? If indigenous, at what time and for how long?

The concepts of water integrity and water pollution are relative, difficult to define, and involve both science and policy. Take, for example, nutrients. Nutrients support biological growth; without nutrients, there would be no fish in the streams, lakes, or oceans. But a superabundance of nutrients causes algae blooms, which in turn die and exert oxygen demand in water to decompose, and if the oxygen demand from the algae decomposition is sufficient, there will be no oxygen left for fish to survive. There is a dead zone in the Gulf of Mexico caused by nutrients flushed down the Mississippi River from farmland, cities, and factories throughout the Midwest.[160] No nutrients equates to no fish; but too many nutrients also equates to no

155. *See* National Cotton Council of Am. v. U.S. EPA, 553 F.3d 927, 938, 39 ELR 20006 (6th Cir. 2009).

156. The dictionary definition of "integrity" is "unimpaired or unmarred state . . . soundness, purity." Webster's New Dictionary of the English Language 1290 (1958). Use of the word in the CWA "is intended to convey a concept that refers to a condition in which the natural structure and function of ecosystems is maintained." H. Rep. No. 92-911, at76-77 (1972), *reprinted in* 3 Legis. History 753, 763-64. Query how much help that is. Which ecosystem are we protecting? The one that existed before the advent of the industrial revolution, or before the introduction of trans-Atlantic human immigrants, or before the introduction of Native American immigrants, or before the last Ice Age? Ecosystems are constantly evolving even without human intervention.

157. One case from the U.S. Court of Appeals for the Eighth Circuit suggests that water with integrity means pre-industrial revolution or even pre-human water. *See* Minnehaha Creek Watershed Dist.v. Hoffman, 597 F.2d 617, 625, 9 ELR 20334 (8th Cir. 1979). Self-evidently, the physical, chemical, and biological composition of a pre-human river may be purely speculative.

158. CWA §101(a)(2).

159. Section 316(a), 33 U.S.C. §1326(a), may provide some help here by specifying that thermal discharges shall be limited to protect a "balanced, indigenous population of fish, shellfish and wildlife." To be sure, indigenous populations change over time.

160. Friends of the Everglades v. South Florida Water Mgmt. Dist., 570 F.3d 1210, 1227 (11th Cir. 2009).

fish. Whether a substance causes pollution is often a situational issue and is always a scientific question. To be sure, there are some substances that cause harm to health or the environment under any circumstances and therefore are pollutants under any circumstances. Examples include PCBs or dioxin.

Congress dealt with the ambiguity of water integrity by establishing the water quality standards regimen, in which states designate the uses they desire to be made of particular water bodies (a policy determination), and states together with EPA establish the maximum levels or criteria for various pollutants allowed in each water body to achieve the designated use (a scientific determination).[161] Congress provided in the CWA that point sources cannot discharge pollutants into a water body that will interfere with the designated use of the water by causing an exceedance of the criteria for pollutants in the water body.[162]

Congress' decoupling of "pollutant" and "pollution" in the definitions of the two words reflects its recognition that whether the discharge of a pollutant causes pollution depends in part on a political judgment as to what use should be made of a particular water body, and in part on a scientific expert judgment as to the extent to which discharges of pollutants must be curtailed to achieve that use. These are the very determinations required to establish water quality standards and effluent limitations based on them. This was the pre-1972 strategy of federal water pollution control. But Congress found the water quality approach was cumbersome, resource-intensive for regulators, and slow. Congress therefore enacted the CWA in 1972 to simplify and expedite regulation of pollution through point sources by the initial substitution, in most cases, of technology-based standard regulation for water quality-based standard regulation.[163] Congress hoped that application of the best available water pollution control technology to point sources would achieve water quality standards in most waterways. But it required permit writers to establish effluent limitations requiring further treatment where necessary to achieve water quality standards. Thus, for courts to ask whether a substance causes environmental harm before hold-

161. For a discussion of water quality standards and how they are implemented, see MILLER ET AL., *supra* note 5, at Chap. V.

162. Of course, limiting the pollution levels of point sources may not achieve the criteria. In that situation, states may limit nonpoint source discharges to achieve the criteria. If states refuse to do so, however, the CWA has reached the limits of its effectiveness, for it authorizes no federal controls on nonpoint sources and no mechanisms to force states to act on them. For a discussion of the CWA's limited nonpoint source program, see MILLER ET AL., *supra* note 5, at Chap. XIII.

163. CWA §301(b). For a quick review of the events leading up to the enactment of the CWA, see *Environmental Prot. Agency v. California ex rel. State Water Resources Control Bd.*, 426 U.S. 200, 203-06, 6 ELR 20563 (1976).

ing that it is a pollutant arrogates to courts a task that Congress delegated to EPA and the states,[164] pushing judicial authority beyond its separation-of-powers limits.

Why do some courts nevertheless continue to examine whether substances cause pollution before finding they are pollutants? Perhaps, because some judges think it unfair to enmesh defendants in the pollution-control regulatory system unless their discharges are actually harmful. If so, these judges misperceive both the statutory process and their role in it. Holding that a substance is a pollutant does not enmesh the substance in the pollution-control system. That occurs only if all of the other elements of §301(a) are also met. Moreover, holding that a substance is a pollutant does not mean that it will be subject to pollution control even if the other elements are met. Water quality standards do not require treatment of harmless substances for which no criteria exist. And the technology-based standards are established, in part, based on a cost-benefit analysis or at least a consideration of cost.[165] If a substance does not cause harm, the costs of treatment for its removal may not be justified.[166] To be sure, someone must determine whether the substance causes sufficient environmental harm to warrant treatment costs. But that is a role that Congress assigned to EPA and its counterpart state agencies, not to the courts.[167]

In conclusion, the plain meaning of the definition of pollutant includes almost all substances that may be added to water, making other interpretive devices unnecessary in almost all cases. For the few substances for which the outcome is ambiguous, the broad general purpose of the CWA to attain and maintain water integrity may be a useful supplemental interpretive device in determining whether an environmentally harmful material is a pollutant. If a material does not cause pollution or interfere with water integrity, however, it is still a pollutant if it fits within the statutory definition.

---

164. As the Eighth Circuit said in *Minnehaha Creek Watershed Dist.v. Hoffman*, 597 F.2d 617, 627, 9 ELR 20334 (8th Cir. 1979):

> Congress has, by the inclusion of certain substances in the definition of "pollutant" . . . determined that the discharge of those substances in navigable waters is subject to the Act's control requirements. The Act contains no provision that the listed substances are to be classified as pollutants and, thus subject to the Act's control requirements, only upon a further administrative or judicial finding that their addition to navigable waters results in a significant decrease in water quality. Nor does the fact that the listed substances may not in themselves be commonly considered "toxic" or "contaminating" change this result.

165. CWA §304(b)(1)(B), (b)(2)(B), & (b)(3)(B), 33 U.S.C. §1314(b)(1)(B), (b)(2)(B), & (b)(3)(B).

166. *See* Miller et al., *supra* note 5, at 278-305.

167. An exception exists in judicial review of EPA's promulgation of technology-based standards, approval of state water quality standards, or issuance of a permit, where courts give substantial deference to EPA's determinations.

E. *Must a Pollutant Be of Human Origin?*

The Ninth Circuit's *Hammersley* decision held that pollutants must be wastes of human or industrial processes. The notion that a pollutant must be a waste was discussed, and dispelled, above. The notion that a pollutant must be the product of human or industrial processes is similarly wrong. The definition of pollutant contains no qualification that a substance be the product of human or industrial activity. While the definition of "pollution" is limited to "man-made or man-induced alteration" of water quality,[168] pollution is not a jurisdictional limitation, and Congress decoupled the definitions of pollutant and pollution. The definitional list of pollutants includes substances that are not man-made (for example, rock and sand), although those substances may be added to navigable water as a result of human activity. But it is not necessary to contort the definition of pollutant to limit the CWA prohibition to the results of human activity, because the offense already includes "by any person" as a separate element. Insisting that pollutant in §502(6) requires human activity merely creates an unnecessary redundancy with the phrase "by any person" in §301(a).

F. *Potential Conflicts With Other CWA Sections*

The CWA authorizes regulatory programs other than the §402 permit program. The §404 permit program for dredged and fill material has already been mentioned and is discussed immediately below in greater detail. Section 311[169] provides a comprehensive program to prevent spills of oil and hazardous materials, to remediate them when they occur, and to recover the government's spill cleanup costs. The Oil Pollution Act,[170] passed in the wake of the *Exxon Valdez* oil spill, has largely supplanted §311. Section 312 authorizes the U.S. Coast Guard to regulate the disposal of untreated or inadequately treated sewage from vessels into navigable waters. Nonpoint pollution sources are regulated, if at all, by states, under their own authorities, as encouraged by §319. For the most part, these other programs do not conflict with §402 or raise particular issues with regard to the definition of pollutant.

168. CWA §502(19).
169. 33 U.S.C. §1321.
170. Oil Pollution Act (OPA), 33 U.S.C. §§2701-2761, ELR Stat. OPA §§1001-7001.

I. Section 311: Spills of Oil and Hazardous Materials

In *United States v. Hamel*,[171] the defendant appealed from a criminal conviction for spilling gasoline into Lake St. Claire in violation of §301(a), arguing that gasoline was not a pollutant under §502(6) because that statutory section did not list petroleum products. He further argued, applying the canon that a specific provision of a statute governs over a general provision of the same statute,[172] that because §311 regulated oil spills and even defined "oil" in §311(a)(1), oil was not a pollutant under §502(6) and oil spills were not included as violations of §301(a). The government argued, and the district court held, that oil and gasoline fell within the category of "biological material" listed as a pollutant in §502(6).[173] The Sixth Circuit could easily have upheld the conviction on that basis. Instead, it undertook a longer but more specific analysis, finding that when Congress enacted the CWA, it intended to include within the §402 permit program all discharges covered by the Refuse Act plus liquid waste from streets and sewers,[174] and that the Supreme Court had held earlier, in *United States v. Standard Oil Co.*,[175]that oil spills were discharges of refuse under the Refuse Act. Although those findings answered both of the defendant's arguments, the court proceeded to demonstrate that there was no conflict between §§402 and 311 in this regard, because they served different purposes. The purpose of the §402 permit program is to require pollution reduction from ongoing discharges of pollutants, while the purpose of §311 is to prevent spills of pollutants and to remediate them when they occur.[176] The court could have performed the same analysis, with the same result, if §311 had been a different statute (for example, the OPA) rather than a different section in the same statute.[177]

---

171. 551 F.2d 107, 7 ELR 20253 (6th Cir. 1977).
172. *See* Eskridge, *supra* note 70, at 324, citing Green v. Bock Laundry Mach. Co., 490 U.S. 504, 524-25 (1989); Scalia & Garner, *supra* note 11, at 183-89.
173. *Hamel*, 551 F.2d at 110.
174. *Id.* at 110-11.
175. 384 U.S. 492 (1966).
176. *Hamel*, 551 F.2d. 111-13.
177. Why would the court have taken such a round-about path to its conclusion, when it would have been far easier to hold that oil is a biological material, a listed category of pollutant? It might be explained by the fact that the decision was written during the heyday of using legislative history as an interpretive device. Eskridge, *supra* note 70, at 207-38; Scalia & Garner, *supra* note 11, at 369-90. *See, e.g.*, Train v. Colorado PIRG, 426 U.S. 1, 6 ELR 20549 (1976). The court may also have been hesitant to start down the biological material path, knowing full well that live fish and perhaps people lay ahead as pollutants.

2. Section 404: Dredged and Fill Material

Section 301(a) prohibits the discharge of a pollutant, except in compliance with a §402 or §404 permit. Section 402 authorizes EPA to issue permits for the discharge of a pollutant, and §404 authorizes the U.S. Army Corps of Engineers (the Corps) to issue permits for the discharge of "dredged or fill material." Many, if not most, courts conclude from this that dredged or fill material is a pollutant, in conclusory formulations such as that §301(a) "prohibits the discharge of any pollutant, including dredged or fill material."[178] A few courts mention §502(6) in support of the conclusion. These decisions, however, are in part either wrong or misleading. Section 301(a) does not define pollutant, much less define it to include dredged or fill material. Section 502(6) defines pollutant to include "dredged spoil"; it does not define pollutant to include either "dredged or fill material" or "fill material." The unstated reasoning of these decisions may be that because §301(a) prohibits the discharge of pollutants except in compliance with permits issued under §402 or §404, and because §404 authorizes the Corps to issue permits for the discharge of dredged or fill material, then dredged or fill material must be a pollutant. This is circular reasoning. There would be no reason for dischargers of dredged or fill materials to apply for §404 permits unless dredged or fill materials are pollutants, because their discharges without permit are not otherwise illegal under §301(a), §404, or any other provision of the CWA. But for the absence of dredged or fill material from the definition of pollutant in §502(6), the courts' logic might be persuasive. But because the definition in §502(6) is exclusive, its failure to explicitly include "dredged or fill material" means that such material is not a pollutant unless it or its constituents are listed in the definition or fall within a category of materials listed in the definition.

Fortunately, the §502(6) definition of pollutant includes "dredged spoil," a term practically synonymous with "dredged material."[179] Although there

178. Bersani v. EPA, 850 F.2d 36, 39, 18 ELR 20874 (2d Cir. 1988). *See also* Leslie Salt Co. v. United States, 55 F.3d 1388, 1391, 25 ELR 21046 (9th Cir. 1995); Town of Norfolk v. Corps of Eng'rs, 968 F.2d 1438, 1445, 22 ELR 21337 (1st Cir. 1992) (citing §502(6)); United States v. Brink, 795 F. Supp. 2d 565, 575 (S.D. Tex. 2011); Gouger v. Corps of Eng'rs, 779 F. Supp. 2d 588, 603 (S.D. Tex. 2011); Alabama Rivers Alliance, Inc. v. Corps of Eng'rs, 697 F. Supp. 2d 1251, 1258 (N.D. Ala. 2009); D'Olive Bay Restoration and Preservation Committee, Inc. v. Corps of Eng'rs, 513 F. Supp. 2d 1261, 1268 (S.D. Ala. 2007); Florida Wildlife Fed'n v. U.S. Army Corps of Eng'rs, 401 F. Supp. 2d 1298, 1308 (S.D. Fla. 2005) (under §502(6) "dredged or fill materials are pollutants"); City of Shoreacres v. Waterworth, 332 F Supp. 2d 992, 1015 (S.D. Tex. 2004); National Wildlife Federation v. Norton, 332 F. Supp. 2d 170, 185 (D.D.C. 2004). *Contra, see* Kentuckians for Commonwealth, Inc. v. Rivenburgh, 204 F. Supp. 2d 927, 932, n.5 (S.D.W. Va. 2002), labeling this as circular reasoning, *rev'd on other grounds*, 317 F.3d 425 (4th Cir. 2003).

179. In *United States v. Wilson*, 133 F.3d 251, 259, 28 ELR 20299 (4th Cir. 1997), the court commented that while "the statutory term 'dredged spoil' carries with it a more pejorative connotation than does

is no hint of "fill material" in the §502(6) definition, most fill material consists at least in part of one of the specific substances included in that definition (rock and sand), or a substance that is within one of the categories included in that definition (demolition debris that falls within the solid, industrial or municipal waste categories).[180] Assuming that fill material is composed at least in part of pollutants, which agency should issue a permit for the discharge: EPA for the discharge of the included pollutants or the Corps for the discharge of the fill material?[181] The interpretive canon that the specific statutory provision governs over the general provision[182] suggests that when a broadly defined pollutant is also a narrowly defined "fill material," then the more specific §404 governs, authorizing the Corps to issue a permit for the discharge of the fill material, rather than authorizing EPA to issue a permit for the discharge of a pollutant included in the fill material. Section 402(a) confirms this by granting EPA authority to issue permits for the discharge of pollutants, "[e]xcept as provided in" §404. This leaves the Corps as the agency issuing permits for the discharge of dredged and fill material, and EPA as the Agency issuing permits for the discharge of all other pollutants.

That was the conclusion reached by courts considering the matter, including the Supreme Court in *Coeur Alaska, Inc. v. Southeast Alaska Conservation Council.*[183] In *Coeur Alaska*, environmental petitioners sought judicial review of a §404 permit authorizing the discharge of a slurry of 30% crushed rock and 70% water from a gold-mining operation[184] into a 23-acre lake that all parties agreed was navigable water. The operation would allow the solids to separate from the slurry by settling to the bottom of the lake, eventually filling all of the natural lake and more than doubling its surface. The operation would build a dam across the outlet from the lake, preventing the slurry water from flowing downstream without treatment. EPA issued a §402 permit imposing effluent limitations requiring the treatment of water flowing downstream from the dam. The Corps issued a §404 permit to fill the lake.

---

the term the court used [in its jury instructions], 'dredged material,' the two are not sufficiently different to constitute error."

180. United States v. Pozsgai, 999 F.2d 719, 23 ELR 21012 (3d Cir. 1993).

181. Justice Stephen Breyer explores this issue in his concurring opinion in *Coeur Alaska, Inc. v. Southeast Alaska Conservation Council*, 557 U.S. 261, 292-93 (2009) (Breyer, J., concurring).

182. Eskridge, *supra* note 70, at 324, citing Green v. Bock Laundry Mach. Co., 490 U.S. 504, 524-25 (1989); Crawford Fitting Co. v. J.T. Gibbons, Inc., 482 U.S. 437 (1987); Scalia & Garner, *supra* note 11, at 183-89.

183. 557 U.S. 261 (2009). *See also* Resource Invs., Inc. v. U.S. Army Corps of Eng'rs, 151 F.3d 1162, 28 ELR 21407 (9th Cir. 1998); Greenfield Mills, Inc. v. Macklin, 361 F.3d 934, 946 (7th Cir. 2004); United States v. United Homes, Inc., 1999 WL 117701 (N.D. Ill. 1999).

184. The operation re-refined previously processed ore by crushing it and treating it with a chemical bath to float and recover remaining gold.

Environmental petitioners contended that discharging pollutants into the lake required a §402 permit, not a §404 permit. The Ninth Circuit agreed, but the Supreme Court reversed.

Section 402 permits issued by EPA or states with approved programs, and §404 permits issued by the Corps or states with approved programs, are subject to very different conditions and limitations. Section 402 permits impose limitations on the discharge of pollutants requiring pollution reduction reflecting the application of the best available treatment technology and meeting state-developed water quality standards. Section 404 permits do not require pollution reduction, but may impose conditions on authorized projects to protect the environment. The Corps may also condition or deny permits when a public interest review determines that their negative environmental impacts outweigh any social or economic benefit. Thus, whether particular pollutants are "dredged or fill material" makes a tremendous difference in the viability of the operation producing the pollutants and the environmental protection afforded the receiving water. Covering streams and filling valleys with mining overburden may be authorized by a Corps §404 permit if the overburden is "fill material." But the same operation would probably be stymied if the pollutants discharged to the streams first had to meet technology-based and water quality standards under an EPA-issued §402 permit.[185] In *Coeur Alaska*, for instance, if a §402 permit was required, the particular mining operation would have been subject to a new source performance standard of zero discharge of pollutants.[186] Under the circumstances, however, the tailings pond technology on which the §402 standard was justified would have resulted in a greater loss of wetlands and natural habitat.[187]

The CWA does not define dredged material. Although CWA §502(6) defines pollutant to include dredged spoil, the statute does not define dredged spoil. EPA and the Corps both define dredged material to mean "material that is excavated or dredged from the waters of the United States."[188] As the concept of navigable waters expands to include wetlands, the question arises whether digging in a wetland requires a §404 permit. Because §404 requires a permit for discharging dredged material into a wetland, not for removing it from a wetland,[189] however, those issues revolve around whether particular

185. *See Coeur Alaska*, 557 U.S. 261; Kentuckians for Commonwealth, Inc. v. Rivenburgh, 317 F.3d 425 (4th Cir. 2003).
186. *Coeur Alaska*, 557 U.S. at 266.
187. *Id.* at 269-70.
188. 33 C.F.R. §323.2(c) and 40 C.F.R. §232.2.
189. Save Our Community v. U.S. EPA, 971 F.2d 1995 (5th Cir. 2002); Orleans Audubon Society v. Lee, 742 F.2d 901, 15 ELR 20030 (5th Cir. 1984).

movements of soil and vegetation in wetlands during landclearing operations constitute addition.[190]

The CWA does not define fill material. EPA and the Corps both define the term to mean "material placed in the waters of the United States where the material has the effect of: (i) Replacing any portion of a water body with dry land; or (ii) Changing the bottom elevation of any portion of the water of the United States."[191] The definitions give inclusive examples of "rock, sand, soil, clay, plastics, construction debris, wood chips, overburden from mining or other excavation activities, and materials used to create any structure or infrastructure in the waters of the United States." Both definitions specifically exclude "trash or garbage," which neither EPA nor the Corps defines. The Corps earlier defined fill material to mean "any material used for the primary purpose of replacing an aquatic area with dry land of or changing the bottom elevation of a [ ] water body." That definition excluded pollutants discharged "primarily to dispose of waste" (not just trash and garbage), an activity it acknowledged was regulated under CWA §402.[192] The newer rule has the virtue of being objective rather than depending on the discharger's state of mind. However, the newer definition creates a giant loophole in §402, allowing all sorts of waste (excepting only trash and garbage) disposal into water under the guise of filling the water or changing its bottom elevation under §404.[193]

While typical municipal and industrial wastewaters are primarily liquid, they also have solid components that can settle out on the bottom of the receiving water and change its elevation, although that is more likely to occur in quiescent rather than in turbulent water. Indeed, in *Rapanos v. United States*, the Supreme Court toyed with the idea that *liquid* effluents are pollutants subject to §402 permits, while *solids* are fill material subject to §404 permits.[194] The Court stated that "'dredged or fill material' . . . unlike tra-

190. For a discussion of the "addition" element, see *supra* Chapter 1.

191. 33 C.F.R. §323.2(e) and 40 C.F.R. §232.2.

192. 42 Fed. Reg. 37122, 37145 (July 19, 1977); 33 C.F.R. §323.2(m) (1978).

193. Neither the older nor the newer version of the definition deals comfortably with the construction in a wetland of a sanitary landfill for the disposal of municipal solid waste. Under either definition, the placement of trash and garbage is excluded from fill material, so that a §402 permit rather than a §404 permit would be required for the operation of the facility. But the construction of the bottom liners and leachate collection systems of the landfill would seem to fit within the definition of fill material, requiring a §404 permit for its construction. Because construction of municipal landfilling wetlands is directly regulated by Subtitle D of RCRA, however, the Ninth Circuit has ruled that it is not regulated by CWA §404. *See* Resource Invs., Inc. v. U.S. Army Corps of Eng'rs, 151 F.3d 1162, 28 ELR 21407 (9th Cir. 1998).

194. "[T]he deposit of *mobile* pollutants into upstream ephemeral channels is naturally described as an 'addition . . . to navigable waters,' 33 U.S.C §1362(12), while the deposit of *stationary* fill material generally is not." Rapanos v. United States, 547 U.S. 715, 745 n.11, 36 ELR 20116 (2006).

ditional water pollution, are solids that do not readily wash downstream."[195] The plaintiffs in *Coeur Alaska* argued that this created a huge loophole in the CWA, and the Supreme Court dissent agreed.[196] Any industrial wastewater with solids in it could be classified as "fill material" if the solids can accumulate on and eventually raise the level of the water bottom. That could effectively remove large segments of industry from the water pollution abatement requirements of the CWA §402 program.[197] The argument was weakened in *Coeur Alaska*, however, because EPA and the Corps agreed that the fill project at issue was governed by §404, and their regulations provided that mining overburden was fill material. Moreover, EPA and the Corps treated the altered lake as a treatment facility by requiring a §402 permit at the outlet of the lake into a navigable river. The Court majority acknowledged the potential loophole, but left the question for another day, if and when EPA and the Corps agreed that other, more obnoxious industrial waste was fill material.[198] In any event, the Court pointed out that EPA could veto Corps permits under §404(c).

The distinction between whether material discharged to water is subject to a §402 or a §404 permit is easy in most cases. The discharge of effluent containing pollutants, such as PCBs, is regulated under §402 because the main impact of the discharge is on public health and water quality. The discharge of soil to extend a building lot into a river is regulated under §404 because its impact is on the navigability of the river. However, if the PCBs are attached to solids in a liquid effluent, under the current regulatory definition, the discharge is arguably regulated under §404, because if the solids accumulate, they can change the bottom elevation of the receiving water. Treating toxic solids as fill material subject to a §404 permit, however, ignores the differences between the purposes and operations of the two sections. The confusion may be the inevitable consequence of mixing water pollution control regulation with water projects regulation, but EPA and the Corps

195. *Id.* at 723.

196. Coeur Alaska, Inc. v. Southeast Alaska Conservation Council, 557 U.S. 261, 302-03 (2009).

197. EPA commonly regulates total suspended solids. *See, e.g.*, best practicable treatment technology for the Rayon Fibers Subcategory of the Organic Chemicals, Plastics and Synthetic Fibers Effluent Guidelines, 40 C.F.R. §414.21.

198. *Coeur Alaska*, 557 U.S. at 275-76. The mining company in that case proposed to treat mining waste by turning a wilderness lake into a settling basin. It applied to the Corps for a §404 permit to fill the lake and to EPA for a §402 permit to discharge water from the lake to the stream originally draining the lake. At the termination of mining activity, the company would restore the lake. The alternative was to build a settlement basin of considerable height and extent that would not have to be deconstructed at the termination of operations. EPA, the Corps, and the Supreme Court majority seemed to think the proposal was the better alternative. *Id.* at 269-70. It goes without saying that the best environmental alternative would have been to have neither the settlement basin nor the flooded lake.

could eliminate confusion and ambiguity by a more careful phrasing of their regulations. Indeed, EPA and the Corps should return to their earlier regulatory definitions of fill material, because those earlier definitions better track differences between the intended purposes of §402 and §404.

G. *Potential Conflicts With Other Statutes*

Special care is needed to interpret a statute in cases where another statute may apply to the same fact pattern, especially if an interpretation of one statute may interfere with achieving the purposes of the other statute. In these cases, common wisdom cautions to read the statutes harmoniously with each other, and certainly canons of statutory construction require a harmonious reading.[199] Pollution control statutes create this type of problem because they normally protect a single environmental medium (air, land, or water) from contamination, whereas pollutants tend to migrate between environmental media. For instance, a power plant's air pollution, which is regulated by the Clean Air Act (CAA),[200] not only fouls the air, but may also be redeposited by gravity or precipitation onto land or water. Indeed, air emissions from coal-burning power plants have long been known as the primary sources of acid deposition in the northeastern United States and neighboring Canada, causing both acid rain and acid runoff into lakes and other surface water.[201] Air pollution from such sources could theoretically be regulated by land and water pollution statutes as well as by an air pollution statute. Pollution control statutes that regulate particular substances (for example, PCBs) create the same sorts of problems.[202] Such substances are ubiquitous in the environment and potentially fall within the purview of the multiple environmental media-oriented statutes. Because EPA administers most of these statutes, it can often implement or interpret them to avoid significant conflicts. Conflicts between EPA-administered statutes may reach the courts when envi-

199. Statutes are to be interpreted *in pari materia* to avoid conflicts. *See* Eskridge, *supra* note 70, at 327, citing Morales v. TWA, Inc., 112 S. Ct. 2031 (1992); TWA, Inc. v. Independent Fed'n of Flight Attendants, 489 U.S. 426 (1989); Scalia & Garner, *supra* note 11, at 252-55.

200. Clean Air Act (CAA), 42 U.S.C. §§7401-7671q, ELR Stat. CAA §§101-618.

201. The problem led to the amendment of the CAA to include the Acid Deposition Program, creating a national cap-and-trade program for the reduction of sulfur oxide emissions from large sources. *See* CAA §§402-416, 42 U.S.C. §§7651a-7651o. Acid deposition is not the only such interstate transport problem. "Burning coal is the main source of mercury generation in our [Great Lakes] region and atmospheric deposition of mercury into the Great Lakes from coal accounts for seventy five percent of those highly toxic compounds." Kendra Fogarty et al., *Emerging Legal Issues in the Great Lakes Such as the Public Trust Doctrine, Subterranean Rights, and Municipal Regulatory Arrangements.*, 34 Can.-U.S. L.J. 279, 309 (2010).

202. Regulated by the Toxic Substances Control Act (TSCA), 15 U.S.C. §§2601-2692, ELR Stat. TSCA §§2-412. *See* 15 U.S.C. §2605(e).

ronmental advocates either challenge an EPA regulation attempting to resolve an interstatute conflict or sue a violator of one environmental statute who defends by alleging that another environmental statute authorized its actions.[203] Statutes administered by other agencies lead to the same types of conflict, with less opportunity or incentive for EPA to avoid the conflict.

1. Atomic Energy Act

The first such conflict under the definition of pollutant arose because Congress defined pollutant in CWA §502(6) to include "radioactive materials," while Congress had earlier established a "pervasive regulatory scheme"[204] for source, byproducts, and special nuclear materials under the Atomic Energy Act (AEA),[205] administered by the Atomic Energy Commission (AEC) or later the Nuclear Energy Regulatory Commission and the U.S. Department of Energy. If the CWA authorized EPA to regulate radioactive discharges from nuclear power plants, the AEC would have lost significant authority over the development of nuclear power for peaceful use. EPA defined pollutant to exclude material "regulated under the Atomic Energy Act of 1954."[206] An environmental group challenged this exclusion in *Colorado PIRG*.[207] The Supreme Court ruled that the CWA's legislative history made explicit that Congress did not intend EPA to regulate source, byproduct, and special nuclear materials,[208] although this did not rule out EPA authority over other water pollutants discharged by nuclear power plants, such as heat. Another environmental group later filed suit against a mining company for its discharge of uranium mill tailings into navigable water, with the same outcome in the Ninth Circuit for the same reasoning.[209] These decisions both ignore

203. Environmental plaintiffs have attempted to bring CWA citizen suits against air pollution sources when their air pollutants eventually enter water. The courts have summarily held that air pollution is regulated by the CAA, not by the CWA. *See, e.g.*, Chemical Weapons Working Grp. v. Dep't of Army, 111 F.3d 1485, 1490-91, 27 ELR 21130 (10th Cir. 1997); *see also* U.S. EPA ex rel. McKeown v. Port Authority of New York and New Jersey, 162 F. Supp. 2d 173, 189 (S.D.N.Y. 2001). Environmental plaintiffs also challenged EPA's attempts to reconcile the CWA with RCRA in *Chemical Waste Mgmt. v. U.S. EPA*, 976 F.2d 2, 23 ELR 20024 (D.C. Cir. 1992), and RCRA with the Safe Drinking Water Act, 42 U.S.C. §§300f to 300j-26, ELR Stat. SDWA §§1401-1465, in *NRDC v. U.S. EPA*, 907 F.2d 1146, 20 ELR 21274 (D.C. Cir. 1990).
204. Train v. Colorado PIRG, 426 U.S. 1, 24, 6 ELR 20549 (1976).
205. Atomic Energy Act (AEA) of 1954, 42 U.S.C. §§2114.
206. 40 C.F.R. §22.2.
207. 426 U.S. 1. The challenge took the form of a citizen suit against EPA for failing to carry out a mandatory duty. The challenge, however, should have been for judicial review of EPA's exclusionary regulation. EPA had performed its mandatory duty of promulgating regulations establishing the §402 regulations; the plaintiff objected to the terms of those regulations.
208. *Colorado PIRG*, 426 U.S. at 10-23.
209. Waste Action Project v. Dawn Mining Corp., 137 F.3d 1426, 28 ELR 21035 (9th Cir. 1998). The Ninth Circuit also considered the applicability of the Uranium Mill Tailings Regulation Control Act.

the canon that the later statute governs over the earlier statute;[210] Congress enacted the CWA in 1972, 18 years after the 1954 enactment of the AEA.[211] The decisions could have relied on the canon that the specific governs over the general, for the AEA's jurisdiction over "source, by-product and special nuclear materials"[212] is more specific than the CWA's jurisdiction over pollutants or over radioactive materials generally. All of these canons, however, can be seen as shortcuts to determine legislative intent, and in this case, legislative intent on the issue was directly addressed by persuasive legislative history.[213] It is worth noting that the *Colorado PIRG* decision was written in the heyday of judicial use of legislative history as an interpretive method, a practice that has been much curtailed in more recent years[214] in favor of an emphasis on plain meaning.

2. FIFRA

The most widespread and considered cases of conflict between EPA-administered statutes are between the CWA and FIFRA. FIFRA pervasively regulates consumer products; pesticides cannot be manufactured, sold, or used until they are registered by EPA. Registration establishes the uses to which pesticides may be put and the means by which they can be applied for those uses. The uses allowed and the application directions are calculated to avoid danger to public health and unintended[215] danger to the environment. Thus,

---

210. Where a later statute conflicts with an earlier statute, it impliedly repeals the earlier statute to the extent of the conflict. *See, e.g.*, Dorsey v. United States, 132 S. Ct. 2321, 2332 (2012).

211. Oct. 18, 1972, Pub. L. No. 92-500 (CWA); Aug. 30, 1954, Pub. L. No. 85-256 (AEA).

212. The Court in *Colorado PIRG* reviews the legislative history, including explicit statements in the reports on the bill by both the responsible Senate and U.S. House of Representatives Committees, colloquies on the floors of both chambers, and the defeat in the House of an amendment to give states authority to regulate radioactive discharges. *See* 426 U.S. at 11-24. The Court found particularly persuasive an extensive dialogue on the Senate floor between Senator Muskie, chief sponsor and author of the CWA, and Sen. John Pastore (D-R.I.), Chair of the Joint Committee on Atomic Energy.

213. The Court held that the legislative history supported the "intent to preserve the pre-existing regulatory plan." 426 U.S. at 24. Textualists argue that interpretation is not directed at discerning legislative intent, but only at discerning what the statute means. *See* Scalia & Garner, *supra* note 11, at 391-96 (discovering legislative intent included as one of treatise's "Thirteen Falsities Exposed").

214. Courts used legislative history in 5 of the 11 decisions interpreting pollutant, or 45%, dated 1982 or earlier, but only in 7 of the 59 decisions, or 12%, dated later. Count the number of decisions dated 1982 or earlier and the number of decisions dated later in Column B of Table B for Pollutant, *infra*. Count the number of times that the earlier set of decisions and the later set of decisions cite device 16 in Column E. Divide the number of times the earlier and later sets of decisions cite device 16 by the number of decisions in the earlier and later sets. Much of the reason for the shift away from reliance on legislative history was the influence of the new textualists on the courts and in academia. One treatise traces the ebb and flow of the Supreme Court's use of legislative history to support or escape the plain meaning of a statute. *See* Eskridge, *supra* note 70, at 207-38.

215. It goes without saying that the very purpose of pesticides is to damage specific living parts of the environment.

FIFRA registration establishes whether pesticides can be applied over or near water and, if so, the FIFRA-approved label establishes how they must be applied over or near water.

As discussed above, several CWA decisions concern the application of pesticides on, in, or near water. *Headwaters* was the first of the court of appeals decisions and the only one directly considering a potential conflict between FIFRA and the CWA and the consequent need to read the statutes harmoniously. *Headwaters* concluded that the statutes did not conflict because they served different purposes. The Ninth Circuit noted that FIFRA established a "nationally uniform labeling system to regulate pesticide use," based on a national cost/benefit analysis weighing the benefits of using the pesticide against the adverse environmental effects of doing so.[216] Congress made no effort in FIFRA to ensure that individual applications of pesticides were compatible with local conditions. On the other hand, CWA requires that §402 permits contain effluent limitations designed to achieve local water quality standards, preventing local environmental harm from individual dischargers, as well as to achieve national technology-based standards. The *Headwaters* court concluded that these were compatible goals and that FIFRA registration did not foreclose requiring a CWA permit for discharging chemical wastes.[217]

The issue of a conflict between FIFRA and the CWA did not arise in the Ninth Circuit's subsequent *Fairhurst* decision[218] because the pesticide applied in that case was a chemical leaving no waste or excess in the water, so there was no chemical waste and hence no pollutant. The court erroneously noted, however, that the necessity to read the CWA and FIFRA harmoniously was required by the Ninth Circuit's earlier holding in *Headwaters*. The Sixth Circuit did not reconsider the issue in *National Cotton Council.*[219]

3. Refuse Act

The CWA §402 permit program was based on the Refuse Act Permit Program, developed by EPA and the Corps under the Refuse Act.[220] Congress took care in the CWA to reconcile the two statutes. Sections 402(a)(4) and (5) provide that no Refuse Act permits for discharges of refuse into navigable waters can be issued after the enactment of the CWA, but that appli-

216. Headwaters, Inc. v. Talent Irrigation Dist., 243 F.3d 526, 532, 31 ELR 20535 (9th Cir. 2001).
217. *Id.* at 531-32.
218. 422 F.3d 1146 (9th Cir. 2005).
219. National Cotton Council of Am. v. U.S. EPA, 553 F.3d 927, 39 ELR 20006 (6th Cir. 2009).
220. 33 U.S.C. §407.

cations for Refuse Act permits filed before the date of that enactment were deemed to be applications for CWA §402 permits. Moreover, Refuse Act permits issued before that date were deemed to be CWA §402 permits, and CWA §402 permits were deemed to be Refuse Act permits. Finally, CWA §402(k) provided a grace period until the earlier of: (1) the end of 1974 for a permit applicant filing a timely application under §402 or the Refuse Act; or (2) until a permit was issued or denied. Although these measures eliminated conflict between CWA §402 and the Refuse Act, defendants from time to time have attempted unsuccessfully to obfuscate the applicability of the CWA by invoking the Refuse Act.[221]

The Refuse Act is one section of the Rivers and Harbors Act of 1899, which includes §403[222] requiring a permit from the Corps to dredge in navigable waters. Dredging produces dredged material or dredged spoil that must be disposed. If dredged material was disposed elsewhere in navigable water prior to 1972, a Refuse Act permit was required. In 1972, CWA §404 authorized the Corps to issue permits for the discharge of dredged and fill material into navigable waters, a function it performed earlier under the Refuse Act. Oddly, §404 did not contain the same sort of provisions as §402 for meshing the new permitting authority with the earlier Refuse Act. CWA §511(a), however, provided generally that the CWA did not supersede the authority of the Corps "to maintain navigation" and specifically that a §404 permit "shall be conclusive as to the effect on water quality of any discharge resulting from any activity subject to section 403 (ocean discharge) of this title."

III. Conclusion

"Pollutant" is not a limiting element of the water pollution offense. Congress intended the element to reach broadly, and it does; its statutory definition encompasses virtually all residuals and byproducts of human activity as well as biological materials. Courts have only held that seven substances or types of substances discharged to water by human activity do not fall within the statutory definition of pollutant, and there are contrary opinions concerning four of the seven. Moreover, the rulings that most of the seven substances are not pollutants were predicated on the argument that their addition to water was governed by another statute rather than on the argument that they did not fit within the definition of pollutant.

221. *See, e.g.*, United States v. Hamel, 551 F.2d 107, 112, 7 ELR 20253 (6th Cir. 1977).
222. 33 U.S.C. §403.

Perhaps, because of the broad plain meaning of the statutory definition of pollutant, there were relatively few judicial challenges to the interpretation of the element. Courts resolved those challenges fairly easily by reference to the plain meaning of its statutory definition, or, after time, by reference to precedents based on the plain meaning of the definition of the element. Courts relied on other canons of statutory construction only to a minor extent to interpret pollutant. Indeed, courts have used only 10 canons beyond plain meaning and precedent to interpret "pollutant," and the courts cited plain meaning and precedent slightly more than they cited the remaining canons in the aggregate.[223] Most courts recognized that Congress decoupled the meanings of pollutant and pollution in the CWA and that a pollutant under the statute does not have to cause pollution. Moreover, a substance does not have to result from human activity to be a pollutant.

The main problem with the definition of pollutant is that its plain meaning sometimes seems to reach too far. Most questionable results from that, however, are blunted because the finding that a substance is a pollutant has no legal effect unless all of the other elements of the water pollution offense are also found. Even so, there are a few specific areas in which congressional or EPA action is warranted. Most follow from the definition listing categories of substances denominated both "material" and "waste," with no apparent reason for the distinction. Why did Congress include all biological material, but only chemical, solid, industrial, and municipal waste? The inclusion of all biological material leads to the absurd result that propagating native fish in a hatchery and adding them to a fishing stream can violate the CWA without a permit. The same problem would occur for other biological material; for example, for planting wild rice or eel grass where they have been depleted by human activity. Congress could easily deal with this issue by a surgical amendment changing "biological material" in CWA §502(6) to "biological wastes." Alternatively, EPA could address the issue by defining "biological material" to exclude "indigenous fish or biota," or by issuing a general permit or a permit by regulation authorizing such discharges under appropriate conditions; for example, where the fish are native species rather than introduced species. Similar solutions are possible for analogous *reductio ad absurdum* situations; for example, people conveyed to navigable waters from diving boards or water slides.

223. *See infra* Table B for Pollutant. Count the number of interpretive devices, other than 22 and 23, cited in Column E. Count the number of times that those devices are cited, 84, and the number of times devices 22 and 23 are cited, 87.

More difficult are the statute's and EPA's failures to define waste, in particular the failure to address the issue of whether consumer products designed for use on or in water become wastes when their useful life is spent. Positioning a shotgun to fire every half hour over (and therefore into) a body of water is a classical addition of pollutants into navigable water from a point source without a permit. However, if that shotgun is fired only for the purpose of killing ducks when they fly into its trajectory, it may not be an addition of pollutants even when some or all of the shot misses the ducks and falls into the water. Variants of this fact pattern are limited only by the imagination: Examples include spraying pesticide on mosquito larvae in water or spraying paint on a bridge.[224] EPA's reaction when faced with such situations is to promulgate a rule exempting the discharges resulting from such activities from requiring §402 permits, a strategy that courts have rejected as contrary to the CWA. EPA might be more successful promulgating a regulatory definition of waste, as it has done under RCRA generally,[225] or for particular substances, such as military munitions.[226] Alternatively, it could issue a permit by rule or authorize states with approved programs to issue permits by rule for such categories of discharges.

The distinction between pollutants subject to §402 permits and dredged and fill material subject to §404 permits is also problematic because "fill material" does not appear in the statutory list of substances that are "pollutants," although most fill material does fall within one of the categories of substances that are listed in those statutory categories. EPA and the Corps have created a loophole in the §402 program for requiring pollution reduction by defining fill material as anything that can change the bottom elevation of a water body. Because toxic substances may be solids or be attached to suspended solids in liquid waste, such substances may be fill material regulated by §404 rather than by §402. The Supreme Court acknowledged this loophole in *Coeur Alaska*.[227] The agencies could easily deal with the loophole by reverting to an earlier version of their definitions.

224. These two examples illustrate the difficulty in this task, because pesticides cannot be sprayed on aquatic pests without spraying waste or residue pesticides in the water, while properly placed plastic barriers can collect sprayed paint that misses the bridge and prevent it from entering the water.
225. 40 C.F.R. §261.2.
226. 40 C.F.R. §266.202, which survived judicial review in *Military Toxics Project v. EPA*, 146 F.3d 948, 28 ELR 21350 (D.C. Cir. 1998).
227. Coeur Alaska, Inc. v. Southeast Alaska Conservation Council, 557 U.S. 261 (2009).

Chapter 3. "Navigable Waters"

This chapter examines the interpretation of the third element of the water pollution offence: "navigable waters." The courts had long interpreted that phrase as a traditional touchstone of jurisdiction under the Commerce Clause, denoting waters associated with interstate and foreign commerce.[1] Congress defined "navigable waters" in the CWA to mean "the waters of the United States,"[2] a phrase having nothing to do with waterborne transportation. EPA and the courts have struggled to reconcile the traditional notion of navigability with Congress' broader definition of the element. This chapter reviews the latest round in this struggle, EPA's detailed and exhaustive amendments to its definition in 2015 (which, under the Trump Administration, EPA proposed to rescind as this book was in production). The chapter also examines EPA's unitary navigable waters theory, concludes the theory is inconsistent with the CWA, and argues EPA should abandon the theory and the courts should strike it down.

In §502(7), the CWA defines navigable waters to mean "the waters of the United States, including the territorial seas." Of all the elements in the CWA's core command, only "navigable waters" had a developed legal meaning before enactment of the statute. The U.S. Supreme Court developed that legal meaning in half a dozen decisions over the preceding century and a half,[3] establishing the extent of the U.S. Congress' constitutional authority to develop and regulate

1. *See* The *Daniel Ball*, 77 U.S. 557 (1870).
2. 33 U.S.C. §1362(7).
3. The Court has interpreted navigable waters in six decisions: *Los Angeles County Flood Control Dist. v. Natural Res. Def. Council*, 133 S. Ct. 710, 43 ELR 20004 (2013); *Rapanos v. United States*, 547 U.S. 715, 36 ELR 20116 (2006); *South Fla. Water Mgmt. Dist. v. Miccosukee Tribe of Indians*, 541 U.S. 95, 34 ELR 20021 (2004); *Solid Waste Agency of N. Cook Cnty. v. U.S. Army Corps of Eng'rs (SWANCC)*, 531 U.S. 159, 31 ELR 20382 (2001); *International Paper Co. v. Ouellette*, 479 U.S. 481, 17 ELR 20327 (1987); and *United States v. Riverside Bayview Homes, Inc.*, 474 U.S. 121, 16 ELR 20086 (1985). It has interpreted "addition" in two decisions, "pollutant" in three decisions, and "point source" in one decision. *See infra* Table B under "Supreme Court Decisions" for each element. The lower courts interpreted navigable waters in 142 decisions; addition in 68 decisions, pollutant in 70 decisions, and point source in 72 decisions, *see infra* Table B. The greater number of navigable waters decisions also results from parties challenging interpretations of that element more than the other elements, but that probably results from the same factors.

   Chapter 5, Section II.B.3., *infra*, however, demonstrates that the predominant reason there are more decisions interpreting "navigable waters" than interpreting any of the other elements is that there are far more §404 decisions interpreting "navigable waters" than any of the other elements.

waterways under the Commerce Clause. The statutory definition of navigable waters attempts to broaden the element's meaning, while the statutory definitions of the other elements narrow their meanings. (For example, the statutory definition of "pollutant" limits it to specifically listed materials and categories of materials.[4]) Congress also devoted substantial legislative history to the intended broad reach of navigable waters, while providing far less legislative history to the meanings of the other elements.[5]

The courts' historical familiarity with interpreting navigable waters suggests they may be more comfortable interpreting the term under the CWA than interpreting other elements of the CWA offense. Indeed, this is the case; the Supreme Court and lower courts have interpreted navigable waters far more often than they have interpreted any of the other elements. The historical role of the courts in shaping the meaning of the term navigable waters also suggests that they should be more willing to disregard or overrule an Agency interpretation of the term than to disregard or overrule Agency interpretations of other elements of the CWA offense. Indeed, this too is the case; the Court has twice overruled EPA and U.S. Army Corps of Engineers (the Corps) interpretations of navigable waters, but has not overruled their interpretations of other elements in the CWA offense.[6] Finally, the availability of one century and a half of judicial decisions interpreting navigable waters prior to the CWA makes it likely that courts will use precedent more often than other interpretive devices such as plain meaning to determine the meaning of the term in the CWA. And indeed they have done so: Precedent is by far the most commonly used device for judicial interpretation of navigable waters,[7] while plain meaning is the most commonly used device for judicial interpretation of some of the other elements.

The major tension in interpreting the CWA's navigable waters element is the inherent conflict between judicial interpretation of Congress' authority

---

4. CWA §502(6), 33 U.S.C. §1352(6). The listed categories of materials, however, cover a vast territory.
5. Compare the discussion of the legislative history of "navigable waters" in this chapter *infra* at Section II.B., with the discussions of legislative history in the chapter on "Addition," *supra* Chapter 1, Section II., on "Pollutant," *supra* at Chapter 2, Section II.B., and on "Point Source," *infra* Chapter 4, Section II.B.
6. *Rapanos*, 547 U.S. 715; *SWANCC*, 531 U.S. 159.
7. In the 142 court decisions the author found interpreting navigable waters, precedent was used in 129 decisions. Courts used the second most popular interpretive device, broad interpretation to achieve statutory goals, in only 56 decisions. By contrast, in the 68 decisions interpreting pollutant, courts used plain meaning in 60 decisions, but precedent in only 30 decisions. *See infra* Column E in Table B for both Navigable Waters and Pollutant, counting interpretive devices 4, broad interpretation, 22, plain meaning, and 23, precedent.

to develop and regulate navigable waters for promoting interstate and foreign commerce, and Congress' subsequent use of the term to establish expansive EPA authority for improving and maintaining water quality. The importance of a waterway for transportation may have little connection to its water quality. Congress appeared to have understood this disconnect and to distance CWA jurisdiction from waterborne commerce by defining navigable waters as "the waters of the United States," a more expansive term having nothing to do with transportation. But because Congress used "navigable waters" as an element of the CWA offense, it invited judicial focus on the term rather than on its definition, because courts were familiar with that term, having interpreted it for over one century and a half. Courts understood the historical meaning of navigable waters far more than they understood the meanings of "waters of the United States" or of the CWA's other elements. And because of the Supreme Court's current fascination with textualism, it has great difficulty divorcing navigable waters entirely from waterborne transportation; "navigable" must mean something.

If the CWA dealt only with pollution control under the §402 permit program,[8] the disconnect between promotion of interstate commerce and protection of water quality might not have ripened into a jurisdictional crisis. The primary issue, most of the time, would have been whether particular tributaries of navigable waters were within the jurisdiction of the CWA, when those tributaries themselves had never and could never support waterborne transportation. Because tributaries, even remote ones,[9] contribute both water and pollution to the navigable waters into which they flow, they directly affect those navigable waters. The water volumes they contribute provide or increase the navigability of the receiving waters, and the water pollution they contribute may interfere with the navigability of the receiving waters. The Court has recognized congressional power to develop flood control projects in watersheds of navigable waters to protect against floods on the navigable waters into which the tributaries directly and indirectly flowed.[10] The same rationale applies to controlling discharges of pollutants into tributaries of navigable waters to protect the water quality of the navigable waters into which the tributaries flow.

But the CWA did not deal only with pollution control under the §402 permit program; it also dealt with filling wetlands under the §404 permit

8. 33 U.S.C. §1342.
9. Headwater streams, for instance, contribute 60% of the total flow to northeastern streams and rivers. *See* 79 Fed. Reg. 22188, 22224 (proposed Apr. 21, 2014) (scientific support for EPA's proposed amendments).
10. *See* Oklahoma ex rel. Phillips v. Atkinson, 313 U.S. 508 (1941).

program.[11] While wetlands contribute both water and pollutants to the navigable waters into which they flow and protect those navigable waters from flooding and pollution,[12] their contributions to adjacent or downstream navigation and water quality may not be apparent to the uninitiated. Moreover, wetlands may not exhibit surface water for months at a time, when to the untrained eye they may be indistinguishable from the surrounding dry land. Wetlands pose two problems for the Court's historical view or even for the plain meaning of navigable waters: Wetlands do not directly support waterborne navigation and, at least when they are dry, are not waters at all. Because a majority of CWA judicial decisions interpreting navigable waters are in §404 cases, 82 of 142,[13] the apparent incongruity to some between navigability and wetlands, particularly when they are dry, threatens to overflow into interpreting navigability in other CWA contexts.

Unfortunately, narrowing interpretations of navigable waters to defeat §404 jurisdiction over wetlands may also defeat §402 jurisdiction over tributaries of navigable waters, because terms are to be interpreted in the same manner throughout a statute,[14] unless the statute explicitly indicates otherwise. The CWA reiterates this interpretive canon by beginning its definitional §502 "[e]xcept as otherwise specifically provided, when used in this chapter . . . 'navigable waters' . . . 'means' . . . ." And it does not provide in §402, §404, or elsewhere that the term has different meanings in §§402 and 404.

Aside from the Corps' initial, interim regulatory definition of navigable waters, which merely repeated the Court's traditional interpretation of the term, EPA and the Corps' regulatory definitions have followed repeated legislative history to interpret the term broadly. After recent setbacks from the Supreme Court, however, the two agencies amended their definitions of "waters of the United States" to preserve much of their earlier regulations' breadth, but to retrench it where the Court's recent decisions give the agencies no room to do otherwise. The preambles to the rulemaking and of the proposed rulemaking offer detailed justifications for the controversial portions of the definitions, much of it scientific.[15]

---

11. 33 U.S.C. §1344.
12. The Court briefly surveyed the benefits of wetlands in *United States v. Riverside Bayview Homes, Inc.*, 474 U.S. 121, 134-35, 16 ELR 20086 (1985). For a more extensive treatment of the subject, see William L. Want, Law of Wetlands Regulation §2.01[1] (1993).
13. *See infra* Column H of Table B for Navigable Waters.
14. William N. Eskridge Jr., Dynamic Statutory Interpretation 324 (1994); Antonin Scalia & Bryan A. Garner, Reading Law 170-73 (2012).
15. *See* 80 Fed. Reg. 37054 (June 29, 2015) and 79 Fed. Reg. 22188 (proposed Apr. 21, 2014). The scientific support is provided at 80 Fed. Reg. 37061-65 and 79 Fed. Reg. 22222-52, but is referenced

These dynamics frame the interpretive battles examined in this chapter. We begin by tracing the development of the traditional judicial interpretation of navigable waters and how Congress and EPA attempted to extend its meaning. The chapter then examines the primary battlefields for interpreting navigable waters: tributaries and wetlands, isolated waters, groundwater, and EPA's unitary theory of navigable waters.[16] Along the way, the chapter takes sidelong glances at a wild card in these interpretive battles, the development of the CWA's §404 program regulating the filling of wetlands. The §404 program has provided proponents of narrowly interpreting "navigable waters" with the rhetorical argument that EPA and the Corps' broad interpretations of the term metaphorically make water of dry land.

I. The Traditional Legal Meaning of Navigable Waters

The legal concept of navigable waters originated in medieval England. The Crown owned the land beneath waters used for navigation and the public had a common-law right to use navigable waters for fishing and transport.[17] We know this public right today as the navigable servitude. Upon independence, the American colonies succeeded to ownership of the beds beneath navigable waters, subject to the common-law navigable servitude. As the United States acquired new territories, it took ownership of the beds beneath their navigable waters, again subject to the common-law navigable servitude. When new states were formed from those territories, ownership of the beds beneath the navigable waters transferred to those states, again subject to the navigable servitude.[18]

At the time that the 13 colonies formed the United States, only waters subject to the ebb and flow of the tide were considered navigable waters.[19] This limited concept of navigability expanded to meet a growing and industrializing nation's needs for federal improvement and regulation of waterborne commerce and transportation. In *Propeller Genesee v. Fitzhugh*,[20] the

throughout the preamble. EPA and the Corps, under the Trump Administration, proposed to rescind the 2015 rule as this book was in production. 82 Fed. Reg. 34899 (July 27, 2017).

16. For more information on EPA's unitary theory of navigable waters and its exemption of water transfers and diversions from §402, see *infra* Chapter 1, Section III.A.3.
17. Martin v. Waddell's Lessee, 41 U.S. 367, 407-14 (1842), traces this history back to the Magna Carta.
18. Pollard v. Hagen, 44 U.S. 212 (1843). The U.S. Constitution did not grant the shores of navigable waters and the soil beneath them to the United States, impliedly reserving them to the states. New states have the same rights, sovereignty, and jurisdiction over navigable waters as the original states. This is known as the "equal footing" doctrine: New states are admitted to the Union on an equal footing with the original states. *See also* Illinois Cent. R.R. v. Illinois, 146 U.S. 387 (1892), in which the navigable servitude gave rise to the public trust doctrine.
19. Wilson v. Blackbird Creek Marsh Co., 27 U.S. 245 (1845).
20. 53 U.S. 443 (1851).

Court held that navigable waters included freshwater. Not long after, it held in *Gibbons v. Ogden*[21] that the U.S. Constitution's Commerce Clause[22] conferred on Congress authority to regulate interstate commerce, including the authority to regulate navigation. Finally, in *The* Daniel Ball,[23] it held that waters forming a highway or part of a continuous highway of foreign or interstate commerce were subject to Commerce Clause jurisdiction.

The Court has subsequently broadened Commerce Clause jurisdiction to include waters that were once navigable, *Economy Power & Light Co. v. United States*[24]; waters that presently are capable of use for navigation, *The* Monticello[25]; and waters that could be made navigable with reasonable improvements, *United States v. Appalachian Electric Power Co.*[26] The Court has also recognized that Commerce Clause jurisdiction extends to tributaries of navigable waters or even to entire watersheds, where necessary to accomplish Commerce Clause-justified purposes; for example, flood control, as in *Oklahoma ex rel. Phillips v. Atkinson.*[27]

Traditional federal navigable water jurisdiction, therefore, extends to waters subject to the ebb and flow of the tide; waters that are navigable in fact, or once were navigable in fact, or could be made navigable in fact with reasonable improvements; and, for some purposes, to their tributaries or watersheds. But this is somewhat simplified, because it reflects an amalgam of decisions interpreting "navigable waters" for different purposes: ownership of land beneath navigable water; public rights to use navigable waters for fishing and transportation; admiralty jurisdiction; and Commerce Clause jurisdiction.[28] Of these uses, the definition of navigable waters for the purpose of delineating land ownership is, predictably, the most conservative, while the definition for Commerce Clause jurisdiction is more flexible.[29] Those differences should not be significant for interpreting navigable waters in the CWA, however, because, as discussed below, Congress intended that term to include all waters within Congress' constitutional authority.

---

21. 9 Wheat. 1; 21 U.S. 1 (1866).
22. U.S. CONST., art. I, §8, cl. 3.
23. 10 Wall. 557; 77 U.S. 557 (1871).
24. 256 U.S. 113, 123 (1921).
25. 20 Wall. 430, 441-42; 87 U.S. 430, 441-42 (1874).
26. 311 U.S. 377, 406 (1940).
27. 313 U.S. 508 (1941).
28. For more information on these uses of navigable waters, see A. DAN TARLOCK, THE LAW OF WATER RIGHTS AND RESOURCES (Clark Boardman 1988).
29. *See, e.g.*, Kaiser Aetna v. United States, 444 U.S. 164, 10 ELR 20042 (1979), in which a privately owned pond separated from an ocean bay by a barrier beach became navigable water for purposes of Commerce Clause navigational regulation under 33 U.S.C. §403 when the owners constructed a channel between the pond and the bay, but the private property did not become subject to the navigational servitude for public access without just compensation.

II. Congressional Action: Statutory Definition and Legislative History of Navigable Waters

A. *Statutory Definition*

Section 502(6) of the CWA defines navigable waters to mean "the waters of the United States, including the territorial seas."[30] As discussed above, "navigable waters" has a well-established legal meaning. "The waters of the United States," on the other hand, has no well-established meaning. Congress defined the limited "navigable waters" with the expansive "waters of the United States" to extend the CWA's jurisdiction to the constitutional limits of congressional authority for water pollution control.

But wording the underlying element to suggest traditional linkage with waterborne commerce was an odd way to accomplish expansiveness. The navigable waters formulation of the element has caused and continues to cause interpretive conflict. EPA and the Corps have promulgated detailed regulatory definitions of waters of the United States, claiming a broad reach for jurisdiction under the CWA.[31] Indeed, EPA ultimately abandoned reference to navigable waters, using waters of the United States throughout its regulations, perhaps hoping that others would forget the origin of the statute's jurisdiction in waterborne transportation. If so, it has not fooled the Supreme Court, which has twice ruled that the government's actions under §404 exceeded its statutory jurisdiction over navigable waters.[32]

B. *Legislative History*

1. Section 402

The CWA draws on two distinct statutory lineages, with different pollution control strategies and different jurisdictional bases. These lineages have their origins in the Federal Water Pollution Control Act (FWPCA) and the Refuse Act. Congress enacted the FWPCA in 1948[33] and has since amended it repeatedly.[34] The FWPCA initially encouraged and provided technical and

30. 33 U.S.C. §1362(6).
31. 40 C.F.R. §122.2 (EPA); 33 C.F.R. §328.3 (Corps).
32. *See* Rapanos v. United States, 547 U.S. 715, 36 ELR 20116 (2006); Solid Waste Agency of N. Cook Cnty. v. U.S. Army Corps of Eng'rs (*SWANCC*), 531 U.S. 159, 31 ELR 20382 (2001).
33. Federal Water Pollution Control Act (FWPCA), 62 Stat. 1155 (1948).
34. FWPCA Amendment Acts: 66 Stat. 755 (1952); 70 Stat. 498 (1956); Pub. L. No. 86-90, 73 Stat. 141 (1959); Pub. L. No. 86-624, 74 Stat. 411 (1960); Pub. L. No. 87-88, 75 Stat. 204 (1960); Pub. L. No. 89-234, 79 Stat. 903 (1965); Pub. L. No. 89-753, 80 Stat. 1246 (1966); Pub. L. No. 91-224, 84

financial support for state water pollution control programs, as well as partial funding for the construction of municipal sewage treatment facilities. Its goal was to achieve state-established water quality standards.[35] It evolved from providing a passive federal role to establishing an active, although secondary, federal role. Its strongest federal role was to abate interstate pollution, assisting states in achieving their water quality standards when they were unable to do so themselves because of pollution originating beyond their jurisdictions in other states. Not surprisingly, the FWPCA's jurisdictional basis was "interstate waters."[36]

On the other hand, the Refuse Act[37] was enacted as part of a consolidation and revision of navigation laws in the Rivers and Harbors Act of 1899.[38] The Refuse Act protected navigation by prohibiting the discharge of refuse into navigable waters or their tributaries, except for flow from streets and sewers (that is, regulating industrial discharges but excluding municipal discharges) without a permit from the Secretary of the Army. Pursuant to a 1970 Executive Order,[39] the Corps and EPA adapted this statute for water pollution control. The Corps processed applications for and issued Refuse Act permits to industries discharging wastes to navigable water, conditioned on the industries' treatment of the wastewater to achieve standards established by EPA based on the technological treatment capabilities of different industrial categories. Not surprisingly, the Refuse Act's jurisdictional basis was "navigable waters."

Stat. 91 (1970); Pub. L. No. 91-611, 84 Stat. 1818 (1970); Pub. L. No. 92-50, 85 Stat. 124 (1971); Pub. L. No. 92-240, 85 Stat. 47 (1972).

35. For a description of the development of the program, see William L. Andreen, *The Evolution of Water Pollution Control in the United States: State, Local and Federal Efforts, 1789-1972; Part II*, 22 Stan. Envtl. L.J. 215 (2003); N. William Hines, *Nor Any Drop to Drink: Public Regulation of Water Quality*, 52 Iowa L. Rev. 186, 432, 799 (1966/1967).

36. Water Quality Act of 1965, Pub. L. No. 89-234, 79 Stat. 903 (1965). Although EPA continues to include interstate waters in its definition of waters of the United States, see 40 C.F.R. §122.1 (2013), and its recent revision of that definition, see 80 Fed. Reg. 37114 (June 29, 2015), to be codified at 40 C.F.R. §122.2(1). There is no explicit constitutional authority for Congress to regulate interstate waters unless they are also parts of the highways of interstate commerce. Indeed, EPA identifies no such authority in its discussion of jurisdiction over interstate waters in the preamble to its proposed amended definition of navigable waters. *See* 79 Fed. Reg. at 22254-55. It has been argued that such jurisdiction does not exist. *See* William Funk, 1 Law of Environmental Protection §13:117 (2014). EPA, however, argues that the Court inherently recognized Congress' authority to regulate pollution of interstate waters when it held in *Milwaukee v. Illinois*, 451 U.S. 304, 11 ELR 20406 (1981), that the CWA preempted the federal common law of interstate water pollution. *See* EPA's legal justification for its proposed amendment of its definition of waters of the United States, 79 Fed. Reg. 22188, 22256-57.

37. Refuse Act, 33 U.S.C. §407.

38. Rivers and Harbors Act of 1899, 33 U.S.C. §§401-148.

39. Exec. Order No. 11574, 3 C.F.R. §1966-1970, 35 Fed. Reg. 19627 (Dec. 23, 1970). For a description of the program, see 42 Fed. Reg. 37122-23 (July 19, 1977).

Environmentalists brought the Refuse Act Permit Program to a halt by winning a suit against the Corps to require it to comply with the National Environmental Policy Act (NEPA)[40] when issuing each of the thousands of anticipated pollution control permits.[41] In comprehensive amendments to the FWPCA in 1972, Congress adopted the permit program from the Refuse Act for all industrial and municipal point sources of pollution. CWA permits were to be conditioned on the permit holders meeting both state-established water quality standards, originally established under the FWPCA, and EPA-established technology-based standards under the CWA, much like those EPA had established under the Corps/EPA Refuse Act permit program.[42] The new statute exempted the issuance of CWA §402 permits from compliance with NEPA, except for permits to new sources,[43] on the theory that for existing sources CWA permits require reduction of existing pollution, to the benefit of the environment, while new sources would add to existing pollution.

The legislative history of the CWA is replete with statements that Congress intended the statute's jurisdiction to be expansive, indeed to reach the outer limits of congressional jurisdiction under the Constitution. Both houses of Congress, however, used "navigable waters" as a jurisdictional element for their respective bills. The U.S. Senate bill defined navigable waters to mean "the navigable waters of the United States, portions thereof, and the tributaries thereof, including the territorial seas and the Great Lakes."[44] The U.S. House of Representatives bill defined it to mean "the navigable waters of the United States, including the territorial seas."[45] Insofar as both bills defined "navigable waters" as "navigable waters," the definitions were circular. The Committee Report accompanying the Senate bill explained the rationale for the shift from the model of the earlier FWPCA to the model of the Refuse Act:

> Through a narrow interpretation of interstate waters the implementation of the 1965 Act was severely limited. Water moves in hydrological cycles and it is essential that discharge of pollutants be controlled at the source. Therefore reference to the control requirements must be made to the navigable waters, portions thereof, and their tributaries.[46]

40. 42 U.S.C. §§4321-4370f, ELR Stat. NEPA §§2-209.
41. *See* Kalur v. Resor, 335 F. Supp. 1, 1 ELR 20637 (D.D.C. 1971).
42. 33 U.S.C. §§1311, 1342, CWA §§301, 402.
43. CWA §511(c)(1), 33 U.S.C. §1371(c)(1).
44. S. 2770, 92nd Cong. §502(h) (1971), *reprinted in* 2 A Legislative History of the Water Pollution Control Act Amendments of 1972 [hereinafter Legis. History), at 1534, 1698.
45. H.R. 11896, 92nd Cong. §502(8) (1971), *reprinted in* 1 Legis. History, *supra* note 44, at 1069.
46. S. Rep. No. 92-414, at 77 (1971), *reprinted in* 2 Legis. History, *supra* note 44, at 1419, 1495. The explanations that water moves in hydrological cycles and that pollution must be treated at its source are not helpful. The hydrological cycle goes far beyond either navigable or interstate waters,

The Conference Committee changed the circular definitions of navigable waters in both bills to the present "waters of the United States" and explained that "[t]he conferees fully intend that the term 'navigable waters' be given the broadest possible constitutional interpretation unencumbered by agency determinations which have been made or may be made for administrative purposes."[47] Sen. Edmund Muskie (D-Me.), widely credited as the author of the legislation, explained the Conference Committee bill to the Senate as containing the "broadest possible constitutional interpretation" of navigable waters.[48] Rep. John Dingell (D-Mich.), chief sponsor of the legislation in the House, explained that the

> conference bill defines the term "navigable waters" broadly for water quality purposes . . . . It means "all of the waters of the United States" in a geographical sense. It does not mean "navigable waters of the United States" in a technical sense as we sometimes see it in some laws. This new definition clearly encompasses all water bodies, including main streams and their tributaries for water quality purposes. No longer are the old, narrow definitions of navigability, as determined by the Corps of Engineers, going to govern matters covered by this bill.[49]

The repeated references to old, narrow determinations of navigable waters referred to formal determinations the Corps had made that particular waterways were navigable and therefore eligible for spending federal funds to improve them.

This legislative history suggests that Congress made a deliberate choice to base the CWA's jurisdiction on navigable waters rather than on interstate waters because it believed navigable waters jurisdiction to be broader.[50] It also indicates that Congress intended the CWA's jurisdiction to be as broad as constitutionally possible, encompassing "all water bodies . . . including their

---

encompassing groundwater beneath and water vapor in the clouds above Kansas. Congress did not intend the CWA to regulate discharges to groundwater and there is no hint in the statute itself or in its legislative history that Congress intended the CWA to reach water vapor in the clouds above Kansas.

47. S. Rep. No. 92-1236, at 144 (1972) (Conf. Rep.), *reprinted in* 1 Legis. History, *supra* note 44, at 327.

48. 118 Cong. Rec. S16876 (daily ed. Oct. 4, 1972) (statement of Sen. Muskie), *reprinted in* 1 Legis. History, *supra* note 44, at 166.

49. 118 Cong. Rec. H33756-57 (daily ed. Oct. 4, 1972) (statement of Rep. Dingell), *reprinted in* 1 Legis. History, *supra* note 44, at 250.

50. Congress was correct in this regard. The traditional definition of navigable waters under the Commerce Clause does not distinguish between interstate and intrastate waterways, as long as an intrastate waterway is part of a highway of foreign or interstate commerce. On the other hand, none of the enumerated powers in the Constitution explicitly or inferentially grant Congress the power to regulate interstate waters that are not highways or parts of a highway of interstate or foreign commerce.

tributaries."[51] Whether the congressional choice of navigable waters jurisdiction accomplished its expansive goals is questionable. Representative Dingell himself explained the reach of navigable waters jurisdiction by describing judicial interpretation of a term denoting waterborne commerce.[52] And while the Senate Committee Report urged that the "Committee fully intends that the term 'navigable waters' be given the broadest possible interpretation," it prefaced that statement by confessing that "[o]ne term that the Committee was reluctant to define was the term 'navigable waters.' The reluctance was based on the fear that any interpretation would be read narrowly."[53] The Committee's observation was not initially justified, but ultimately proved prescient.

Congress' choice of navigable waters jurisdiction for the CWA appears not to extend to the full limits of Commerce Clause jurisdiction, because navigable waters evokes only the first of the three prongs of Commerce Clause jurisdiction the Supreme Court enunciated in *United States v. Lopez*[54]: (1) channels of interstate commerce; (2) instrumentalities of interstate commerce; and (3) activities substantially affecting interstate commerce. With regard to the CWA's jurisdiction to regulate water pollution sources under the §402 permit program, many polluting discharges are not to traditional navigable waters, but instead to waters that are directly or indirectly tributary to traditional navigable waters. An industrial discharge of pollutants, for instance, may be to a ditch flowing into a small stream, which in turn flows into a larger stream that eventually flows into a navigable-in-fact river. An ordinary ditch[55] is not a channel of interstate commerce in the traditional sense, although a discharge to the ditch might

51. It could be argued, however, that abandoning "the tributaries" in the Senate bill's definition of navigable waters narrowed the definition to exclude tributaries. The more expansive definition from the Conference Committee, "waters of the United States," coupled with the expansive language of the Conference Committee Report and the expansive meanings attributed to it by the sponsors on the House and Senate floors, however, rebut this negative inference.

52. 118 Cong. Rec. H9114-35 (daily ed. Oct. 4, 1972) (House Consideration of the Report of the Conference Committee), *reprinted in* 1 Legis. History, *supra* note 44, at 250.

53. S. Comm. on Pub. Works, Federal Water Pollution Control Act Amendments of 1972, S. Rep. No. 92-911 (1972), *reprinted in* 1 Legis. History, *supra* note 44, at 818.

54. 514 U.S. 549, 558-59 (1995).

55. *See infra* Chapter 4, Section III.F.1. Writing the plurality opinion in *Rapanos v. United States*, 547 U.S. 715, 735-36, 36 ELR 20116 (2006), Justice Antonin Scalia argues that a ditch cannot be navigable water because a ditch is a "point source." The §502(14) definition of point source is a "discernible, confined and discrete conveyance," and incorporates a list of inclusive examples, including "ditch." He argues that a ditch cannot be "navigable water" because it is already a "point source" and the same thing cannot be two elements. Justice Scalia's argument is a non sequitur, however. True, a ditch is a conveyance; it conveys water, whether or not navigable, but often polluted. A ditch is not water and therefore cannot be navigable water, although it can *convey* navigable water. Justice Scalia has confused the issue by conflating a ditch with the water it conveys, and conflating "point source" with "navigable waters." Navigable waters are distinct from their beds, a distinction that underlies the doctrine that

be considered an indirect discharge to navigable waters, because the tributaries to a navigable water in the aggregate contribute most of the flow of that water as well as most of the pollutants discharged to it, invoking either the first or third prong of *Lopez*.[56]

Municipal and industrial pollutants also might be considered instrumentalities of interstate commerce, analogous to liquid and solid wastes that the Court has held are items in interstate commerce when they are transported on interstate roadways, invoking the second prong of *Lopez*.[57] Moreover, the discharge of pollutants to non-navigable water, aggregated with other discharges to the same or similar smaller waterways, can have adverse impact on the navigable waters they eventually flow into, again implicating the *Lopez* third prong of Commerce Clause jurisdiction and the Necessary and Proper Clause.[58] Those arguments, however, are beyond the scope of this chapter, because it interprets navigable waters as a statutory term in the CWA, not as a constitutional term under the Commerce Clause jurisdiction.[59]

Congressional choice of navigable waters as the CWA's jurisdictional basis implicitly limits its claim of jurisdiction only to the first of the three *Lopez* prongs of Commerce Clause jurisdiction, and that prong alone does not reach the outer limits of Commerce Clause jurisdiction. Congress attempted to reach the outer limits of Commerce Clause jurisdiction by defining "navigable waters" as the expansive "the waters of the United States" in CWA §502(7) and by repeated pronouncements in the legislative history to the effect that it intended to confer jurisdiction as broad as constitutionally possible. Despite the initial willingness of courts to interpret the term expansively, however, the jurisdictional term is still "navigable

the public may use navigable water for fishing or transport, regardless of who owns the land beneath the waters. See *infra* the discussion of ditches in Chapter 4, Section III.F.1.

56. *See* Rapanos v. United States, 547 U.S. 715, 743, 36 ELR 20116 (2006).

57. C&A Carbone, Inc. v. Clarkstown, 511 U.S. 383, 24 ELR 20815 (1994); Chemical Waste Mgmt., Inc. v. Hunt, 504 U.S. 334, 22 ELR 20909 (1992); Philadelphia v. New Jersey, 437 U.S. 617, 8 ELR 20540 (1978). The analogy is not perfect, because the transportation of municipal garbage to and the disposal of it in landfills are compensable services, and hence traditional interstate commerce, while dumping liquid wastes in navigable waters or their tributaries is a free substitute for otherwise compensable services. That implicates *Wickard v. Filburn*, 317 U.S. 111 (1942), extending Commerce Clause jurisdiction to activities that would in themselves escape Commerce Clause jurisdiction but for their aggregate effect on interstate commerce.

58. U.S. Const., art. I, §8, cl. 18.

59. Justice John Paul Stevens' dissenting opinion in *Solid Waste Agency of N. Cook Cnty. v. U.S. Army Corps of Eng'rs (SWANCC)*, 531 U.S. 159, 192-96, 31 ELR 20382 (2001) (Stevens, J., dissenting), joined by three other Justices, outlined the case for Commerce Clause jurisdiction over isolated ponds as habitat for migratory birds. Circuit courts had earlier reached similar conclusions. *See, e.g.*, United States v. Byrd, 609 F.2d 1204, 1209-11, 9 ELR 20757 (7th Cir. 1979); United States v. Ashland Oil & Transp. Co., 504 F.2d 1317, 1325-28, 4 ELR 20784 (6th Cir. 1974).

waters" and textualist jurists insist that "navigable" must have something to do with floating commerce.[60]

The CWA's definition of navigable waters as the waters of the United States does not solve the problem, because while "navigable waters" has an established meaning, "waters of the United States" has no commonly understood meaning. Representative Dingell's explanation that waters of the United States is used in a geographical sense is ambiguous. The term could include the water in the glasses on our desks, the water vapor in the clouds above Kansas, water 500 feet below the surface that never flows into surface water, or water on federal lands. "Waters of the United States," standing alone, might be interpreted to break out of the first prong of the *Lopez* test to incorporate all three prongs. But for textualist jurists, defining navigable waters as waters of the United States does not abandon entirely the concept of waterborne transportation in navigable waters.

The issue is exacerbated by §404's protection of wetlands, which EPA and the Corps interpret to include areas that might be dry land part of the year. Dry land does not appear to most people to be water, let alone navigable water. The Court has fixated on this issue.[61] Three of the six Supreme Court decisions interpreting navigable water were §404 cases,[62] and the Court reached narrow interpretations of the term in two of those three decisions. These narrow interpretations affect §402 pollution control cases, where they raise the question of whether §402 governs additions of pollutants to remote or intermittent tributaries of navigable waters.[63] If the CWA does not provide the agencies with jurisdiction to stop additions of pollutants to remote or intermittent tributaries of navigable waters, it cannot stop those pollutants from flowing into navigable waters, thwarting admitted congressional jurisdiction to control pollution of navigable waters.

60. *See Rapanos*, 547 U.S. at 734 ("The plain language of the statute simply does not authorize the 'Land is Water' approach to federal jurisdiction."); *SWANCC*, 531 U.S. at 172 ("We cannot agree that Congress' separate definitional use of the phrase 'waters of the United States' constitutes a basis for reading the term 'navigable waters' out of the Statute.").

61. *Id.*

62. The six decisions are *Los Angeles Cnty. Flood Cont. Dist. v. Natural Res. Def. Council, Inc.*, 133 S. Ct. 710, 43 ELR 20004 (2013); *Rapanos*, 547 U.S. 715; *South Fla. Water Mgmt. Dist. v. Miccosukee Tribe of Indians*, 541 U.S. 95, 34 ELR 20021 (2004); *SWANCC*, 531 U.S. 159; *International Paper Co. v. Ouellette*, 479 U.S. 481, 17 ELR 20327 (1987); and *United States v. Riverside Bayview Homes, Inc.*, 474 U.S. 121, 16 ELR 20086 (1985). Of these six, *Rapanos*, *SWANCC*, and *Riverside Bayview* were §404 decisions, and form the focus of the chapter's discussion.

63. *Compare* United States v. Robinson, 505 F.3d 1208 (11th Cir. 2007), decided after *Rapanos*, *with* United States v. Texas Pipe Line Co., 611 F.2d 345, 10 ELR 20184 (10th Cir. 1979); United States v. Ashland Oil & Transp. Co., F.2d 1317, 4 ELR 20784 (6th Cir. 1974); and Sun Enters., Inc. v. Train, 394 F. Supp. 1212 (C.D.N.Y. 1975), all decided before *Riverside Bayview*, the first Supreme Court CWA decision interpreting navigable water. All four circuit court decisions involved discharges to remote tributaries, flowing through several intermediary tributaries to navigable water.

2. Section 404

The role of §404 in wetlands protection is obvious today, but it was not obvious when Congress enacted the CWA in 1972. The initial §404 contained only §404(a)-(c), which have scarcely been altered since.[64] The word "wetland" did not then and does not now appear in §404(a)-(c). Indeed, the word wetland did not appear in the descriptions and explanations of §404 in any of the committee reports accompanying the House, Senate, or Conference Committee bills or in any of the congressional debates over the bills leading to enactment of the 1972 legislation. A close reading of §404(a)-(c) and its legislative history suggests Congress intended to establish a program regulating the disposal in open water of spoil from dredging rivers and harbors for navigation maintenance, rather than a program regulating the filling of wetlands. Indeed, the legislative history demonstrates that is exactly what Congress intended it to be.

Dredging is essential to waterborne commerce. Harbors and navigational channels must be dredged or they will fill with silt and become impassable. Construction in navigable waters also requires dredging bottom sediments to make way for the foundations of bridges, jetties, and other water-related structures. To carry out such activities, vast quantities of bottom sediment, commonly known as dredged spoil, must be removed and disposed. Existing legislation prohibited dredging in navigable waters without a Corps permit.[65] As part of a permit for a dredging project, the Corps specified where and how the dredged spoil generated by the project would be disposed.

As a result of NEPA[66] and the Fish and Wildlife Coordination Act,[67] the Corps increasingly took environmental concerns into account in dredged spoil disposal decisions.[68] Rather than designating individual disposal sites for each project, the Corps began to designate regional sites that could be used for multiple dredging projects in an area, thereby limiting the number

64. *Compare* Pub. L. No. 92-500, §404, 86 Stat. 816, 884 (1972), *with* 33 U.S.C. §1344 (2006).

65. 33 U.S.C. §403.

66. *See* 42 U.S.C. §§4321-4370h. NEPA requires federal agencies to consider environmental impacts when taking actions. Indeed, when an agency takes a major federal action, it must produce an environmental impact statement disclosing the environmental consequences of proposed action and alternative actions.

67. Fish and Wildlife Coordination Act, 16 U.S. C. §§661-666. This statute requires federal agencies taking actions that will alter a water body to consult with the U.S. Fish & Wildlife Service (FWS) to preserve wildlife resources.

68. *See* Want, *supra* note 12, at §2.02[1]. The Corps' "public interest review" began in 1968, when it began to take environmental and public interest factors into account in issuing permits. *See* 33 C.F.R. §209.120(d). For a description of one such disposal project, see *American Dredging Co. v. Dutchshyn*, 480 F. Supp. 957 (E.D. Pa. 1979).

of environmental reviews required for disposal areas. Hence, the Corps has authority to issue discharge permits "at specified disposal sites" in §404(a). Although dredged spoil could be disposed on land, it was more feasible to dispose of this semi-solid ooze in open water, including the ocean.[69] Hence, EPA has authority in §404(b) to develop guidelines for the Corps to use in designating "each such disposal site." While environmental advocates were only becoming aware of environmental concerns from the disposal of dredged spoil in open waters, maritime interests were acutely aware that delays in the use of disposal sites would interrupt dredging necessary for maritime commerce.

The legislative history of §404 reflected a policy conflict between prevention of water pollution and promotion of waterborne commerce. The Senate bill, S. 2770, the primary model for much of the CWA, contained no §404, but instead treated dredged spoil as just another pollutant subject to the §402 permit program.[70] Senators proposed an amendment in the Public Works Committee to authorize the Corps rather than EPA to regulate the disposal of dredged spoil, but the Committee rejected the amendment.[71] In the Senate debate on S. 2770, Sen. Allen Ellender (D-La.) asserted that "strict adherence to the published standards [for pollution control] would result in 90 per cent of the ports and harbors of the United States being closed, until such time as land disposal areas are provided. This would create a catastrophic situation with respect to our foreign and domestic commerce."[72] Senator Muskie, tacitly acknowledging the problem,[73] offered an amendment to the Senate bill adding §402(m), which basically required EPA to issue permits for the discharge of dredged material certified by the Corps unless EPA found that the disposal would adversely affect water supplies, fisheries, shellfish beds, and so forth. The Senate adopted this amendment.[74]

Maritime interests had greater success in the House. The House bill authorized the Corps to issue permits for the discharge of "dredged or fill material" where it would not unreasonably affect human health or the environment.[75] Although it required the Corps to apply EPA guidelines for such discharge,

69. *See* 1 Legis. History, *supra* note 44, at 177-78.
70. S. 2770, 92nd Cong. §§402, 504(f) (1971), *reprinted in* 2 Legis. History, *supra* note 44, at 1534, 1685, 1697.
71. S. Rep. No. 92-414, at 92 (rollcall votes during Committee consideration), *id.* at 1415, 1509.
72. 117 Cong. Rec. S. 2770 (daily ed. Nov. 2, 1971) (statement of Sen. Ellender), *id.* at 1386-88.
73. After all, he was a senator from Maine, where Portland is one of the most active ports on the East Coast because it receives oil tankers offloading cargo for transfer by pipeline to Canada.
74. 117Cong. Rec. S. 2770 (daily ed. Nov. 2, 1971), *reprinted in* 2 Legis. History, *supra* note 44, at 1392-93.
75. H.R. 11896, *reprinted in* 1 Legis. History, *supra* note 44, at 1063.

it allowed the Corps to disregard the guidelines if there was no economically feasible alternative reasonably available.[76] Again, the concern was that "until such time as economic and feasible alternative methods are available, no arbitrary or unreasonable restrictions shall be imposed on dredging activities essential for the maintenance of interstate and foreign commerce."[77] The wording of the Conference Committee Report underlined congressional intention to deal with spoil from dredging to maintain navigation. Thus "specific spoil" was to be deposited at a site; the section dealt with the "disposal of dredged spoil"; and advancing technology would eventually end the need for "dumping dredged spoil" in water.[78]

The relationship between CWA §§403 and 404 underscores that §404 was aimed at open-water disposal of spoil from dredging rivers and harbors. Section 403(c) required EPA to develop guidelines to protect human health and welfare and *marine life* and diversity, from the discharge of pollutants into the territorial seas, contiguous zone, and ocean. Section 404(b) forbade the issuance of a §404 permit not meeting the §403 guidelines unless the "economic impact of the site on *navigation and anchorage*" (emphasis added) outweighs compliance with the guidelines. The §403 criteria are designed to protect marine waters and are irrelevant to filling inland wetlands, again suggesting that Congress intended §404 to regulate the open-water disposal of spoil from dredging rivers and harbors rather than filling wetlands. The Corps' authority to override EPA's criteria because of their economic impact on navigation and anchorage reiterates the section's purpose to address spoil from dredging for navigation purposes.

Nevertheless, §404(a)-(c) authorized the Corps to regulate the discharge of "dredged *or fill* material" (emphasis added) into navigable waters. The entire wetlands protection tilt of CWA §404 derives from the inclusion of fill material in §404. That term, however, is at best ambiguous in addressing protection of wetlands from landfilling activities. Fill material could mean material to fill abutments, jetties, and other marine structures, but it has come to include material removed from high elevations in a wetland and then redeposited at lower elevations in the same wetland, thus converting it to fast land. The term originated in H.R. 11896, §404, which provided that discharges of dredged or fill material "will not unreasonably degrade or endanger . . . the *marine* environment" (emphasis added), evidencing legislative intent to deal with dredged spoil disposed in marine

76. *Id.*
77. H. Comm. on Pub. Works, Federal Water Pollution Control Act Amendments of 1972, H.R. Rep. No. 92-911, at 130 (1972), *reprinted in* 1 Legis. History, *supra* note 44, at 817.
78. 1 Legis. History, *supra* note 44, at 177.

waters rather than filling wetlands. The Report accompanying the bill expressed the intent of the House Committee on Public Works that the Corps "shall act promptly on the dredging permits essential for the maintenance of interstate commerce,"[79] emphasizing the same intent to regulate the dredging of navigable channels. The absence of congressional intent to regulate filling of wetlands in §404 is also evident from the disconnect between "fill" and "pollutant" in the CWA; the definition of pollutant in §502(6) includes dredged spoil, but does not mention fill or dredged or fill material.[80]

Even when the Corps first promulgated regulations to administer §404, the program it described was for disposing of dredged spoil from maintaining navigational waterways. Although those regulations acknowledged the importance of wetlands, the only wetland protections the regulations anticipated was care in discharging dredged spoil from navigational maintenance into wetlands. Indeed, the Corps did not include wetlands in its definitions of navigable waters.[81] EPA's initial regulatory definition of navigable water did not mention wetlands either.[82]

Section 404 is so associated with wetlands protection today that its original goal to facilitate navigational dredging is forgotten or altogether unknown. The considerable legislative history expressing concern that §404 not interfere with the maintenance of navigation or anchorage and the complete absence of legislative concern that §404 not interfere with agriculture and development, points directly at disposal of spoil from navigational dredging, rather than filling wetlands, as the focus of §404. Even the story told by the Corps' general counsel when the agency began to administer the section to

79. H. Comm. on Pub. Works, Federal Water Pollution Control Act Amendments of 1972, H.R. Rep. No. 92-911, at 129-30 (1972), *reprinted in* 1 Legis. History, *supra* note 44, at 816-17.

80. This omission does not affect the reach of the section's jurisdiction, because most fill material is composed wholly or partially of material defined as a pollutant in §502(6); for example, sand or rock. *See supra* Chapter 2, Section III.F.2.

81. The Corps' proposed regulations addressed "policy, practice and procedures . . . in connection with [the Corps'] performance of *Federal dredging projects*." 39 Fed. Reg. 6113 (Feb. 19, 1974) (emphasis added). The final regulations addressed those policies, practices, and procedures "in connection with . . . the review of Federal projects performed by the Corps of Engineers which involve the disposal of *dredged material* in navigable water." 39 Fed. Reg. 26635 (July 22, 1974) (emphasis added). Both required the Corps to consider the "effect on wetlands" of proposed projects, 39 Fed. Reg. 6114, 26637, but neither mentioned wetlands outside of that narrow paragraph. These regulations dealt only with "Federal dredging projects." The Corps traditionally had conducted projects to dredge navigable channels and harbors. Private landclearing activities to convert wetlands to other uses are not "Federal dredging projects."

82. 40 C.F.R. §125.1(o), 38 Fed. Reg. 13528, 13529 (May 22, 1973).

protect wetlands acknowledges that the section was not designed to protect wetlands from landclearing.[83]

While protection of wetlands contributes to the control of pollution in navigable waters and provides other environmental and economic benefits, such a significant expansion of a statute's jurisdiction would not ordinarily be made without notice to both legislative chambers. Yet, there was no such notice in any of the following: the draft legislation; reports of the House Public Works Committee, Senate Public Works Committee, or Conference Committee; debates on the bills in either chamber; or the explanations of the Conference Committee's actions to either the House or the Senate. The omissions in this legislative history raise significant questions on the legitimacy of the entire initial wetlands protection orientation of EPA and the Corps' initial wetlands protection program under §404.

In 1977, Congress enacted a set of "mid-course correction" amendments to the CWA.[84] One of the chief controversies in the congressional deliberations was over the scope of §404. Farming and development interests, asleep at the switch when Congress enacted §404 in 1972, now launched a coordinated effort to scale back the scope of regulations on filling wetlands. While interest regarding §404 during the enactment of the 1972 legislation was focused entirely on allowing disposal of spoil from dredging rivers and harbors to maintain navigation, interest regarding §404 during the 1977 legislation was focused entirely on the scope of regulation on filling wetlands to promote agriculture and development.[85] Although efforts for significant reductions in the jurisdiction of the program were unsuccessful, considerable accommodations to agricultural and development interests were adopted,[86] increasing the section to more than nine times its original length.[87] The depth of attention to §404 in 1977 suggests that if Congress had been aware in 1972 that the section could be used to regulate filling of wetlands, at least some legislative attention would have been devoted to wetlands in 1972.

83. *See* Charles D. Abalard & Charles Boru O'Neill, *Wetlands Protection and Section 404 of the Federal Water Pollution Control Amendments of 1972*, 1 Vt. L. Rev. 51 (1975), at 71-74.

84. *See* Weyerhaeuser Co. v. Costle, 590 F.2d 1011, 43, 9 ELR 20284 (D.C. Cir. 1978).

85. Pub. L. No. 95-217, 91 Stat. 1556 (1977). *See* H.R. Rep. No. 95-139 (1977), *reprinted in* 4 Legis. History, *supra* note 44, at 1195, 1215-20; and additional views of Reps. Robert Edgar and John Myers, *id.* at 1247-51.

86. Pub. L. No. 95-217, §67(a) & (b), 91 Stat. 1556, 1600-06 (1977).

87. Subsections 404(a)-(c) have changed little since their enactment in 1972. All the following subsections were added later. In the 2006 edition of the U.S. Code, §404 occupies nine page-long columns, of which §404(a)-(c) occupies less than one column. *See id.* at 950-54.

III. Administrative Action: Regulatory Definitions of Navigable Waters and Wetlands

A. *Navigable Waters*

Although EPA implements the CWA §402 permit program for the discharge of pollutants, EPA and the Corps share administration of the CWA §404 program for the discharge of dredged or fill material. Pursuant to §404(a) & (s), the Corps issues §404 permits and enforces against violations of the section. EPA, however, supervises the Corps' conduct of the program. Pursuant to §404(b), EPA issues guidelines that the Corps must follow in deciding whether to issue and how to condition permits; under §404(c), EPA may veto Corps-proposed permits as contrary to the requirements of the CWA; and it is authorized by §404(n) to enforce against the discharge of dredged or fill material without a §404 permit or in violation of the terms of a §404 permit. In §404(g), Congress also gave EPA authority to approve state programs to administer §404 in place of the Corps, except in traditionally navigable waters; and in §404(i), Congress authorized the Agency to withdraw approval of a state program if it no longer meets the requirements for approval. Both EPA and the Corps have promulgated regulatory definitions of navigable waters[88]; moreover, each has promulgated two or more definitions of the term, one for use only under §404, and one for use under other programs that the agencies administer.[89]

Prior to the CWA's enactment, the Corps had promulgated a regulatory definition of navigable waters for use in all of its water-related programs, reflecting the traditional judicial interpretation of "navigable waters." A few weeks prior to the statute's enactment, the Corps amended that definition to reflect extensive, up-to-date research on judicial interpretation of the term: "those waters of the United States which are subject to the ebb and flow of the tide, and/or are presently, or have been in the past, or may be in the future susceptible for use for purposes of interstate or foreign commerce."[90] The Corps continued to use that definition under §404.[91]

---

88. *Compare* 40 C.F.R. §122.2 (EPA definition), *with* 33 C.F.R. §328.3 (Corps definition).
89. EPA's 40 C.F.R. §122.2 definition applies to the §402 permit program and its 40 C.F.R. §230.3 definition applies to the §404 permit program. Other EPA definitions of the waters of the United States appear at 40 C.F.R. §§110.1, 112.2, 116.3, 117.1, 232.2, 300.5, 302.2, and 401.11. The Corps' 33 C.F.R. §328 definition applies to the §404 permit program and its 33 C.F.R. §329 definition applies to the programs it administers in aid of navigation.
90. Abalard & O'Neill, *supra* note 83, at 65. *See also* 37 Fed. Reg. 18289 (Sept. 9, 1972).
91. *See* the Corps' initial regulations under §404 at 33 C.F.R. §209.145(d)(1) (1975), 39 Fed. Reg. 26635, 26637 (July 22, 1974). In that definition of navigable waters, the Corps referenced and incorporated the more-detailed definition it had earlier developed for all its water-related programs,

EPA's initial definition of navigable waters, adopted for the §402 permit program, expanded it beyond the traditional judicial interpretation to include: (1) tributaries of navigable waters; (2) interstate waters; and (3) intrastate waters (a) used by interstate travelers for recreation and other purposes, (b) from which fish or shellfish are taken and sold in interstate commerce, or (c) utilized for industrial purposes by industries in interstate commerce.[92] But that definition made no mention of wetlands. Shortly thereafter, EPA promulgated interim final guidelines under §404(b)(1), including a definition of navigable waters that did not mention wetlands or tributaries, but incorporated by reference the Corps' general definition of navigable waters.[93] EPA's guidelines, however, stated that destruction of wetlands was one of the most egregious results of discharges of dredged and fill material.[94] Although the interim final nature of the guidelines invited comments and promised revisions if warranted by the comments, EPA did not revise the guidelines for several years. When the Agency did revise them, EPA adopted the definition it had subsequently developed for "waters of the United States" under §402, including tributaries and adding wetlands adjacent to other waters of the United States.[95]

The significant differences between the Corps and EPA's initial definitions of navigable waters were contrary to the interpretive canon that words are interpreted consistently throughout a statute unless the statute explicitly required otherwise. The introductory phrase in CWA §502, "[e]xcept as otherwise specifically provided, when used in this chapter . . ." a term, e.g., "navigable waters," reiterates the canon. Nothing in the CWA specifically provides that navigable waters or the waters of the United States have different meanings in §§402 and 404. The differences between EPA and the Corps' initial regulatory definitions of navigable water reflected the Corps' historical authority to make improvements in aid of navigation, covering traditional Commerce Clause highways of commerce jurisdiction; and EPA's more recent mission to reduce water pollution in all of the nation's waters, reaching beyond traditional highways of commerce jurisdiction. The difference led to confusion and conflict within the executive branch and between the executive branch and the interested publics.

then contained in 33 C.F.R. §209.260. Neither definition mentioned wetlands adjacent to navigable waters or tributaries of navigable waters.

92. 40 C.F.R. §125.1(p) (1974).
93. 40 C.F.R. §230.2(b), 40 Fed. Reg. 41292-93 (Sept. 5, 1975).
94. 40 C.F.R. §320.4-1(a), 40 Fed. Reg. 41292, 41294 (Sept. 5, 1975).
95. 45 Fed. Reg. 85340, 85346 (Dec. 24, 1980).

Both EPA and the U.S. Department of Justice (DOJ) believed that the Corps' initial definition of navigable waters was too narrow, particularly with regard to its failure to include wetlands, and they unsuccessfully urged the Corps to expand its definition. Environmental advocates won an early case challenging the Corps' regulations as too narrow, particularly because they did not include wetlands.[96] The Corps was dissatisfied with DOJ's representation in *Natural Resources Defense Council v. Callaway* because DOJ agreed with EPA; accordingly, DOJ allowed the Corps to present its views to the district court.[97] Finding with little analysis that Congress intended to confer in the CWA "federal jurisdiction over the nations' waters to the maximum extent permissible under the Commerce Clause of the Constitution," the court ordered the Corps to promulgate a more expansive definition.[98] That, plus some nudging by DOJ,[99] led the Corps to promulgate a definition of navigable waters parallel to EPA's, but including wetlands adjacent to navigable waters.[100] The Corps began a practice of publishing two definitions of navigable waters, one for all of its water-related programs except for §404, and another for §404. It continues that practice today.[101]

EPA's initial regulations defined and used "navigable waters" and did not define or use "waters of the United States."[102] In 1980 amendments to its regulations, EPA ceased defining navigable waters or using that term; instead, the Agency defined and used the term waters of the United States.[103] EPA explained it did so "for the same reason that ["the waters of the United States"] is used in the Clean Water Act: the Act covers much more than waters that are traditionally 'navigable.'"[104] The Corps' regulations under §404 also switched from defining navigable waters to defining waters of the United States.[105] Neither EPA nor the Corps can amend the CWA to eliminate navigable waters from its text. Perhaps, the agencies hoped that if they

---

96. Natural Resources Defense Council v. Callaway, 392 F. Supp. 685, 5 ELR 20285 (D.D.C. 1975). *See also* Abalard & O'Neil, *supra* note 83, at 63, n.68.
97. *Callaway*, 392 F. Supp. 685.
98. *Id.* at 686.
99. DOJ issued an opinion stating that EPA rather than the Corps had authority to interpret the term navigable waters in §404 because EPA administers the entire CWA, which is premised on navigable waters jurisdiction, while the Corps' role is limited to implementing §404, and the CWA authorizes EPA to oversee the Corps' administration of §404. *See* 43 U.S. Op. Atty. Gen. 197 (Sept. 5, 1979).
100. *See* Want, *supra* note 12, at §2.02[3].
101. 33 C.F.R. §328 (definition for the §404 program), 33 C.F.R. §329 (definition for other Corps programs).
102. 45 Fed. Reg. 33290, 33298 (May 19, 1980).
103. *Id.* at 33424.
104. *Id.* at 33298.
105. *Compare* 33 C.F.R. §209.145(d)(1), 39 Fed. Reg. 26635, 26637 (July 22, 1974), *with* 33 C.F.R. §328.3 (2013).

began to articulate the jurisdictional basis of the CWA exclusively as waters of the United States rather than navigable waters, the world would follow. In any event, the Supreme Court did not.

Prior to EPA's amendment of the regulatory definition of the waters of the United States in 2015, that definition meant: (a) traditionally navigable waters; (b) interstate waters, including wetlands; (c) other waters[106] whose "use, degradation or destruction would . . . or could affect interstate or foreign commerce"[107]; (d) impoundments of waters identified in (a)-(c); (e) tributaries of waters identified in (a)-(c); (f) the territorial sea; and (g) wetlands adjacent to waters identified in (a)-(f).[108] The definition excluded "waste treatment systems, including treatments ponds or lagoons designed to meet the requirements of the CWA."[109]

EPA promulgated this definition prior to the Supreme Court's decisions in *SWANCC and Rapanos.* After these decisions, EPA and the Corps issued a guidance document interpreting waters of the United States in light of them and followed it with proposed and final amendments to the regulatory definitions.[110] Although the three documents are similar, they differ in many details.[111] Rather than describing each document, this chapter analyzes the final regulations as EPA's latest interpretation of the jurisdictional term. The definition now includes[112]: (a) traditionally navigable waters; (b) interstate waters, including interstate wetlands; (c) the territorial seas; (d) impoundments of navigable waters identified elsewhere in the definition; (e) tributaries of navigable waters identified in (a) through (c); (f) waters, including

106. Including "intrastate lakes, rivers streams (including intermittent streams), mudflats, sandflats, 'wetlands,' sloughs, prairie potholes, wet meadows, playa lakes, or natural ponds."

107. The definition includes the following uses as potentially affecting interstate commerce: (1) use by interstate or foreign travelers for recreational or other activities; (2) harvesting fish or shellfish for sale in interstate or foreign commerce; and (3) use or industrial purposes by industries in interstate or foreign commerce.

108. *See* 40 C.F.R. §122.2, waters of the United States.

109. But "the exclusion applies only to manmade bodies of water which neither were originally created in waters of the United States [ ] nor resulted from the impoundment of waters of the United States."

110. The guidance document can be found at http://water.epa/gov/lawsregs/guidance/wetlands/CWAwaters.efm. The proposed regulation appears at 79 Fed. Reg. 22188, 22267-68 (proposed Apr. 21, 2014), and the final regulations were published at 80 Fed. Reg. 37054 (June 29, 2015). EPA and the Corps proposed to rescind the rule as this book was in production. 82 Fed. Reg. 34899 (July 27, 2017) (open for comment as this book went to press),

111. Because there were over one million comments on the proposed regulations, it is not surprising that the final regulations differed from the proposal. 80 Fed. Reg. 37057.

112. The rulemaking amended 10 different EPA and Corps regulations with identical or substantially identical definitions of navigable waters or waters of the United States. *See* 33 C.F.R. §328.3, 40 C.F.R. §§110.1, 112.2, 116.3, 117.9(i), 122.2, 230.3(s) & (t), 232.2, 300.5, 300 Appendix E to Part 300, 3.2.3 and 401.11, 79 Fed. Reg. 22188, 22262-74. References in this chapter are to proposed amendments to 40 C.F.R. §122.2, EPA's primary definition of waters of the United States, from which most of the other amendments are derived.

wetlands, adjacent to waters identified in (a) through (e); (g) on a case-specific basis for specified waters, such as prairie potholes, that, alone or aggregated with similarly situated waters, have a "significant nexus" to waters identified elsewhere in the definition; and (h) all waters within 100 year floodplains of waters identified in (a)-(e) and all waters that are within 4,000 feet of the high watermark of such identified waters that are determined on a case-specific basis to have a significant nexus to such identified waters. The definition then excludes over a dozen specific types of water, including puddles, groundwater, and particular types of ditches. Finally, the revision adds definitions for "adjacent," "neighboring," significant nexus," "tributaries," and "wetlands." The preamble to the proposed and final rules support much of the definition by a massive peer-reviewed compendium of scientific literature as well as the agencies' experience and knowledge, especially for its definitions of adjacent, tributaries and significant nexus.[113] (EPA's current and proposed definitions are set forth in Appendix I.)

EPA's definition of waters of the United States occupies two full pages in the *Federal Register* and is several times longer than its earlier definition of the term. The sheer length of the definition makes it difficult to describe or comprehend. Its length derives in part from its illumination of many issues that were ambiguous or unclear in the earlier regulations, such as which ditches contain navigable waters and which do not. It also addresses most of the issues the Court raised with the agencies' jurisdictional claims in *SWANCC* and *Rapanos.* EPA specifically dropped its claim that use of specific waters or similarly situated waters in the aggregate for fishing or recreational use made them navigable under the statute. At the same time, the revisions claim CWA jurisdiction over similar waters that in the aggregate have a significant effect on the quality of other waters identified as navigable.

This chapter will refer to the revised regulatory definition at a number of points in its examination of judicial interpretations of navigable waters and will return to it in Chapter 5.

B. *Wetlands*

Prior to EPA's 2015 revision of its definition of "waters of the United States," EPA defined and, after that revision, continues to define wetlands as "areas that are inundated or saturated by surface or ground water at a frequency and duration to support, and that under normal circumstances do support, a prevalence of vegetation typically adapted for life in saturated soil conditions.

113. 79 Fed. Reg. 22188, 22222-52 (proposed Apr. 21, 2014).

Wetlands generally include swamps, marshes, bogs, and similar areas."[114] The Corps' definition is similar.[115] But the Corps and EPA have gone much further to develop a Wetlands Delineation Manual, detailing the types of soils and vegetation that are typical of saturated soil conditions.[116] *Avoyelles Sportsmen's League, Inc. v. Marsh* provides a good example of a hotly contested delineation.[117]

IV. Judicial Interpretation of Navigable Waters Under the CWA

Because courts have had no difficulty in finding that discharges into traditionally navigable water, such as tidal water,[118] the territorial seas,[119] and major inland waterways[120] are within the jurisdiction of the CWA, this chapter will not discuss such decisions. Instead, it will focus on areas of controversy. Most decisions interpreting navigable waters have arisen in five contexts: (1) remote or dry tributaries; (2) wetlands; (3) isolated waters; (4) groundwater; and (5) unitary water.

The tributary and wetlands decisions are often intertwined and are analyzed together because to be within the CWA's jurisdiction, wetlands must be adjacent to navigable waters, often tributaries of traditionally navigable water. The Corps established the adjacency requirement in its regulations[121] and adjacency is a central feature of the recently revised definition of waters of the United States.[122] Thus, whether a wetland is within CWA jurisdiction often depends on whether an adjacent tributary is navigable; accordingly, many decisions on the navigability of tributaries are §404 decisions. The lower courts' decisions on whether tributaries and wetlands are within the CWA's jurisdiction usually reflect the latest Supreme Court decision on the subject. This chapter, therefore, considers tributary and wetlands decisions together and organizes them by reference to the latest preceding Supreme Court decision on the CWA's navigable water jurisdiction.

---

114. 40 C.F.R. §122.2 (2013) and 80 Fed. Reg. 37115.
115. 33 C.F.R. §328.3.
116. For the history of the Wetlands Delineation Manual, see Want, *supra* note 12, at §4.09.
117. 715 F.2d 897, 13 ELR 20942 (5th Cir. 1983).
118. United States v. Milner, 583 F.3d 1174, 39 ELR 20232 (9th Cir. 2009) (expansion of jurisdiction above mean high tide). *See also* Want, *supra* note 12, at 2-9 & n.1082; Abalard & O'Neill, *supra* note 83, at 85.
119. Natural Res. Def. Council v. EPA, 863 F.2d 1420, 19 ELR 20225 (9th Cir. 1998).
120. Gibbons v. Ogden, 9 Wheat. 1 (1866).
121. 33 C.F.R. §209(d)(2((i)(b), 40 Fed. Reg. 31324 (1975).
122. 80 Fed. Reg. 37114-15 (June 29, 2015), to be codified at 40 C.F.R. §122.2 waters of the United States (3)(i).

Prior to any Supreme Court decision interpreting navigable waters in the CWA, the lower courts followed the admonition of the CWA's legislative history to interpret the jurisdictional reach of the CWA to the outer boundaries of Congress' constitutional jurisdiction over waters of the United States. Lower courts interpreted the term broadly to extend to the following: primary tributaries, flowing directly into traditionally navigable waters; very remote tributaries, flowing eventually into traditionally navigable waters through a series of intermediate tributaries; and their adjacent wetlands. These expansive interpretations by the lower courts continued after *Riverside Bayview*,[123] the Supreme Court's first CWA decision interpreting both §404 and navigable waters under the CWA. In that decision, the Court held that wetlands adjacent to tributaries of traditionally navigable waters were within the jurisdiction of the CWA. The decision did not define which tributaries were within the CWA's jurisdiction, but the tributaries at issue were remote and the decision suggested an expansive interpretation of navigable waters.

The Court's next §404 decision, *SWANCC*,[124] did not deal with wetlands or tributaries, but held that isolated waters were not navigable, signaling that there is a limit to the jurisdictional reach of "navigable waters." After *SWANCC*, a few lower courts began to narrow their interpretations of navigable waters, but only a few. *Rapanos*,[125] the Court's latest §404 decision, dealt directly with wetlands and tributaries of navigable waters and did so in a restrictive manner. Because the Court could not muster a majority opinion, *Rapanos*' precedential meaning is uncertain; as a result, lower courts have spent much of their interpretive energies deciphering how to apply it. Nevertheless, lower courts have narrowed their interpretations of navigable water to varying degrees, sometimes startlingly so.

Courts have fairly consistently held that isolated waters are not navigable. Although courts also have fairly consistently held that groundwater is not navigable, some have held that when a defendant adds pollutants to groundwater that flows into a nearby navigable water or its tributary, the groundwater is navigable, a proposition that has not been retested since *Rapanos* in 2006 and that may have difficulty surviving that decision. EPA's proposed redefinition of waters of the United States excludes groundwater.[126] EPA's "unitary navigable waters" theory is raised at various points, but is particu-

123. United States v. Riverside Bayview Homes, Inc., 474 U.S. 121, 16 ELR 20086 (1985).
124. Solid Waste Agency of N. Cook Cnty. v. U.S. Army Corps of Eng'rs (*SWANCC*), 531 U.S. 159, 31 ELR 20382 (2001).
125. *Rapanos*, 547 U.S. 715.
126. 79 Fed. Reg. 22188, 22268 (proposed Apr. 21, 2014). The proposed rule would be codified at 40 C.F.R. §122.2(b)(5)(vi), *Waters of the United States*.

larly relevant to the Agency's water transfer rule and will be discussed in conjunction with that rule. Water diversions and transfers raise particular problems in the context of navigable waters.[127] How EPA's revised definition of waters of the United States affects these issues will be addressed in this chapter's discussion of them below.

A. *Tributaries and Wetlands*

The Supreme Court's traditional interpretation of Commerce Clause authority to regulate activities on navigable waters did not include tributaries or wetlands unless they independently meet the criteria for navigable waters. But the Court also has recognized that federally authorized activities on navigable waters, such as flood control, often can be accomplished only if they also take place on non-navigable tributaries to navigable waters. Indeed, the Court recognized the necessity of and constitutional permissibility of a watershed approach to flood control,[128] an extension of Commerce Clause jurisdiction authorized by the Necessary and Proper Clause.[129] Congress could not perform its Commerce Clause authority to promote and protect interstate and foreign commerce, including waterborne commerce, from flooding without flood control projects on non-navigable tributaries of navigable water.

1. Tributaries

Because the protection of navigable waters from pollution requires the protection of their tributaries from pollution, it is logical that the CWA's definition of navigable waters includes their tributaries. If the CWA does not control the addition of pollutants to the tributaries of a navigable water, pollutants added to those tributaries will flow into and pollute the navigable water, making it impossible for the CWA to control the pollution of traditionally navigable waters. Congress recognized this when it enacted the Refuse Act, prohibiting the deposit of refuse "into any navigable water of the United States, or into any tributary of any navigable water from which the same shall float or be washed into such navigable water."[130] It also forbade the deposit of

127. *See supra* Chapter 1, Section III.A.3.
128. Oklahoma ex rel. Phillips v. Atkinson, 313 U.S. 508 (1941).
129. U.S. Const., art. I, §8, cl.18.
130. 33 U.S.C. §407. Discharge of refuse into a tributary of navigable water violated the statute even in the absence of proof that the refuse eventually flowed into navigable waters, when it was likely to have done so. *See* United States. v. American Cyanamid Co., 480 F.2d 1132, 3 ELR 20656 (2d Cir. 1973).

refuse on the banks of navigable waters or their tributaries from which it "is liable to be washed into" navigable waters.

The drafters of the CWA were equally aware of the need to make tributaries of navigable waters subject to CWA jurisdiction. The Senate bill did so explicitly. Although the final statute does not mention tributaries in its definition of navigable waters as the waters of the United States, that definition is at least expansive enough to cover tributaries. The congressional explanations of "navigable waters" accompanying the enactment of the CWA describe it as reaching far beyond its traditional meaning, to reach the limits of Congress' jurisdiction over water, and many specify that it includes tributaries.

Assuming that tributaries of navigable waters are within the jurisdiction of the CWA, however, the question becomes *which* tributaries? The primary tributaries, the creeks that flow into the navigable river? The secondary tributaries, the streams that flow into the creeks? The tertiary tributaries? The trickles of water that eventually flow into the most remote stream, but only after rainfall events? Does it make a difference whether an identifiable drop of water from the trickle eventually flows into navigable water or if an identifiable molecule of a pollutant added to that trickle does? Justice Antonin Scalia warned that expanding the interpretation of navigable waters to include those trickles may "engulf entire cities and immense arid wastelands. In fact, the entire land area of the United States lies in some drainage basin, and an endless network of visible channels furrows the entire surface, containing water ephemerally whenever rain falls."[131] He was correct in asserting that this is not what Congress intended. On the other hand, as he acknowledged, Congress did not intend to confine CWA jurisdiction to traditionally navigable waters. Alas, legal decisions are often exercises in line-drawing where there are no obvious places to draw lines.

EPA's revisions of its regulatory definition of "waters of the United States" include a definition of "tributary" as " characterized by the presence of a bed and banks and ordinary high water mark . . . [which] indicators demonstrate there is volume, frequency and duration of flow sufficient to create a bed and banks and an ordinary high water mark." The Agency's new definition specifies that a tributary does not lose its status as such because of natural or man-induced interruptions, such as underground flows, impoundments, or canals.[132] The areas encompassed in the amended definition seem to stop

131. Rapanos v. United States, 547 U.S. 715, 722, 36 ELR 20116 (2006).
132. 80 Fed. Reg. 37114-15 (to be codified at 40 C.F.R. §122.2 waters of the United States (3)(iii).

short of Justice Scalia's parade of horribles, but we will never know whether it does so sufficiently to have satisfied him.

2. Wetlands

The traditional definition of navigable waters did not include wetlands as such. The Corps' first regulatory definition of navigable waters did not include wetlands,[133] nor did EPA's.[134] After *Natural Resources Defense Council v. Callaway*[135] overturned the Corps' regulations as too narrow, the Corps promulgated new regulations substantially expanding its definition of navigable waters, most notably including wetlands adjacent to coastal or inland navigable waters,[136] and EPA followed suit.

3. Pre-*Riverside Bayview* Decisions

Twenty-nine of the 108 (27%) lower court decisions interpreting navigable waters under the CWA in the context of tributaries or wetlands were decided before the first Supreme Court decision on §404 or wetlands, *Riverside Bayview*,[137] issued in 1985. All but one of those 29 decisions (that is, 97%) decided that the waters at issue were navigable; the one negative decision was based on unusual equities rather than reflecting a narrow interpretation of navigable waters.[138]

One of the decisions was in a challenge by environmental groups to the Corps' initial regulations claiming jurisdiction under §404 only for discharges to traditionally navigable waters. The court issued a terse decision ordering the Corps to promulgate a more expansive definition of navigable waters.[139] Once the Corps did so, Wyoming and Puerto Rico unsuccessfully

133. 33 C.F.R. §209.102(d)(1) (1975).
134. 40 C.F.R. §125.1(p) (1975).
135. 392 F. Supp. 685, 5 ELR 20285 (D.D.C. 1975).
136. 33 C.F.R. §209.120(d)(2)(b) & (h), 40 Fed. Reg. 31320, 31324 (July 25, 1975).
137. United States v. Riverside Bayview Homes, Inc., 474 U.S. 121, 16 ELR 20086 (1985). *See* Tables A & B at the end of the book. In using Table B to confirm this and other calculations in the text or to make your own calculations, beware that the groupings of lower court decisions between Supreme Court decisions include more than just tributary and wetlands cases. For instance, the pre-*Riverside Bayview* groupings of circuit and district court decisions include 33 decisions, of which 29 are tributary or wetlands cases.
138. *See* Table B. The one exception is *United States v. City of Fort Pierre*, 747 F.2d 464, 15 ELR 20177 (8th Cir. 1984). The Corps' own dredging and disposal of dredge spoil created a wetland on private property, where there had been no wetland before. The court viewed it as inequitable that the Corps could expand its jurisdiction by its own activities.
139. *Callaway*, 392 F. Supp. 685.

challenged its more expansive regulations as intruding on their jurisdictions over intrastate or intra-commonwealth waters.[140]

a. Tributaries

No decisions during this period held that a stream, lake, or other tributary was not navigable. Six of the decisions held that tributaries were navigable. Three of those decisions held that discharges to small streams, entering navigable water only after flowing through four or five intermediaries, were discharges to waters of the United States.[141] A fourth held that discharges to a normally dry arroyo, flowing eventually to navigable waters only after heavy rainfall, were to waters of the United States.[142] Two more held that discharge to canals leading to navigable water were to waters of the United States.[143] Although a 2,000- to 3,000-acre lake was not connected to other navigable water, the U.S. Court of Appeals for the Seventh Circuit held it to be navigable because interstate travelers used it for recreational boating.[144] Finally, another decision held that a small stream that began and ended in a single county was a water of the United States because it was used to produce agricultural crops sold in interstate commerce.[145]

b. Wetlands

Most of the decisions during this period were in §404 cases where landowners argued that their wetlands were not navigable. Only one decision focused solely (and unsuccessfully) on whether the landowner's property factually met EPA and the Corps' definition of wetlands.[146] The remaining decisions determined whether wetlands were sufficiently connected to navigable waters or their tributaries to be waters of the United States. Nine of them held that various tidal wetlands were navigable.[147] Three recited that the wetland at

140. Wyoming v. Hoffman, 437 F. Supp. 114, 8 ELR 20001 (D. Wyo. 1977); Commonwealth of Puerto Rico v. Alexander, 438 F. Supp. 90, 7 ELR 20751 (D.D.C. 1977).
141. United States v. Texas Pipe Line Co., 611 F.2d 345, 10 ELR 20184 (10th Cir. 1980); United States v. Ashland Oil & Transp. Co., 504 F.2d 1317, 4 ELR 20784 (6th Cir. 1974); Sun Enters., Inc. v. Train, 394 F. Supp. 211 (C.D.N.Y. 1975).
142. United States v. Phelps Dodge Corp., 391 F. Supp. 1185, 5 ELR 20308 (D. Ariz. 1975).
143. United States v. St. Bernard Parish, 589 F. Supp. 617, 14 ELR 20794 (E.D. La. 1984); Bayou des Familles Dev. Corp. v. U.S. Army Corps of Eng'rs, 541 F. Supp. 1025, 13 ELR 20055 (E.D. La. 1982).
144. United States v. Byrd, 609 F.2d 1204, 9 ELR 20757 (7th Cir. 1979).
145. United States v. Earth Sci., Inc., 599 F.2d 368, 9 ELR 20542 (10th Cir. 1979).
146. Avoyelles Sportsmen's League, Inc. v. Marsh, 715 F.2d 897, 13 ELR 20942 (5th Cir. 1983).
147. Leslie Salt Co. v. Froehlke, 578 F.2d 742, 8 ELR 20480 (9th Cir. 1978); United States v. Ciampitti, 583 F. Supp. 483 (D.N.J. 1984); United States v. Robinson, 570 F. Supp. 1157, 14 ELR 20056 (M.D.

issue was navigable because it was adjacent to navigable water, but did not explain what they meant by "adjacent."[148] Others defined "adjacent" variously as bordering[149]; having a hydrological connection[150]; being in close proximity[151]; being contiguous[152]; or neighboring.[153] Some held that a wetland was navigable because it was adjacent to navigable water even though it was never flooded by that navigable water[154] or was separated from it by a 30-foot-wide berm[155] or another tract of land.[156] They held that navigable waters could include an artificial wetland.[157]

The government was a party to the case in all but two of these 29 decisions.[158] The decisions used 10 interpretive devices: precedent (26 decisions); broad interpretation to serve statutory purposes (19 decisions); legislative history (15 decisions); deference (4 decisions); reading harmoniously with other statutes (4 decisions); plain meaning (3 decisions); interpreting the statute to avoid absurd results (2 decisions); structure of the statute (1 decision); give meaning to every word (1 decision); the exception proves the rule (1 decision); and equity (1 decision)—for an average of 2.7 interpretive devices per decision. Of the 25 decisions citing precedent,[159] 12 cited *United States v. Ashland Oil & Transportation Co.*[160] and 10 cited *United States v. Holland*,[161] the first court of appeals and district court decisions, respectively, interpreting navigable water under the CWA. Many of the decisions not directly citing legislative history did so indirectly by citing *Ashland Oil* or *Holland*, both of which included an analysis of the pertinent legislative history. The decisions cited an average of 3.5 precedents.

Fla. 1983); Bayou des Familles Dev. Corp. v. U.S. Army Corps of Eng'rs, 541 F. Supp. 1025, 13 ELR 20055 (E.D. La. 1982); United States v. Bradshaw, 541 F. Supp. 880, 12 ELR 20629 (D. Md. 1981); United States v. Weisman, 489 F. Supp. 1331, 10 ELR 20698 (M.D. Fla. 1980); Conservation Council of N.C. v. Costanzo, 398 F. Supp. 653, 5 ELR 20666 (E.D.N.C. 1975); P.F.Z. Properties, Inc. v. Train, 393 F. Supp. 1370 (D.D.C. 1975); United States v. Holland, 373 F. Supp. 665, 4 ELR 20710 (M.D. Fla. 1974).

148. *Robinson*, 570 F. Supp. 1157; *Weisman*, 489 F. Supp. 1331; American Dredging Corp. v. Dutchyshyn, 480 F. Supp. 957 (E.D. Pa. 1979).
149. *Byrd*, 609 F.2d 1204; *Bradshaw*, 541 F. Supp. 880.
150. United States v. Lee Wood Contractors, Inc., 529 F. Supp. 119, 12 ELR 20421 (E.D. Mich. 1981).
151. United States v. Lambert, 589 F. Supp. 366, 14 ELR 20588 (D. Md. 1984).
152. *Byrd*, 609 F.2d 1204; *Lee Wood*, 529 F. Supp. 119.
153. *Lee Wood*, 529 F. Supp. 119.
154. *Byrd*, 609 F.2d 1204.
155. United States v. Tilton, 705 F.2d 429, 13 ELR 20583 (11th Cir. 1983).
156. *Lee Wood*, 529 F. Supp. 119.
157. Track 12, Inc. v. District Eng'r, 618 F. Supp. 448, 16 ELR 20163 (D. Minn. 1985).
158. In some of these decisions, an official of the United States is a named party; for example, in *Avoyelles Sportsmen's League, Inc. v. Marsh*, Marsh was the Secretary of the Army.
159. *See* Table B.
160. 504 F.2d 1317, 4 ELR 20784 (6th Cir. 1974).
161. 373 F. Supp. 665, 4 ELR 20710 (M.D. Fla. 1974).

None of these decisions suggested that courts were troubled by interpreting navigable water expansively to include even waters that could not float a boat, despite Congress' use of the term navigable waters as a jurisdictional phrase. The lower courts considered and took at face value the legislative history indicating that Congress was exercising its full constitutional authority to protect the nation's waters from pollution, and did not question the constitutionality of that authority.

4. *Riverside Bayview* and Subsequent Decisions

The Supreme Court's first decision interpreting navigable waters under the CWA was *Riverside Bayview*,[162] in which the Corps sought an injunction against defendants filling wetlands, without a §404 permit, to support the subsequent construction of a housing development. The U.S. Court of Appeals for the Sixth Circuit narrowly interpreted the Corps' regulations, defining navigable water to include only wetlands adjacent to navigable waters and periodically inundated or flooded by those adjacent waters.[163] Its purpose in interpreting the Corps' regulations narrowly was to avoid the question of whether §404 constituted an unconstitutional taking of the wetlands. In a unanimous opinion written by Justice Byron White, the Supreme Court reversed. The Court began its analysis by making clear there was no constitutional issue justifying a narrow interpretation of the regulations.[164] That decided, the Court had no trouble finding that the Corps' regulations covered the wetlands at issue.[165]

The Court then addressed whether §404 authorized the Corps to regulate filling wetlands "adjacent to but not regularly flooded by rivers, streams, and other hydrographic features more conventionally identifiable as 'waters.'"[166] The Court specifically noted that it was not addressing the filling of nonadjacent wetlands,[167] and then commenced a quasi-*Chevron*[168] analysis to determine whether the Corps' interpretation of §404 was subject to deference. Skipping step one of the *Chevron* analysis—determining whether the statute

162. United States v. Riverside Bayview Homes, Inc., 474 U.S. 121, 16 ELR 20086 (1985).
163. United States v. Riverside Bayview Homes, Inc., 729 F.2d 391, 14 ELR 20365 (6th Cir. 1984).
164. *Riverside Bayview*, 474 U.S. at 127-28. Just because the Corps may deny a §404 permit does not mean that it will do so. If the Corps does deny a fill permit, there is no taking of the property unless it is deprived of all economic value. Even if there is a taking, it is not unconstitutional unless no just compensation is given.
165. *Id.* at 129-30.
166. *Id.* at 131.
167. *Id.* at 131, n.8.
168. *Id.* at 131. *See* Chevron, U.S.A., Inc. v. Natural Res. Def. Council, Inc., 467 U.S. 837, 14 ELR 20507 (1984).

is ambiguous—the Court moved directly to *Chevron* step two—determining whether the agency's interpretation was a reasonable one. The Court began its analysis by observing that although "[o]n a purely linguistic level, it may appear unreasonable to classify 'lands,' wet or otherwise, as 'waters,'" the interpretation is "reasonable because there is no fine line dividing the waters of a navigable body of water and the waters in an adjacent wetland, both are intermixed parts of a continuum in the hydrological cycle."[169] The Court concluded that a broad interpretation of navigable waters was supported by the CWA's expansive definition of navigable waters and the unanimous legislative history that the term be interpreted to reach the full constitutional limits of congressional power.[170]

Finally, the Court concluded that Congress' acquiescence in the Corps' broad interpretation of navigable waters during the 1977 amendments to the CWA supported the reasonableness of the Corps' interpretation. Although the Court was mindful that subsequent legislative history or legislative failure to act are not ordinarily helpful in interpreting earlier legislation, it concluded that the unsuccessful efforts in Congress to restrict the Corps' jurisdiction over wetlands in 1977 were pertinent because even the proponents of restricting that jurisdiction would not have divested the Corps of jurisdiction over wetlands adjacent to navigable waters, the very waters at issue in *Riverside Bayview*. Thus, the Congress that amended §404 in 1977 specifically intended that §404 protect such wetlands, even if the Congress that enacted §404 in 1972 had no articulated intent with regard to them.

The Court's recognition that Congress "chose to define the waters covered by the Act broadly,"[171] and the Court's deference to the interpretations of navigable waters by the Corps and EPA[172] were taken by lower courts as signals to continue construing navigable waters broadly in the context of tributaries and wetlands. Nineteen of the 23 (83%) lower court decisions interpreting navigable waters in the context of tributaries or wetlands after *Riverside Bayview*, and prior to the next Supreme Court decision interpreting navigable waters, held that the waters at issue were navigable.[173] The decisions holding the waters at issue not to be navigable did so primarily because the

169. *Riverside Bayview*, 474 U.S. at 132. The Court also deferred to the expertise of the Corps and EPA in dealing with the hydrological cycle, noting that the agencies were more capable than courts in determining whether there are lines between particular navigable waters and their adjacent wetlands. *Id.* at 134.

170. *Id.* at 133-34.

171. *Id.* at 133.

172. *Id.* at 131-34.

173. *See* Table B *infra*. For explanation on calculations, see note 137, *supra*.

plaintiff did not carry its factual burden of proof, rather than because the courts narrowly interpreted the term navigable waters.[174]

a. Tributaries

Nine of the decisions considered only whether tributaries were navigable. Two of these held that creeks and bayous flowing directly into ocean bays or sounds were navigable[175] and one held that an unnamed tributary to interstate waters was navigable.[176] Two held that discharges to intermittently flowing ditches eventually connecting with navigable water were navigable,[177] but one held that a dry arroyo was not navigable, absent proof of a nexus to interstate commerce or tributaries to interstate waters.[178] Three held that discharges to or subsequently flowing through ditches, canals, or other man-made waterways were discharges to navigable waters.[179] In holding that a creek was navigable, one court commented that "virtually any surface water, navigable or not" is within the jurisdiction of the CWA.[180]

b. Wetlands

Three decisions upheld the government's delineation of wetlands,[181] two of them deferring to the agency's expertise.[182] Another court, however, held the wetlands at issue were not within the Corps' jurisdiction, because it had failed to carry the burden of proof on delineation.[183] The remaining decisions considered mixed issues of whether tributaries and wetlands were navigable. Most considered whether wetlands were adjacent to navigable water.

---

174. *Id.*
175. United States v. Gulf Park Water Co., Inc., 972 F. Supp. 1056 (S.D. Miss. 1997) (bayou to Mississippi Sound); United States v. Zanger, 767 F. Supp. 1030, 22 ELR 20231 (N.D. Cal. 1991) (creek to San Francisco Bay).
176. Georgia v. City of East Ridge, Tenn., 949 F. Supp. 1571, 27 ELR 20782 (N.D. Ga. 1996) (sewer overflow to unnamed tributary to interstate creek).
177. Driscoll v. Adams, 181 F.3d 1285, 29 ELR 21387 (11th Cir. 1999); United States v. Eidson, 108 F.3d 1336, 27 ELR 20853 (11th Cir. 1997).
178. Friends of Santa Fe Cnty. v. LAC Minerals, Inc., 892 F. Supp. 1333, 26 ELR 20135 (D.N.M. 1995).
179. *Driscoll*, 181 F.3d 1285; *Eidson*, 108 F.3d 1336; United States v. TGR Corp., 171 F.3d 108, 29 ELR 21059 (2d Cir. 1999) (channelized natural brook, in places carried below ground in pipes).
180. Beartooth Alliance v. Crown Butte Mines, 904 F. Supp. 1168, 26 ELR 20639 (D. Mont. 1996).
181. Golden Gate Audubon Soc'y, Inc. v. U.S. Army Corps of Eng'rs, 796 F. Supp. 1306, 23 ELR 20267 (N.D. Cal. 1992); United States v. Malibu Beach, Inc., 711 F. Supp. 1301, 19 ELR 21247 (D.N.J. 1989); United States v. Larkins, 657 F. Supp. 76, 17 ELR 20783 (W.D. Ky. 1987).
182. *Golden Gate*, 796 F. Supp. 1306; *Larkins*, 657 F. Supp. 76.
183. In *United States v. Hallmark Constr. Co.*, 30 F. Supp. 2d 1033, 28 ELR 21438 (N.D. Ill. 1998), the court characterized the wetland as isolated, but appeared willing to consider the Corps' argument that the pond was within its jurisdiction because it was used by migratory birds. However, the Corps did not carry its burden of proof that the pond actually was used by migratory birds.

Three held that wetlands were adjacent to a tidal lake, a tidal pool, or other navigable water.[184] One held that adjacent meant "bordering, contiguous, or neighboring" and that a wetland was adjacent to navigable water, although one-half mile away, because there was a groundwater connection and a surface water connection during hurricanes.[185] Three held that isolated wetlands were not within the Corps' jurisdiction.[186]

The government was a party in all but four of the 23 decisions on tributaries and wetlands during this period. The decisions used a total of eight interpretive devices: precedent (applied in 19 decisions); broad interpretation to serve legislative purpose (8 decisions); deference (5 decisions); structure of the statute (2 decisions); interpret statutory exemptions narrowly (2 decisions); statutory history (1 decision); avoid constitutional issues (1 decision); and plain meaning (1 decision). None of the decisions cited legislative history. The decisions used an average of 1.8 interpretive devices.[187] Of the 19 decisions citing precedent, nine used *Riverside Bayview*[188] and three used *Ashland Oil*.[189]

What differences are observable in lower court decisions before and after *Riverside Bayview*? While the pre-*Riverside Bayview* set of decisions on tributaries and wetlands used a total of 10 and an average of 2.8 interpretive devices, the set of decisions immediately after *Riverside Bayview* used a total of eight and an average of 1.8 devices to interpret navigable waters. While the earlier set of decisions used an average of 3.4 precedents, the latter set used 1.8. While in the earlier set of decisions, 15 cited legislative history, only one cited it in the latter set of decisions. This diminution in the depth of the courts' analyses of the meaning of "navigable waters," including their almost complete disregard of legislative history, suggests that after *Riverside Bayview* the "navigable waters" status of tributaries of navigable waters, even remote tributaries, and of wetlands adjacent to tributaries had become routine and required little analysis.

184. Hanson v. United States, 710 F. Supp. 1105, 19 ELR 21074 (E.D. Tex. 1989) (tidal lake); *Malibu Beach*, 711 F. Supp. 1301 (tidal pool); Conant v. United States, 786 F.2d 1008, 16 ELR 20453 (11th Cir. 1986) (unspecified navigable water).

185. United States v. Banks, 873 F. Supp. 650, 25 ELR 20776 (S.D. Fla. 1995).

186. Hoffman Homes, Inc. v. Administrator, EPA, 999 F.2d 256, 23 ELR 21139 (7th Cir. 1993); United States v. Wilson, 133 F.3d 251, 28 ELR 20299 (4th Cir. 1997); Town of Norfolk v. U.S. Army Corps of Eng'rs, 968 F.2d 1438, 22 ELR 21337 (1st Cir. 1992).

187. *See supra* note 137 (explanation on calculations).

188. *See* Table B.

189. *Id.*

5. *SWANCC* and Subsequent Decisions

The Supreme Court next interpreted the CWA's "navigable waters" term in *Solid Waste Agency of Northern Cook County v. U.S. Army Corps of Engineers* (*SWANCC*).[190] In that case, local governments challenged the Corps' denial of a §404 permit to fill a series of ponds remaining from a sand and gravel mining operation, to construct a municipal landfill. The Corps had found the 31 acres of relatively shallow ponds to be navigable under its Migratory Bird Rule, which extended its §404 jurisdiction to intrastate waters used as habitat by birds protected by migratory bird treaties to which the United States is a party or that migrated across state lines.[191] The five-to-four majority opinion, written by Chief Justice William Rehnquist, held that isolated waters were not subject to the Corps' jurisdiction. The majority first emphasized that *Riverside Bayview* addressed wetlands adjacent to navigable water rather than ponds wholly isolated from navigable water. It noted that the *Riverside Bayview* opinion specifically disclaimed addressing waters not adjacent to navigable waters.[192] While noting that *Riverside Bayview* had commented "the word 'navigable' in the statute was of 'limited import,'" the *SWANCC* majority retorted that the definition of "navigable waters" as the "waters of the United States," does not "constitute a basis for reading the term 'navigable waters' out of the statute."[193]

The Court then denied *Chevron* deference to the Corps' regulations and interpretation because the Corps' interpretation had been inconsistent over time: first adopting the traditional interpretation of navigable waters in 1974, then adopting a more expansive interpretation the next year,[194] and finally adopting the Migratory Bird Rule in 1986.[195] The Court noted that deference is particularly inappropriate when an agency interpretation extends jurisdiction to the farthest reaches of the Commerce Clause, and alters the traditional federal/state balance that leaves regulation of land use to state and

---

190. 531 U.S. 159, 31 ELR 20382 (2001).
191. 531 U.S. at 164.
192. *SWANCC*, 531 U.S. at 167-68.
193. *Id.* at 172 (quoting United States v. Riverside Bayview Homes, Inc., 474 U.S. 121, 133, 16 ELR 20086 (1985)). Justice Scalia commented similarly in his dissent to *Babbitt v. Sweet Home Chapter of Cmtys. for a Great Oregon*, 515 U.S. 687, 718-19, 25 ELR 21194 (1995) (Scalia, J., dissenting), that "take" in the Endangered Species Act, could not be deprived of its traditional meaning with regard to wildlife by an expansive interpretation of "harm" in the statutory definition of "take."
194. The Court failed to note that, as discussed above, a court ordered the Corps to expand its interpretation of navigable waters, or that EPA adopted a much broader definition than the Corps from the outset.
195. *SWANCC*, 531 U.S. at 172-74. This appears to be contrary to *Riverside Bayview*, which had given deference to the Corps' second interpretation, although that interpretation was far more expansive than its first interpretation. At the time, the Corps' 1986 migratory bird amendment to the regulatory definition had yet to be promulgated and accordingly was not considered in *Riverside Bayview*.

local authorities.[196] The Court also noted that the Commerce Clause did not grant unlimited jurisdiction to Congress, citing *United States v. Morrison*[197] and *United States v. Lopez*,[198] both decided after *Riverside Bayview*. It rejected the relevance of the same legislative history of the CWA 1977 Amendments that was cited in *Riverside Bayview* in support of that decision. While that legislative history may not have been relevant to the Migratory Bird Rule, promulgated a decade later, the Court rejected it on a far broader basis.[199]

The *SWANCC* dissent by Justice John Paul Stevens, joined by three other Justices, reads as if they had used a different translation of *Riverside Bayview* than did the majority. The dissent even interpreted different sections of the statute: Whereas the majority interpreted navigable waters in §404, the dissent interpreted waters of the United States in §502(7).[200] The dissent did not characterize the wetland at issue in the Court's earlier *Riverside Bayview* decision as adjacent to navigable waters, but instead characterized it as "not itself navigable, directly adjacent to navigable water, or even hydrologically connected to navigable water."[201] The dissent also laid out the argument that the Migratory Bird Rule was well within Commerce Clause jurisdiction, because of the aggregate effects of piecemeal elimination of habitats essential to the multibillion-dollar recreational commerce connected to bird-watching and fishing.[202]

---

196. Query whether the CWA significantly interferes with state land use decisions? If states want to control CWA permitting decisions, they may administer the §402 program and much of the §404 program. The §402 program does not involve land use. The permit applicant can make whatever use of the land it wants, as long as it treats resulting water pollution to federal and state standards. The §404 permit program does not prohibit particular uses of land, but it does prevent filling wetlands for any use without a permit. If, as Justice Scalia claims, the Corps interprets §404 to include "ripples of sand in the desert . . . engulf[ing] entire cities and immense arid wastelands," *see* Rapanos v. United States, 547 U.S. 715, 722, 36 ELR 20116 (2006), then §404's interference with state land use decisions could be real. However, as this chapter discusses, that description is overblown.
197. 529 U.S. 598 (2000).
198. 514 U.S. 549 (1995). In the course of invalidating a federal statute prohibiting possession of firearms near a school, the Court explained that the Commerce Clause authorized only federal statutes regulating: (1) highways of interstate or foreign commerce; (2) instrumentalities of interstate commerce; or (3) activities having a substantial effect on interstate commerce. In a 5/4 decision, it held that the effect on interstate commerce of educational disruption caused by gun-related violence near schools was too tangential to support Commerce Clause jurisdiction.
199. *SWANCC*, 531 U.S. at 169-71. The Court generally disavowed the relevance of failed amendments in interpreting a statute, especially failed amendments in a later Congress. This was only part of a broader attack on the relevance and reliability of legislative history, which has suffered a decline in use since the enactment of the CWA. *See infra* Chapter 5, Section I.D.
200. Compare the majority's statement of the case, "We are asked to decide whether the provision of §404(a) may be fairly extended to the isolated ponds," *SWANCC*, 531 U.S. at 161, with the dissent's: "[It] is the *definition* [§502(7)] that is the appropriate focus of our attention," *SWANCC*, 531 U.S. at 182 (Stevens, J., dissenting).
201. *SWANCC*, 531 U.S. at 175.
202. *Id.* at 192-97.

SWANCC did not discuss tributaries or wetlands, but merely held that isolated ponds were not navigable waters. Nevertheless, the decision's refusal to defer to the Corps' interpretation of the statute[203] and its observation that the CWA's broad definition of navigable waters did not read "the term 'navigable waters' out of the statute"[204] signaled to lower courts that there are limits to which tributaries to navigable waters and their adjacent wetlands are navigable. The Court stated: "The term 'navigable' has at least the import of showing us what Congress had in mind as its authority for enacting the CWA: its traditional jurisdiction over waters that were or had been navigable in fact or which could reasonably be so made."[205]

In response to the Court's first restrictive opinion on the CWA's navigable waters jurisdiction, 22 of 25 lower court decisions (88%) held the waters at issue to be navigable,[206] an increase over the 83% decided between *Riverside Bayview* and *SWANCC*. Some lower courts interpreted *SWANCC* as narrowly as they had interpreted navigable waters broadly, reading *SWANCC* only to hold that isolated waters are outside the CWA's jurisdiction.[207] Other courts, however, took *SWANCC*'s reading of *Riverside Bayview* to limit the CWA's jurisdiction to waters that are navigable in fact or are adjacent to navigable-in-fact open waters.[208] Others wrestled with the meaning of adjacency, noting that the Supreme Court did not define it in either of its §404 decisions.[209] One district court sought to reconcile the Court's two rulings by holding that when "a drop of rainwater landing in the [wetlands] is certain to intermingle with water from the [adjacent navigable river]," the relationship is direct and therefore a "significant nexus."[210] *SWANCC*, however, provided more conservative courts with the opportunity to fashion narrower interpretations of "navigable waters."

---

203. *Id.* at 168-69.
204. *Id.* at 172.
205. *Id.* (citing United States v. Appalachian Elec. Power Co., 311 U.S. 377, 407-08 (1940)).
206. *See* Table B.
207. North Carolina Shellfish Growers Ass'n v. Holly Ridge Assocs., LLC, 278 F. Supp. 2d 654 (E.D.N.C. 2003). The U.S. Court of Appeals for the Ninth Circuit in *Borden Ranch P'ship v. U.S. Army Corps of Eng'rs*, 261 F.3d 810, 32 ELR 20011 (9th Cir. 2001), agreed that an isolated vernal pool was not navigable under *SWANCC*, but its other decisions cited in this segment of the chapter continue with an expansive interpretation of navigable waters.
208. In re Needham, 354 F.3d 340 (5th Cir. 2003); Rice v. Harken Exploration Co., 250 F.3d 264, 31 ELR 20599 (5th Cir. 2001); FD & P Enters., Inc. v. U.S. Army Corps of Eng'rs, 239 F. Supp. 2d 509 (D.N.J. 2003); United States v. RGM Corp., 222 F. Supp. 2d 780 (E.D. Va. 2002).
209. United States v. Rueth Dev. Co., 335 F.3d 598 (7th Cir. 2003); United States v. Interstate Gen. Co., 152 F. Supp. 2d 843, 31 ELR 20750 (D. Md. 2001).
210. United States v. Rueth Dev. Co., 189 F. Supp. 2d 874, 877-78 (N.D. Ind. 2002), *aff'd* 335 F.3d 598 (7th Cir. 2003).

a. Tributaries

Narrower interpretations of navigable waters by lower courts were particularly pronounced with regard to tributaries. In three of these decisions, the courts held tributaries not to be within navigable waters jurisdiction,[211] based on facts very similar to both earlier and contemporary decisions that had held tributaries to be navigable. In a fourth decision, the court held the tributary at issue to be navigable, but rejected the government's expansive interpretation of navigable waters.[212]

In *Rice v. Harken Exploration Co.*,[213] one of the three narrow decisions, the plaintiff brought suit under the citizen suit provision of the Oil Pollution Act (OPA),[214] which prohibits the discharge of oil into navigable waters. He alleged that the defendant discharged oil into intermittent streams leading to navigable waters. But there was nothing in the record, the U.S. Court of Appeals for the Fifth Circuit concluded, "that could convince a reasonable trier of fact that . . . any of the . . . intermittent creeks on the ranch are sufficiently linked to an open body of navigable water as to qualify for protection under the OPA."[215] While this perhaps appears to be a case in which the plaintiff failed to carry his burden of proof that the intermittent streams at issue ever reached navigable water, the Fifth Circuit stated with regard to tributaries that "a body of water is protected under the Act only if it is actually navigable or is *adjacent* to an open body of navigable water."[216] This is a narrow interpretation of navigable waters with regard to wetlands, but it is untenable with regard to tributaries, as discussed below. Because the court searched the record for evidence that the "intermittent creeks on the ranch are sufficiently linked to an open body of navigable water as to qualify for protection under the OPA,"[217] its use of "adjacent to . . . navigable water" may not have been an accurate description of the Fifth Circuit's perception of how tributaries must be related to navigable waters to be within the jurisdiction

211. *Rice*, 250 F.3d 264; *RGM Corp.*, 222 F. Supp. 2d 780; United States v. Newdunn Assocs., LLP, 195 F. Supp. 2d 751, 32 ELR 20573 (E.D. Va. 2002). The significance of the two district court decisions, by the same judge, is undercut by their invective against the Corps.
212. In re Needham, 354 F.3d 340.
213. 250 F.3d 264, 31 ELR 20599 (5th Cir. 2001). This opinion does not interpret navigable waters under the CWA; instead, it interprets the same jurisdictional phrase in the Oil Pollution Act. That statute and §311 of the CWA, 33 U.S.C. §1321, both address oil spills into the "navigable waters" and are intimately intertwined.
214. 33 U.S.C. §§2701-2761, ELR Stat. OPA §§1001-7001.
215. *Rice*, 250 F.3d at 271.
216. *Id.* at 270 (emphasis added).
217. *Id.* at 271.

of the statute. Unfortunately, the court repeated this misleading adjacency concept in a later decision.

In re Needham[218] was an appeal from the denial by a bankruptcy court of the U.S. claim for its costs of remediating an oil spill. The government argued that navigable waters included all tributaries "that have any hydrological connection with 'navigable waters.'"[219] In the Fifth Circuit's view, however, CWA jurisdiction did not extend "over 'tributaries' that are neither themselves navigable nor truly adjacent to navigable waters," citing its *Rice* decision.[220] The government "may not simply impose regulations over puddles, sewers, roadside ditches and the like; under *SWANCC* 'a body of water is subject to regulation . . . if the body of water is actually navigable or adjacent to an open body of navigable water,'"[221] the court said. It continued: "[T]he proper inquiry is whether Bayou Folse, the site of the farthest traverse of the spill, is navigable-in-fact or adjacent to an open body of navigable water."[222] Because "Bayou Folse flows directly into the Company Canal," which is navigable in fact, the court allowed the government's claim.[223] Thus, the Fifth Circuit may be using "adjacent" to mean "flowing into," at least for tributaries.

Determining whether tributaries are navigable waters by asking if they are adjacent to navigable-in-fact or open water is misdirected for several reasons. First, the focus on adjacency originates in the Corps' §404 regulations, which use adjacency only to claim jurisdiction over wetlands adjacent to navigable waters, not over tributaries adjacent to navigable waters.[224] Second, the Supreme Court in *Riverside Bayview* held that only wetlands adjacent to navigable waters were subject to §404's jurisdiction, not that tributaries adjacent to navigable waters were subject to §402's jurisdiction.[225] Third, "adjacent" means "having a common border" or "abutting,"[226] which may describe the relationship between wetlands and navigable waters, such as a river, for wetlands are often next to rivers. Non-wetland tributaries of rivers, such as creeks and streams, however, cannot be adjacent to those rivers because those tributaries do not flow *next to*, but rather *into* rivers.

---

218. 354 F.3d 340 (5th Cir. 2003).
219. *Id.* at 345.
220. *Id.*
221. *Id.* at 345-46.
222. *Id.* at 345.
223. *Id.* at 343.
224. 33 C.F.R. §328.3.
225. United States v. Riverside Bayview Homes, Inc., 474 U.S. 121, 126, 16 ELR 20086 (1985).
226. Webster's Seventh New Collegiate Dictionary 11 (1967).

Most lower courts continued to interpret the term navigable waters with regard to tributaries and wetlands as if *SWANCC* had never happened. Courts interpreted navigable waters to include the following: primary tributaries to navigable waters[227]; remote tributaries to navigable waters[228]; intermittent flows[229]; man-made or man-improved flows, including irrigation canals[230]; ditches[231]; and waters channelled[232] or pumped[233] through pipes and culverts under roads.

b. Wetlands

Lower court wetlands decisions were largely confined to decisions regarding tributaries. Decisions held that wetlands adjacent to both primary and remote tributaries were navigable waters.[234] One court held a wetland to be adjacent to navigable water despite being separated from it by a 70-foot-wide berm.[235] The government was a party in only 14 of the 22 cases interpreting navigable waters in the context of tributaries and wetlands. The increased proportion of citizen suits did not affect their outcomes, however, as all three decisions interpreting navigable waters narrowly were in cases in which the government was a party.[236] The decisions used 10 interpretive devices: precedent (applied in 22 decisions); broad interpretation (6 decisions); legislative history (6 decisions); deference (6 decisions); plain meaning (5 decisions); finality (4 decisions); avoid constitutional issues (4 decisions); structure of statute (3 decisions); lenity (1 decision); and read harmoniously with other statutes (1 decision). The decisions used an average of 2.6 interpretive devic-

---

227. United States. v. Hubenka, 438 F.3d 1026 (10th Cir. 2006); In re Needham, 354 F.3d 340 (5th Cir. 2003); Aiello v. Town of Brookhaven, 136 F. Supp. 2d 81 (E.D.N.Y. 2001).
228. Parker v. Scrap Metal Processors, Inc., 386 F.3d 993 (9th Cir. 2004); United States v. Adams Bros. Farming, Inc., 369 F. Supp. 2d 1166 (C.D. Cal. 2003); Idaho Rural Council v. Bosma, 143 F. Supp. 2d 1169 (D. Idaho 2001); United States v. Buday, 138 F. Supp. 2d 1282 (D. Mont. 2001).
229. California Sportfishing Prot. Alliance v. Diablo Grande, Inc., 209 F. Supp. 2d 1059 (E.D. Cal. 2002).
230. Headwaters, Inc. v. Talent Irrigation Dist., 243 F.3d 526, 31 ELR 20535 (9th Cir. 2001); *Adams Bros.*, 369 F. Supp. 2d 1166; *Idaho Rural Council*, 143 F. Supp. 2d 1169.
231. Treacy v. Newdunn Assocs., LLP, 344 F.3d 407 (4th Cir. 2003); United States v. Deaton, 332 F.3d 698 (4th Cir. 2003).
232. *Treacy*, 344 F.3d 407; *Adams Bros.*, 369 F. Supp. 2d 1166; *California Sportfishing*, 209 F. Supp. 2d 1059.
233. United States v. Adams Bros. Farming, Inc., 209 F. Supp. 2d 1059 (E.D. Cal. 2002).
234. *Treacy*, 344 F.3d 407; *Deaton*, 332 F.3d 698; North Carolina Shellfish Growers Ass'n v. Holly Ridge Assocs., LLC, 278 F. Supp. 2d 654 (E.D.N.C. 2003); United States v. Buday, 138 F. Supp. 2d 1282 (D. Mont. 2001) (the discharge was 240 miles from traditionally navigable waters).
235. Baccarat Freemont Devs., LLC v. U.S. Army Corps of Eng'rs, 425 F.3d 1150 (9th Cir. 2005).
236. *See* Tables A & B.

es.[237] All 22 decisions cited precedent; 20 cited *SWANCC* and 10 cited *Riverside Bayview.*[238] The decisions cited an average of 5.2 precedents.

While the lower court decisions dating between *Riverside Bayview* and *SWANCC* used a total of eight and an average of 1.8 interpretive devices, the decisions between *SWANCC* and *Rapanos* used a total of 10 and an average of 2.6 interpretive devices. While the earlier set of decisions cited an average of 1.8 precedents, the lower court decisions issued between *Riverside Bayview* and *SWANCC* cited an average of 6.4. What conclusions can we draw from this quantitative analysis? Decisions on tributaries and wetlands after *SWANCC* undertake their legal and factual analysis in considerably more depth than do decisions issued between *Riverside Bayview* and *SWANCC.* If *SWANCC* did not prompt lower courts to find more frequently that waters were non-navigable, it did remind them that expansive interpretations of navigable waters were not to be made or accepted without careful scrutiny.

6. *Rapanos* and Subsequent Decisions

The 2006 *Rapanos* decision[239] arose out of enforcement actions for filling wetlands without §404 permits. The wetlands at issue were adjacent to man-made drainage ditches that flowed into a succession of other ditches and creeks and ultimately into Lake Michigan. The plurality opinion, written by Justice Scalia and joined by three other Justices, held that the wetlands at issue were not within the jurisdiction of the CWA. In a concurring opinion, Justice Anthony Kennedy agreed with the plurality's result, but not with its reasoning. Chief Justice John Roberts joined the plurality opinion, but filed a short concurrence lamenting that there was no majority opinion and acknowledging the difficulty that would pose to lower courts. Justice Stevens wrote a dissent joined by three Justices. In addition to joining Justice Stevens' dissenting opinion, Justice Stephen Breyer filed his own dissenting opinion.

As discussed above, the lack of a majority opinion in *Rapanos* has required lower courts to spend more of their energy interpreting *Rapanos* than interpreting the CWA. Although the holding of *Rapanos* is opaque, it is clear that the Court's interpretation of navigable waters has become increasingly restrictive since *Riverside Bayview.*

Two distinct but related questions are at issue in *Rapanos.* First, and most important, when is a tributary to a traditionally navigable water within the CWA's jurisdiction (that is, a jurisdictional tributary)? Second, when is a

237. *See supra* note 137 (explanation on calculations).
238. *Id.*
239. Rapanos v. United States, 547 U.S. 715, 36 ELR 20116 (2006).

wetland sufficiently connected to a traditionally navigable water or a jurisdictional tributary for the wetland to be within the CWA's jurisdiction (that is, a jurisdictional wetland)? Although the plurality opinion addressed these questions, its answers are incomplete. One reason is that the plurality's analyses conflated different elements of §301(a). Another reason is that the plurality's analyses are sometimes obscured by vituperative outbursts against the Corps, lower courts, and the dissenting and primary concurring opinions.

Justice Scalia began his plurality opinion with what can only be described as a tirade designed to demonstrate that the Corps "exercises the discretion of an enlightened despot"[240] by claiming §404 jurisdiction over "storm drains, roadside ditches, ripples of sand in the desert that may contain water once a year, and lands that are covered by floodwaters once every 100 years . . . engulf[ing] entire cities and immense arid wastelands."[241] Scalia complained mightily that the Corps and the lower courts have continued to expand §404 jurisdiction "to include 'ephemeral streams' and 'drainage ditches' as 'tributaries' . . . [extending] to virtually any land feature over which rainwater or drainage passes and leaves a visible mark—even if only the presence of litter and debris," despite the Court's narrowing of §404 jurisdiction in *SWANCC*.[242]

---

240. *Id.* at 720.

241. *Id.* at 722. In particular, Justice Scalia's plurality opinion criticized *Treacy v. Newdunn Assocs., LLP*, 344 F.3d 407 (4th Cir. 2003); *United States v. Deaton*, 332 F.3d 698 (4th Cir. 2003); *Community Ass'n for the Restoration of the Env't v. Henry Bosma Dairy*, 305 F.3d 943 (9th Cir. 2002); *Headwaters, Inc. v. Talent Irrigation Dist.*, 234 F.3d 526, 31 ELR 20535 (9th Cir. 2001); *United States v. Eidson*, 108 F.3d 1336, 27 ELR 20853 (11th Cir. 1997); and *Quivira Mining Co. v. EPA*, 765 F.2d 126, 15 ELR 20530 (10th Cir. 1985). *See Rapanos*, 547 U.S. at 725-27. Part of Justice Scalia's ire is probably attributable to the fact that lower court decisions more frequently used expansive interpretations of navigable waters after *SWANCC* than before it. This could also reflect plaintiffs' greater care in bringing and arguing cases after *SWANCC*. See also the Fifth Circuit's diatribe in *In re Needham*, 354 F.3d 340, 345 (5th Cir. 2003), against the Corps for claiming jurisdiction over "puddles, sewers, roadside ditches and the like."

Many of Justice Scalia's outlandish examples of Corps overreach are hyperbole. The Corps does not and never has claimed jurisdiction over mud puddles or ripples of sand in the desert. For one thing, sand ripples in deserts and dunes are created by wind; when precipitation falls on them, it generally percolates down through the sand rather than flowing down the outsides of dunes. But when tempered, Justice Scalia's concerns are understandable. Most land is covered at times with water, however thinly, although Congress did not claim jurisdiction for the CWA over most land: Most land is neither navigable water nor water of the United States. Justice Scalia's inconsistent focus on the permanence of waterways elsewhere in the opinion may be seeking a distinction between stormwater runoff and water bodies. That is analogous to the distinction Congress established between point sources and nonpoint sources. Unfortunately, neither distinction is a bright line, any more than the edge between a river and an adjacent wetland is a bright line. Sheetwater runoff is stormwater and few would claim that it is a tributary of navigable water, although much of it may flow to navigable water. On the other hand, many would claim that a desert gully, roaring with a flash flood after rains, is a tributary of the navigable river into which its flash floods flow.

242. *Rapanos*, 547 U.S. at 724-29.

Justice Scalia acknowledged that Congress intended CWA jurisdiction to extend beyond traditional navigable waters, as the Court had recognized in both *Riverside Bayview* and *SWANCC*. On the other hand, he noted that in *SWANCC*, the Court took pains to emphasize "that the qualifier 'navigable' is not devoid of significance"[243] and that CWA jurisdiction is limited to 'waters,' whether qualified by 'navigable' or 'of the United States."[244] He then launched into a counterintuitive demonstration that the definitions of "water" and "waters" in the 1954 edition of *Webster's New International Dictionary* lead to the conclusion that "*the* waters of the United States" is a subset of "water of the United States."[245] (Query: How can the plural "waters" be a subset of the singular "water"?)

In short, Scalia argues, the CWA's incorporation of the term navigable waters confers jurisdiction "only over relatively *permanent* bodies of water" and "does not authorize [a] 'Land Is Water' approach to federal jurisdiction."[246] (Query: Assuming that the term "the waters of the United States" is a subset of "water of the United States," how does that suggest that "navigable waters" include only permanent bodies of water? That is a non sequitur.) Justice Scalia posits that the statute's definition of another element of the water pollution offense confirms this, because ditches, channels, and conduits, which typically convey transitory waters, are defined as "point sources" and therefore discharge into "navigable waters" rather than carry "navigable waters."[247] (Again, this is a non sequitur. Granted that point sources convey polluted water to navigable waters, how does that establish that the conveyed waters are not themselves navigable?) Scalia supports his analysis with the need to narrowly interpret statutes when their jurisdictions extend to the outer edges of Congress' constitutional authority or intrude on states' traditional authority over land use without an explicit statement by Congress of its intent to do so.[248]

Justice Scalia then summarizes the plurality's conclusions. Jurisdictional waters (presumably including tributaries)

243. *Id.* at 731 (quoting Solid Waste Agency of N. Cook Cnty. v. U.S. Army Corps of Eng'rs (*SWANCC*), 531 U.S. 159, 172, 31 ELR 20382 (2001).
244. *Rapanos*, 547 U.S. at 731-36.
245. *Id.* at 732-33 (emphasis added).
246. *Id.* at 734.
247. *Id.* at 735-36. Ditches, channels, and conduits convey water that is transitory in the sense that the water they convey is in transit. But is the water transitory in a temporal sense, as Justice Scalia seems to be using the term? Roadside ditches normally flow during and after storm events, but in places, they continually convey streamwater that has been rerouted, for instance, to pass under roads in culverts. The English Channel is always wet, as is the main channel of the Mississippi River, and both are navigable. Moreover, Justice Scalia's argument is a non sequitur, as pointed out in the text.
248. *Id.* at 738.

> include[ ] only those relatively permanent, standing or continuously flowing bodies of water "forming geographic features" that are described in ordinary parlance as "streams[,] . . . oceans, rivers [and] lakes." [The jurisdiction] does not include channels, through which water flows intermittently or ephemerally, or channels that periodically provide drainage for rainfall.[249]

He asserts that this narrower definition of navigable waters will not jeopardize the reach of the §402 program, because the discharge of pollutants into an intermittent stream from which they will wash downstream into navigable waters would still be covered: The statute does not require that the discharge be directly into navigable water.[250] Jurisdictional wetlands are "*only* those wetlands with a continuous surface connection to bodies that are 'waters of the United States' in their own right, so that there is no clear demarcation between 'waters' and wetlands."[251] The plurality opinion does not directly address the question of whether tributaries of navigable waters are waters of the United States, but tributaries are easily susceptible to the analysis that the plurality poses for jurisdictional waters. Moreover, the plurality commented that "the discharge into intermittent channels of any pollutant *that naturally washes downstream* likely violates §1311(a), even if the pollutants discharged from a point source to not emit 'directly into' covered waters."[252] Pollutants discharged to tributaries of navigable water naturally wash downstream into those navigable waters.

It is difficult to discern the *Rapanos* plurality's precise conclusion with regard to jurisdictional tributaries, in part because its analysis is abstract and largely divorced from the facts of the cases before it and in part because it does not sharply separate its analysis of whether a tributary is jurisdictional from its analysis of whether a wetland is jurisdictional. Finally, the plurality opinion is inconsistent on what constitutes a navigable tributary. It begins by stating that navigability connotes "continuously present, fixed bodies of water, as opposed to ordinarily dry channels through which water occasionally or intermittently flows . . . [and does not] encompass[ ] transitory puddles or ephemeral flows of water."[253] The opinion does not define what it means by "ephemeral flows." The plurality continues:

> [A]pplying the definition [of "navigable waters"] to "ephemeral streams," "wet meadows," storm sewers and culverts, "directional sheet flow during storm

249. *Id.* at 739.
250. *Id.* at 742-45. *See also* Amil Anthony, *Shotguns, Spray, and Smoke: Regulating Atmospheric Deposition of Pollution Under the Clean Water Act*, 29 UCLA J. Envtl. L. & Pol'y 215, 235-37 (2011).
251. *Rapanos*, 547 U.S. at 742.
252. *Id.* at 743.
253. *Id.* at 733.

> events," drain tiles, man-made drainage ditches, and dry arroyos in the middle of the desert, the Corps has stretched the term "waters of the United States" beyond parody. The plain language of the statute simply does not authorize this "Land is Waters" approach to federal jurisdiction.[254]

While some of these examples are admittedly far-fetched to deem as tributaries to navigable waters, others are not. Storm sewers, culverts, and drainage ditches, for example, often replace or channel existing natural streams that otherwise might be easily categorized as tributaries. The opinion dwells on the necessity of "at bare minimum, the ordinary presence of water."[255] While the plurality acknowledges that ditches, channels, and conduits can carry constant flows, it asserts that when they do, they are called rivers, creeks, streams, or moats, implying they may be navigable waters under those circumstances.[256] (Query: What about the English Channel, which carries a constant flow; should we call it the English River or English moat?)

Finally, the opinion states that "waters of the United States" "does not include channels through which water flows intermittently or ephemerally, or channels that periodically provide drainage for rainfall."[257] The plurality only examined two aspects of whether a tributary is navigable: whether it forms a "geographic feature"; and whether its flow is permanent. But in a footnote, the plurality admitted that it did not "necessarily exclude *seasonal* rivers, which contain continuous flow during some months of the year but no flow during dry months."[258] Indeed, the opinion did not distinguish between the intermittency of flow that would include or exclude a water body from being "waters of the United States," beyond "[c]ommon sense and common usage."[259]

The *Rapanos* plurality did not address what constitutes a wetland, but limited itself to addressing what types of wetlands are jurisdictional wetlands. A jurisdictional wetland must be adjacent to a navigable water or a jurisdictional tributary of a navigable water. Wetlands adjacent to such a water are "*only* those wetlands with a continuous surface connection to . . . 'waters of the United States' in their own right, so that there is no clear demarcation

---

254. *Id.* at 734.
255. *Id.*
256. *Id.* at 736 n.7. Justice Scalia's plurality opinion seeks to dismiss this issue by commenting that when ditches do flow constantly, they are called streams, a dubious proposition for which he offers no authority.
257. *Id.* at 739.
258. *Id.* at 733 n.5.
259. *Id.*

between 'waters' and wetlands."[260] The plurality's test applies to whether a wetland is a jurisdictional wetland, not to whether a tributary is a jurisdictional tributary. At the same time, the plurality's concern with whether a water is permanent, seasonal, or ephemeral was in the context of whether tributaries were jurisdictional tributaries, not to whether a wetland is permanent, seasonal, or ephemeral, although it is not much of an additional step to consider the issue in that context, given the plurality's concern that dry land not be classified as water.

The analysis in the concurring opinion by Justice Kennedy focused almost entirely on wetlands. His reading of *Riverside Bayview* and *SWANCC* is that the Corps' "jurisdiction over wetlands depends upon the existence of a significant nexus between the wetlands in question and navigable waters in the traditional sense."[261] Wetlands have a significant nexus with navigable waters "if the wetlands, either alone or in combination with similarly situated lands in the region, significantly affect the chemical, physical, and biological integrity of other covered waters more readily understood as 'navigable.'"[262] Adjacency of a wetland to navigable-in-fact waters is sufficient to establish a significant nexus, and adjacency to major tributaries may be sufficient. But adjacency to lesser tributaries requires the Corps to "establish a significant nexus on a case-by-case basis."[263]

Justice Kennedy would have remanded the cases for a determination of whether the wetlands in question had a significant nexus to navigable-in-fact waters. His opinion does not indicate when a tributary is sufficiently "major" for adjacency to satisfy his "significant nexus" test. He rejected the plurality's conclusions that tributaries must be permanently flowing and have a continuous surface connection to be waters of the United States,[264] pointing out that the requirement of permanent flow "makes little practical sense in a statute concerned with downstream water quality."[265]

The dissenting opinion of Justice Stevens and three other Justices accorded *Chevron* deference to the Corps' regulations defining navigable waters to include wetlands adjacent to navigable waters and their tributaries. Like Justice Kennedy's concurring opinion, the dissenting opinion provides a point-by-point refutation of the plurality opinion. Not surprisingly, the focus of the plurality is on the meaning of navigable waters, while the focus of both

260. *Id.* at 742.
261. *Id.* at 779 (Kennedy, J., concurring).
262. *Id.* at 780.
263. *Id.* at 782.
264. *Id.* at 769-71.
265. *Id.*, at 769.

Justice Kennedy's concurrence and the dissent is on the meaning of waters of the United States.[266]

Thus, in regard to tributaries, the plurality in *Rapanos* developed a test under which most tributaries of navigable waters, even remote tributaries, would be navigable waters, but inconsistently drew a line at intermittent tributaries. With regard to intermittent tributaries, it suggested that the discharge of pollutants into intermittent waters from which the pollutants naturally wash downstream into navigable waters are indirect discharges to navigable water, violating §301(a). The concurrence did not focus on tributaries. The dissent concluded that tributaries of navigable water, major or remote, are navigable water, and argued that even intermittent tributaries could be navigable waters. None of the Justices argued that tributaries to navigable waters were not themselves waters of the United States. As to intermittent tributaries, the plurality's guidance is negative, but also ambiguous and in the minority.

Post-*Rapanos* lower court decisions perceive *Rapanos* as game-changing for the interpretation of the terms navigable waters and waters of the United States, and focus their interpretations of those terms on *Rapanos.* Of the 40 lower court decisions issued after *Rapanos*, 27 of those decisions (68%) held the waters at issue to be navigable, a decrease from the findings in 88% of the decisions dating between *SWANCC* and *Rapanos.* Unfortunately, a large part of the lower courts' attention to *Rapanos* is devoted to determining what its holding means in the absence of a majority opinion. Chief Justice Robert's short concurring opinion lamented that without a majority opinion, the "[l]ower courts and regulated entities will now have to feel their way on a case-by-case basis," citing *Marks v. United States.*[267] *Marks* directed lower courts that "[w]hen a fragmented Court decides a case and no single rationale explaining the result enjoys the assent of five Justices, the holding . . . may be viewed as that position taken by those Members who concurred in the judgments on the narrowest grounds."[268]

Lower courts struggled mightily to determine whether to use the plurality test, the concurrence test, or both. The U.S. Court of Appeals for the Eleventh Circuit has followed the concurrence's significant nexus test.[269] The U.S. Courts of Appeals for the First, Third, and Eighth Circuits have

266. *Compare Rapanos*, 547 U.S. at 723 (plurality), *with Rapanos*, 547 U.S. at 759 (Kennedy, J., concurring), *and Rapanos*, 547 U.S. at 787 (Stevens, J., dissenting).
267. 430 U.S. 188 (1977).
268. *Id.* at 193.
269. United States v. Robinson, 505 F.3d 1208, 1219-21 (11th Cir. 2011).

applied alternately either the plurality or the concurrence tests.[270] The Fifth and Sixth Circuits apparently require plaintiffs to prove both tests.[271] And the Seventh Circuit and U.S. Court of Appeals for the Ninth Circuit have preferentially employed the concurrence test, but if that is not met will look to the plurality test.[272]

These courts misperceive the Kennedy concurrence's significant nexus test, however, when they apply it to determine whether a tributary is a water of the United States. Justice Kennedy developed and used that test only to determine whether a wetland is a water of the United States. As to tributaries, his concurrence only commented that the plurality's insistence that tributaries must be permanent to be waters of the United States was ill-founded. On that issue, he joined the four dissenting Justices. A few courts have recognized the inapplicability of the significant nexus test to tributaries.[273] Justice Kennedy's significant nexus test requires a jurisdictional wetland to affect the water quality of a traditionally navigable water and presumes such an effect if the wetland is *adjacent* to traditionally navigable water or a major tributary thereof. Because tributaries flow directly into navigable-in-fact waters, tributaries inevitably affect the water quality of navigable waters; in the aggregate, even remote tributaries do so, echoing Justice Kennedy's conclusion that it "makes little practical sense in a statute concerned with downstream water quality"[274] to require a nexus showing for tributaries.[275] In this light, EPA's scientific study[276] demonstrating that 60% of the flow of rivers and streams in the northeast originate in headwaters, confining the presumption of a nexus to primary tributaries, does not make sense.

---

270. United States v. Bailey, 571 F.3d 791, 799 (8th Cir. 2009); United States v. Donovan, 661 F.3d 174, 76, 42 ELR 20328 (3d Cir. 2011); United States v. Johnson, 467 F.3d 56, 66 (1st Cir. 2006).

271. United States v. Cundiff, 535 F.3d 200, 210, 39 ELR 20025 (6th Cir. 2009); United States v. Lucas, 516 F.3d 316, 24 (5th Cir. 2008).

272. Northern Cal. River Watch v. City of Healdsburg, 496 F.3d 993, 1000 (9th Cir. 2007); United States v. Gerke Excavating, Inc., 464 F.3d 723, 725 (7th Cir. 2006).

273. United States v. Vierstra, 2011 WL 1064526, *5 (D. Idaho); Benjamin v. Douglas Ridge Rifle Club, 673 F. Supp. 2d 1210, 1215 (D. Or. 2009).

274. Rapanos v. United States, 547 U.S. 715, 769, 36 ELR 20116 (2006) (Kennedy, J., concurring).

275. While Justice Kennedy made that comment with regard to the plurality's test of permanence and continuing surface connection, the comment applies as well in this context.

276. *See* U.S. EPA, Connectivity of Streams and Wetlands to Downstream Waters: A Review and Synthesis of the Scientific Evidence (2013). The draft version of this study is summarized in Appendix A to the preamble of EPA's proposed rule, 79 Fed. Reg. 22188, 22222-52 (Apr. 21, 2014). The document may be found at http://yosemite.epa.gov/sab/sabproduct.nsf/fedrgstr_activities/Watershed%20Connectivity%20Report?. The study was conducted by EPA's Office of Research and Development. It underwent peer review under the supervision of EPA's Science Advisory Board (SAB). SAB review was completed in October 2014, and EPA released the final study in January 2015. 80 Fed. Reg. 2100 (Jan. 15, 2015).

The government was a party in 20 of the 34 lower court tributary and wetland decisions after *Rapanos*.[277] All 34 decisions used precedent and all cited *Rapanos*, the first time that all precedent-citing decisions cited the same decision. Fourteen of the decisions cited *Riverside Baywiew* and 12 of them cited *SWANCC*. The decisions cited an average of 5.9 precedents. The number of precedents cited by the post-*Rapanos* decisions was comparable to the number of decisions cited by lower court decisions issued between *SWANCC* and *Rapanos*, but the importance of precedent as an interpretive device was far greater in the post-*Rapanos* decisions. Although 34 post-*Rapanos* lower-court decisions cited precedent, only 11 used deference,[278] 10 employed a broad interpretation to achieve the statute's objective,[279] six each used the structure of the statute,[280] five used its plain meaning,[281] three used reading harmoniously with another statute,[282] two interpreted exceptions narrowly,[283] and one used avoiding constitutional issues.[284] While this set of decisions used precedent as an interpretive device 34 times, they used all other interpretive devices in the aggregate only 38 times, compared to the decisions issued between *SWANCC* and *Rapanos*, which used precedent as an interpretive device 22 times and all the other interpretive devices in the aggregate 36 times.[285] Similarly, 19 of the post-*Rapanos* decisions used precedent as their sole interpretive device, while only four of the earlier decisions did so.[286]

Why did precedent become preeminent as an interpretive device after *Rapanos*? In part, it is because there were by then four Supreme Court decisions interpreting navigable waters under the CWA (three of them under §404), twice as many as interpreted any of the other elements of the water pollution offense. Unfortunately, it is also in part because the principal focus of the lower courts is how *Rapanos* applies to the facts of their cases, a question whose answer is far from clear due to the lack of a majority decision and the opaqueness of the plurality opinion.

---

277. *See supra* note 137 (explanation on calculations).
278. *Id.*
279. *Id.*
280. *Id.*
281. *Id.*
282. *Id.*
283. *Id.*
284. *Id.*
285. *Id.*
286. *Id.*

7. The Agencies' Response to *SWANCC* and *Rapanos*

In the wake of *SWANCC* and *Rapanos*, EPA and the Corps jointly issued a guidance document on their regulatory definitions of waters of the United States,[287] proposed amendments to their regulatory definitions, and promulgated final revisions of their definitions. This chapter focuses on the final amendments.[288]

The amended regulatory definition continues virtually the earlier regulatory definition's claim of jurisdiction over traditionally navigable waters, interstate, waters, the territorial seas, and impoundments of the same. The new definition also continues the earlier definition's claim of jurisdiction over tributaries to other waters of the United States, but in a sharper, more detailed and less inclusive manner set forth in a new definition of "tributary." While the earlier regulations asserted jurisdiction over wetlands adjacent to other waters of the United States, the amended regulations assert jurisdiction over all adjacent waters, including wetlands, but define "adjacent" in a detailed manner, less inclusive in some respects than earlier case law. The new regulations extend jurisdiction over specified types of waters that are not adjacent to but that are determined on a case-specific basis to have a significant nexus with traditionally navigable waters, interstate waters or the territorial seas. Those types of waters are: prairie potholes, Carolina or Delmarva bays, pocosins, western vernal pools, and Texas coastal prairie wetlands. The amended regulations also extend jurisdiction over waters within the 100-year floodplain of traditionally navigable waters, interstate waters or the territorial seas and over waters within 4,000 feet of the high tide line or ordinary high watermark of traditionally navigable waters, interstate waters, the territorial seas, impoundments thereof or tributaries thereof, where those waters are determined on a case-specific basis to have a significant nexus with a traditionally navigable waters, interstate waters or the territorial seas.

The revised definition abandons the earlier definition's extension to other waters that could affect interstate commerce in fishing, recreation or industry, potentially justified by the third prong of *Lopez* for Commerce Clause jurisdiction. This appears to follow from the conclusion of the plurality in *SWANCC* that "The term 'navigable' has at least the import of showing us what Congress had in mind as its authority for enacting the CWA: its tra-

287. U.S. EPA and U.S. Army Corps of Engineers, Clean Water Act Jurisdiction Following the U. S. Supreme Court Decision in *Rapanos v. United States* & *Carabell v. United States* (June 2007 Legal Memorandum), *available at* http://www2.epa.gov/cleanwaterrule/documents-related-proposed-definition-waters-united-states-under-clean-water-act.

288. 79 Fed. Reg. 22188 (Apr. 21, 2014) and 80 Fed. Reg. 37054.

ditional jurisdiction over waters that were or had been navigable in fact or which reasonably could be made so."[289] Finally, the amended revision makes explicit that it does not include many of the waters that had apparently troubled some courts and Justices, such as ephemeral erosion features and puddles.

How do the EPA/Corps amended definitions square with the Supreme Court's two recent decisions? In some respects, they follow the decisions. For instance, the Court in *SWANCC* held that the CWA did not extend to waters solely because they are habitats for migratory birds. In response, the amendments drop the present claim over other waters "the use, degradation, or destruction of which would affect or could affect interstate or foreign commerce including . . . waters which are or could be used by interstate or foreign travelers for recreational or other purposes . . . ." Although there are strong arguments that Congress has the authority to cover such waters, under either the Commerce Clause or the Treaty Powers Clause, the Court concluded that Congress did not exercise such powers in the CWA. That leaves it open for Congress to explicitly exercise such powers over these waters, an action that Congress is unlikely to take in the present political climate. For these waters, EPA's amended definition substitutes jurisdiction over a specified range of other waters that the Agency determines on a case-specific basis have a significant nexus to traditionally navigable waters, following Justice Kennedy's suggestion. Ironically, that may result in a claim of jurisdiction over a larger set of waters, although that is difficult to quantify.

The plurality in *Rapanos* repeatedly demanded that jurisdictional waters have continuous surface connections to traditionally navigable waters. In response, the EPA/Corps amendments specifically exclude groundwater, although tributaries will not lose their status merely because they flow underground for parts of their courses due to natural or human-induced conditions. The Court did not consider such situations, but they are in keeping with the nature of tributaries. The amendments also dropped EPA's earlier claim over intermittent streams and playa lakes affecting interstate commerce, but they did not exclude intermittent streams meeting the definition of tributaries. On the other hand, the amendments claim limited jurisdiction over waters within the 100-year floodplains, high tide line or high watermark of traditionally navigable waters, prairie potholes, and similar waters, determined on a case-specific basis to have a significant nexus with traditionally navigable waters. Such waters are not confined to those having a surface connection

289. 531 U.S. at 172.

with traditionally navigable waters, courting disapproval by the plurality, but probably not by the majority of justices.[290]

The plurality in *Rapanos* expressed great concern with the Corps' purported claim of jurisdiction over mud puddles or standing water after rains in farmers' fields on floodplains. It's concern was more rhetorical than real, however. The author is not aware of any such claims by the COE. The new definition's inclusion of puddles and other ephemeral features may disarm the plurality on this and some related issues. The amended definition's limited claim of jurisdiction over waters within floodplains, high tide lines and high watermarks, however, may raise more questions with the plurality than it answers. It does establish that in the general case, the agencies claim no jurisdiction over such waters. Their claim is limited to special situations in which they make significant nexus determinations, which courts can be expected to scrutinize with some care. Ditches, unless specifically excluded, also runs afoul of the *Rapanos* plurality's observation that ditches are point sources and therefore cannot be navigable waters. The plurality further observed that ditch flows are intermittent and are called streams if their flows are permanent. The Court did not deal with the fact that many ditches are rerouted streams, no less tributaries than in their original natural beds. All of this suggests further litigation on the status of ditches. Many ditches fall within the proposed definition or the plain meaning of tributaries, especially those that are rerouted streams or creeks.

The amendments to EPA and the Corps' definitions of waters of the United States claim jurisdiction over "all waters adjacent to" traditionally navigable waters, interstate waters, the territorial seas, and impoundments of and tributaries to such waters, "including wetlands, ponds, oxbows, impoundments and, similar waters."[291] This changes the tone of the definition from focusing on wetlands to focusing on all adjacent waters, including wetlands. The amendments use both the plurality's direct connection test and Justice Kennedy's significant nexus tests for adjacency, although the definition is closer to Justice Kennedy's test than it is to the plurality's test.[292] The plurality test for adjacency is whether the surface waters from the wetland and the adjacent navigable water or tributary intermingle to the extent that they cannot be separated. The amended definition of adjacent is "bordering, contiguous or neighboring," picking up on many lower court definitions of the term. It provides that separation by natural or human-made dikes, berms, "and the

290. *Id.*, codified at 40 C.F.R. §122.2(a)(6), (c)(1), (2) & (4).
291. *Id.*, codified at 40 C.F.R. §122.2(a)(6).
292. *Id.*, codified at 40 C.F.R. §122.2(c)(6).

like" does not affect adjacency, again drawing on many lower court definitions of adjacency.

B. *Isolated Waters*

Neither the CWA nor its legislative history state explicitly whether the statute's jurisdiction includes discharges to any isolated water body. EPA and Corps regulations do not explicitly claim jurisdiction over isolated waters, but EPA's regulations before their 2015 revisions included the following: natural ponds, which could be isolated; playa lakes, which are usually isolated; and "other waters" which "could affect interstate commerce."[293] Both EPA and Corps regulatory definitions before their 2015 revisions included intrastate waters that are used by interstate travelers for recreation, which could describe many isolated waters. But EPA's and the Corps' current regulations do not include such waters, dropping their claims to jurisdiction over discharges to all but a few specified isolated waters that alone or in combination with similarly situated waters have a significant nexus to navigable waters. A significant nexus to navigable waters, of course, means that they are not really isolated. While isolated waters could affect interstate commerce under the third prong of *Lopez* sufficiently to provide Commerce Clause jurisdiction, the CWA's use of the jurisdictional term "navigable waters," however, suggests to the courts that Congress relied on the first prong of *Lopez*, in which jurisdiction rarely extends to isolated waters because few are highways or parts of highways of interstate or foreign commerce.[294]

The isolation of a water body alone does not always defeat its status as navigable water under the first prong of *Lopez*. For instance, the Court held in *Utah v. United States* that the Great Salt Lake is navigable, despite its isolation from other waters, because it had been used as a highway or part of a highway of interstate commerce, squarely within the first prong of *Lopez*.[295] The Seventh Circuit held in *United States v. Byrd*[296] that a 2,000- to 3,000-acre isolated lake in Indiana was subject to CWA navigable water jurisdiction because interstate travelers used it for recreational boating, also placing it within the first prong of *Lopez*, but not as clearly so because the use was recreational rather than commercial.[297]

---

293. 33 C.F.R. §323.2(a)(5); 40 C.F.R. §122.2. Playa lakes are usually dry and without outlets.

294. For an argument to the contrary, see Jon Devine et al., *The Intended Scope of Clean Water Act Jurisdiction*, 41 ELR 11118, 11124 (Dec. 2011).

295. Utah v. United States, 403 U.S. 9, 1 ELR 20250 (1971).

296. 609 F.2d 1204, 9 ELR 20757 (7th Cir. 1979).

297. Query: How compelling a distinction is this, when recreational boating is a multibillion-dollar business and some recreational boaters rent their vessels from commercial mariners?

Such claims cannot be made for most isolated waters, however, especially small isolated water bodies. Isolated waters, even smaller ones, might be within the third *Lopez* prong of Commerce Clause jurisdiction, but the Court has not recognized that Congress intended to use such jurisdiction for the CWA. An example of such a claim is the U.S. Court of Appeals for the Tenth Circuit's decision in *United States v. Earth Sciences, Inc.*,[298] holding that an intrastate stream, located totally within one county and unconnected with other waters, was navigable based on the use of its waters for agricultural irrigation from which products were sold in interstate commerce. Isolated water bodies might have an aggregate effect on recreational or other interstate commerce, exemplified by fishing from boats.

The Supreme Court's decision in *SWANCC* on the issue of isolated water bodies is not unique, but it is determinative. The facts of the case involved a site proposed for a landfill that included an aggregate of 31 acres of shallow ponds remaining from a sand and gravel operation, with no connections to other waters. The Corps did not claim jurisdiction over them as natural ponds, presumably because they were created by the former mining operations. The Corps' only jurisdictional claim was its Migratory Bird Rule, which the Court held was beyond the authority Congress delegated to the Corps in §404. The Court implicitly interpreted the CWA to limit Congress' claim to Commerce Clause jurisdiction to be within the first prong of the *Lopez* test,[299] while the dissent would have extended the claim to the third prong of the *Lopez* test and would have found that claim to be constitutional.[300]

Several lower courts have held that the CWA's jurisdiction does not extend to isolated waters. After *Rapanos*, the Ninth Circuit held in *San Francisco Baykeeper v.Cargill Salt Division*[301] that a 17-acre artificial pond was not navigable although it was adjacent to admittedly navigable water, because the adjacency test applied only to wetlands, not to ponds. In *Borden Ranch Partnership v. U.S. Army Corps of Engineers*,[302] the Corps, after *SWANCC*, withdrew the portion of its action that had claimed filling a vernal pond violated §404, and the Ninth Circuit reversed the lower court's ruling that the vernal pond was navigable water.

Before *SWANCC*, the Seventh Circuit in *Village of Oconomowoc Lake v. Dayton Hudson Corp.*[303] held that a six-acre retention pond built to collect

298. 599 F.2d 368, 9 ELR 20542 (10th Cir. 1979).
299. Solid Waste Agency of N. Cook Cnty. v. U.S. Army Corps of Eng'rs (*SWANCC*), 531 U.S. 159, 171-74, 31 ELR 20382 (2001).
300. *Id.* at 192-96.
301. 481 F.3d 700 (9th Cir. 2007).
302. 261 F.3d 810, 816, 32 ELR 20011 (9th Cir. 2001).
303. 24 F.3d 962, 24 ELR 21080 (7th Cir. 1994).

stormwater runoff from a shopping mall parking lot, with no connection to other surface water, was not navigable water. The court noted that EPA's definition of navigable water included natural ponds and therefore by implication excluded artificial ponds such as the pond at issue. The tenor of the opinion, however, suggests that the court would have reached the same conclusion even if the pond had been natural. In *Hoffman Homes, Inc. v. Administrator, EPA*,[304] the Seventh Circuit held there was no CWA jurisdiction over an isolated wetland. Although the opinion, issued prior to *SWANCC*, appeared willing to accept a jurisdictional claim under the Corps' Migratory Bird Rule, the government made no showing that the wetland was actually used by migratory birds or had other effects on interstate commerce. In *United States v. Hallmark Construction Co.*,[305] a district court held that an isolated five-acre wetland was beyond the Corps' jurisdiction in the absence of proof of a connection with interstate commerce.[306]

EPA's current regulatory definition of waters of the United States does not explicitly include or exclude isolated waters. But it no longer includes the existing claim to jurisdiction over

> other waters such as intrastate . . . lakes, prairie potholes, wet meadows, playa lakes or natural ponds the use, degradation, or destruction of which would affect or could affect interstate or foreign commerce; . . . [including use] by interstate or foreign travelers for recreational or other purposes; . . . [use for taking] fish or shellfish . . . [for sale] in interstate or foreign commerce; . . . [or use] for industrial purposes by industries in interstate commerce.[307]

Moreover, it specifically excludes the sand and gravel pits at issue in *SWANCC*.[308] Thus, EPA is abandoning the argument that in enacting the CWA, Congress exercised its Commerce Clause jurisdiction under the third prong of *Lopez*, the jurisdictional claim most likely to support regulating some isolated waters. Although the proposed regulation noted that ponds can be tributaries,[309] EPA dropped that claim in its final amendments. In any event, isolated ponds would not benefit from the definition of tributaries.

On the other hand, the current regulations authorize EPA to determine, on a case-by-case basis, that a particular water falls within enumerated categories and alone or in conjunction with similarly situated waters has a sig-

---

304. 999 F.2d 256, 23 ELR 21139 (7th Cir. 1993).
305. 30 F. Supp. 2d 1033, 28 ELR 21438 (N.D. Ill. 1998).
306. *See also* United States v. Wilson, 133 F.3d 251, 28 ELR 20299 (4th Cir. 1997), in which the U.S. Court of Appeals for the Fourth Circuit held that an isolated wetland was not navigable water.
307. 40 C.F.R. §122.2, *Waters of the United States* (c) (2013).
308. 80 Fed. Reg. 37114, to be codified at 40 C.F.R. §122.2 waters of the United States (2)(E).
309. 79 Fed. Reg. 22268, to be codified at 40 C.F.R. §122.2(1)(vi).

nificant nexus with a water of the United States.[310] The categories of waters include: prairie potholes, Carolina and Delmarva bays, pocosins, western vernal pools, Texas coastal prairie wetlands, and 100-year floodplains or water within 4,000 feet of the high tide line or ordinary high watermark of a water of the United States.[311] EPA's scientific justification for that part of the regulation includes studies suggesting that many "unidirectional wetlands," such as prairie potholes, vernal pools, and playa lakes with subsurface connections to waters of the United States, may be shown to have a significant nexus to waters of the United States.[312] Because such waters lack surface connections to navigable waters, the *Rapanos* plurality is unlikely to agree with the Agency's reasoning. However, Justice Kennedy and the *Rapanos* minority (who together form a majority of the Justices) may well agree with it.

The decisions discussed above in this section deal with both natural and artificial bodies of water. They suggest that, even absent *SWANCC*, courts have not been receptive to holding isolated waters to be within the CWA's jurisdiction, unless they are used as parts of a traditional highway of commerce. In any event, *SWANCC* now requires that result. In the unusual case where isolated waters have been used, may currently be used, or with reasonable improvements could be used in the future as parts of highways of interstate commerce, those isolated waters are arguably navigable. But showing that uses of isolated water bodies have an aggregate impact on interstate commerce will not make them navigable because the Court has in effect ruled that Congress did not intend to exercise the third *Lopez* prong of its Commerce Clause jurisdiction in the CWA. Moreover, EPA has impliedly acceded to this conclusion in its recently revised regulatory definition of waters of the United States.

Notice that in its current regulatory definition, EPA adopted from the aggregate effects on interstate commerce test, the third prong of *Lopez*, for constitutional jurisdiction under the Commerce Clause, the aggregate effects on water quality test for statutory jurisdiction under the CWA. Perhaps EPA did so hoping for a favorable resonance with the Court for the new regulatory definition. That strategy may work with Justice Kennedy, whose concurrence in *Rapanos* was oriented toward protecting water quality. But it is not likely to resonate with the plurality.

Congress could claim expanded jurisdiction by eliminating the use of navigable waters as a jurisdictional term in the CWA, perhaps substituting

310. 80 Fed. Reg. 37114, to be codified at 40 C.F.R. §122.2 waters of the United States (1)(vii).
311. *Id* §122.2 waters of the United States (1)(vii)(A)-(E) and (1)(viii).
312. *Id.* at 22225-26.

waters of the United States, and making findings that water pollution control of isolated waters, or at least some classes of isolated waters, meet the second or third prong of the *Lopez* test. Congress could also claim expanded jurisdiction over some isolated waters under the Constitution's federal treaty powers and the implementing Migratory Bird Treaty Act to protect migratory birds, including protection of their habitats.[313] Indeed, even the U.S. Department of the Interior could claim jurisdiction to protect isolated waters that are habitats to birds in interstate or foreign migrations. In the meantime, the question of whether transporting boats or fishermen across state lines to fish, hunt, or engage in other recreational activities on isolated waters makes those waters part of the highways of interstate commerce subject to Commerce Clause jurisdiction remains undecided.[314] Although the Court interprets the CWA not to claim such jurisdiction, the Court has not addressed, and has not needed to address, whether Congress has authority to claim it.

C. *Groundwater*

Neither the CWA's general prohibition against the discharge of pollutants to navigable waters in §301(a) nor its definition of navigable waters in §502(7) mention groundwater. The statute, however, is replete with references to groundwater.[315] These frequent references indicate that Congress was aware

313. U.S. Const. art. II, §2, cl. [2]; Migratory Bird Treaty Act, 16 U.S.C. §§703-711. *See* Scott Finet, *Habitat Protection and the Migratory Bird Protection Act*, 10 Tul. Envtl. L.J. 1 (2003).

314. One commenter lays out the argument that such recreational use of isolated waters will not render them navigable under any prong of *Lopez. See* Funk, *supra* note 36, at §13:117. Indeed, he argues that Congress may not have the authority to regulate pollution on traditionally navigable waters unless it has a negative impact on navigation because the Commerce Clause does not give Congress plenary power to regulate navigable waters, but only power to protect and promote interstate commerce on navigable waters. For a contrary perspective, see *FPL Energy Marine Hydro, LLC v. Federal Energy Reg. Comm'n*, 287 F.3d 1151, 1157 (D.C. Cir. 2002) (three canoe trips establish potential for navigational use).

315. For example, CWA §§102(a), 104(a)(5), & 106(e)(1), 33 U.S.C. §§1252(a), 1254(a)(5), & 1256(e)(1), authorize plans for improving water quality in groundwater, monitoring water quality in groundwater, and making grants to state agencies to do the same. CWA §§202(a)(2) & 208(b)(2)(K), 33 U.S.C. §§1282(a)(2) & 1288(b)(2)(K) authorize grants to state agencies for programs to protect groundwater quality. CWA §304(a)(1) & (e), 33 U.S.C. §1314(a)(1) & (e), authorize EPA to establish water criteria for groundwater, factors to restore groundwater quality, and methods to control pollution from disposal of pollutant in wells.

> Rep. Les Aspin (D-Wis.) stated on the House floor that groundwater appears in the bill in every section, in every title, except Title IV. It is under the title which provides EPA can study ground water. It is under the title dealing with definitions. But when it comes to enforcement, Title IV, the section on permits and licenses, then ground water is suddenly missing.

92 Cong. Rec. 10,666 (1972) (statement of Rep. Aspin), *reprinted in* 1 Legis. History, *supra* note 44, at 727. The author cannot find groundwater mentioned in Title V, in which the definitional §502 is located. Most lawyers would label §309, 33 U.S.C. §1319, the enforcement section rather than the sections authorizing permit issuance in Title IV. And many sections in Titles I, II, and III do not mention groundwater, e.g., §§102A, 103, 206, 207, 307, & 308, 33 U.S.C. §§1252a, 1253,

of groundwater and groundwater pollution when it drafted the statute, suggesting that its failure to mention groundwater in either §301(a) or the definition of navigable waters was an intentional choice not to regulate discharges of pollutants to groundwater. Moreover, the statute's frequent coupling of "navigable waters and ground waters" indicates that Congress viewed them as distinct rather than overlapping categories of water.[316] Finally, repeated legislative history explicitly stated that the CWA did not regulate discharges of pollutants to groundwater, and Congress rejected amendments that would have included regulation of discharges of pollutants to groundwater.[317] Legislative history did not distinguish between isolated groundwater and tributary groundwater,[318] a distinction courts subsequently developed to justify holding discharges to tributary groundwater to be within the jurisdiction of the CWA, while discharges to isolated groundwater are not.

On the other hand, the legislative history is clear that Congress intended that the term navigable waters be given its full constitutional reach. The

---

1286, 1287, 1317, & 1318. Although the statement is demonstrably wrong in several respects, it has been uncritically repeated. *See, e.g.*, Anna Makowski, *Beneath the Surface of the Clean Water Act: Exploring the Depth of the Act's Jurisdictional Scope of Ground Water Pollution*," 91 Or. L. Rev. 495, 513 (2012); Thomas L. Casey, *Reevaluating "Isolated Waters": Is Hydrologically Connected Ground Water "Navigable Water" Under the Clean Water Act*, 54 Ala. L. Rev. 159, 171 (2002).

316. All of the statutory references immediately above, except §208(b)(2)(K), contain either the phrase "navigable waters and ground waters" or "navigable waters or groundwaters."

317. In the report accompanying S. 2770 from the Committee on Public Works to the Senate, the Committee stated that it recognized that the distinction between surface and groundwaters was artificial. Although the Committee was concerned with groundwater pollution, it rejected several proposals to establish federal standards because "the jurisdiction regarding groundwaters is so complex and varied from State to State." S. Rep. No. 414, at 73 (1971), *reprinted in* 2 Legis. History, *supra* note 44, at 1491. The Committee also stated that the problem of groundwater pollution, although serious, was not as acute as the problem of surface water pollution. *Id.* It did so after hearings on the effect of deep well injection on groundwater pollution. *See Hearing on S. 2770 Before the Subcomm. on Air and Water Pollution of the S. Comm. on Pub. Works*, 92nd Cong. pt. 7 (1971). *See also* Exxon Corp. v. Train, 554 F.2d 1310, 1329, 7 ELR 20594 (5th Cir. 1977).

After passage of S. 2770, the House Committee on Public Works held hearings on H.R. 11896, the House counterpart to S. 2770, but differing in some respects. During the hearings, Representative Aspin proposed amendments adding groundwater control to the bill, because he understood both the House and Senate bills not to authorize regulation of groundwater pollution. *See Hearing on H.R. 11896 Before the H. Comm. on Pub. Works*, 92nd Cong. 727-28 (1972) (statement of Rep. Aspin), *reprinted in* 1 Legis. History, *supra* note 44, at 597. Those amendments were not adopted by the Committee. The only mention of groundwater in the Committee Report accompanying H.R. 11896 to the House floor is its expectation that EPA will be diligent in gathering information on deep well injection. *See* H.R. Rep. No. 911, at 109 (1972). When the bill reached the House floor, Representative Aspin again introduced his amendment to include groundwater pollution control in the legislation. *See* 118 Cong. Rec. 10666 (1972). Members supporting and opposing the amendment held a lively debate, but all agreed that the bill as written did not regulate discharges of pollutants to groundwater. Opposition to the amendment focused primarily on the lack of information on which to base a regulatory program. After the debate, the House rejected the proposed amendment. *See* 118 Cong. Rec. 10666-69 (1972), *id.* at 597.

318. "Tributary groundwater" is groundwater that flows into surface water; "isolated groundwater" is groundwater that does not flow into surface water.

few courts considering whether groundwater can be within federal Commerce Clause jurisdiction concluded that it could be,[319] although not under the first prong of the *Lopez* test, which Congress apparently chose as the jurisdictional basis of the CWA by using "navigable waters" in the CWA as its jurisdictional claim. Highways of interstate commerce are inherently surface waters. One twist in the CWA suggests congressional intent to regulate some discharges to groundwater. Section 402(b)(1)(D) provides that EPA can approve a state program for administering and enforcing the §402 permit program only if the state has authority to issue permits to "control the disposal of pollutants into wells." In the absence of an approved state §402 program, §402(a)(3) provides that EPA is to administer the program in the state under the same terms and conditions as an approved state program. This provision implies that if EPA is administering the §402 program, it also has authority to issue permits for disposing of pollutants into wells. Two references to "well" in the definitions of elements of the CWA offense support this argument,[320] suggesting that Congress intended CWA §402 to regulate discharges through wells. Because wells lead to groundwater rather than to surface water, congressional intent to regulate discharges through wells necessarily includes congressional intent to regulate discharges to groundwater.

Predictably, the issue of whether the CWA regulates discharges to groundwater was first raised with regard to discharges of pollutants into groundwater through waste disposal wells, a waste disposal method known as deep well injection. In the first reported case, *United States v. GAF Corp.*, EPA sought injunctive relief against the construction of deep wells for chemical waste disposal without a §402 permit.[321] EPA relied on the inclusion of discharges through wells in §§402 and 502, while the defendant relied on the legislative history. The district court held that "[d]isposal of chemical

319. Village of Oconomowoc Lake v. Dayton Hudson Corp., 24 F.3d 962, 965, 24 ELR 21080 (7th Cir. 1994); Inland Steel Co. v. EPA, 901 F.2d 1419, 1422, 20 ELR 20889 (7th Cir. 1990). Indeed, groundwater may be interstate. Many aquifers underlie and flow between multiple states. The best known is the Ogallala aquifer, underlying eight states: Colorado, Kansas, Nebraska, New Mexico, Oklahoma, South Dakota, Texas, and Wyoming. Groundwater may not ordinarily be a highway of interstate commerce, but it is used in interstate commerce. The Ogallala aquifer, for example, provides irrigation water that is a key component in the region's agriculture. The Supreme Court held in *Sporhase v. Nebraska*, 458 U.S. 941, 12 ELR 20749 (1982), that groundwater also can be an article in interstate commerce.

320. The definition of "point source" gives "well" as an example of conveyances included in that term, CWA §502(14). The definition of "pollutant" excludes "material which is injected into a well to facilitate the production of oil or gas" under stated conditions, CWA §502(6). There would be no reason to exclude the injection of material into wells if injection of materials into groundwater was not otherwise prohibited by the CWA.

321. United States v. GAF Corp., 389 F. Supp. 1379, 5 ELR 20581 (S.D. Tex. 1975).

wastes into underground waters which have not been alleged to flow into or otherwise affect surface waters does not constitute a 'discharge of a pollutant' within the meaning of" §301(a).[322] It so held because of explicit legislative history that the CWA did not regulate discharges to groundwater, while EPA's §§402 and 502 arguments were "back-door."[323] "Congress could not possibly have meant to achieve in roundabout fashion what it expressly declined to accomplish straightforwardly,"[324] the court wrote. While legislative history may not trump statutory content as easily today,[325] it was quite the fashion at the time.[326] It is significant, however, that the district court in *GAF* noted the polluted groundwater at issue was not tributary groundwater, suggesting that the court might have been open to a different result if the groundwater was tributary.

In the next deep well decision, *U.S. Steel Corp. v. Train*,[327] the petitioner sought judicial review of EPA's permit for discharges of pollutants to surface water and for discharges to groundwater from a deep injection well at the same facility. Upholding EPA's action, the Seventh Circuit distinguished *GAF* as an action against discharge to groundwater alone, not in conjunction with the control of discharges to surface water.[328] The Seventh Circuit was convinced by EPA's argument that §402(a)(1) authorizes the Agency, in the absence of an approved state permit program, to issue permits "subject to the same terms, conditions and requirements" as apply to approved state permit programs, including the authority to regulate pollutants discharged into wells in §402(b)(1)(D).[329] The court found support for that interpretation of the statute in the definition of pollutant in §502(6), which excludes material injected into wells in association with the production of oil and gas, an exclusion that would have no meaning unless the material would be a pollutant when discharged into wells absent the exclusion.[330] Finally, the court found support for that interpretation in the limited CWA legislative history

322. *Id.* at 1383.
323. *Id.* at 1384. In his majority opinion in *Whitman v. American Trucking Ass'ns, Inc.*, 531 U.S. 457, 468, 31 ELR 20512 (2001), Justice Scalia made a similar point more colorfully, saying that Congress does not "hide elephants in mouseholes."
324. *GAF*, 389 F. Supp. at 1385.
325. *See* Eskridge, *supra* note 14, at 207-38; Scalia & Garner, *supra* note 14, at 369-90 (discussing one of their Thirteen Falsities Exposed).
326. *See* Train v. Colorado Pub. Interest Research Grp., 426 U.S. 1, 10-23, 6 ELR 20549 (1976), in which the Court held that the inclusion of radioactive materials in the definition of pollutant in CWA §502(6) did not authorize EPA to regulate the discharge of "radioactive materials from nuclear power plants," largely because explicit legislative history trumped the plain meaning of the statute.
327. 556 F.2d 822, 7 ELR 20419 (7th Cir. 1977).
328. *Id.* at 851 n.60.
329. *Id.* at 852.
330. *Id.*

referring to deep well injection, although it ignored substantial contrary legislative history regarding discharges to groundwater generally.[331]

The Fifth Circuit came to the opposite conclusion in *Exxon Corp. v. Train*,[332] when it reviewed EPA's claim of authority to regulate discharges of wastes through deep wells to groundwater as part of a §402 permit that the Agency issued to regulate discharges of pollutants to surface waters from the same facility. EPA did not argue that §301(a) prohibited discharges to groundwater, but only that the Agency had ancillary authority to issue permits for deep well injection by facilities also requiring CWA permits to discharge into surface navigable water under §402(a)(3) and (b)(1)(D).[333] This limited claim of authority was articulated in an opinion by EPA's Office of General Counsel.[334] While the court commented that EPA's interpretation was "not implausible," it concluded that the distinction the Agency drew between deep well injection at a facility with a surface water discharge and injection at a facility with no surface water discharge had "the strange result of dividing jurisdiction over deep-well injections between federal and state authorities, based on the apparently fortuitous factor of whether the person engaging in the deep-well injection also happens to be engaging in 'associated' surface discharges."[335]

Consequently, the Fifth Circuit proceeded with an exhaustive examination of the statute's structure and legislative history. It stated that references to groundwater in the CWA establish a "pattern . . . of federal information gathering and encouragement of state efforts to control groundwater pollution but not of direct federal control over groundwater pollution."[336] Moreover, the court said that the legislative history "demonstrates conclusively that Congress believed it was not granting the Administrator any power to control disposals into groundwater."[337] Acknowledging the logic of EPA's argument based on §402(a)(3) and (b)(1)(D), the court concluded that "[w]e cannot attribute to Congress an intention to achieve silently and by indirec-

331. *Id.* at 852-53. This legislative history was from House and Senate Committee Reports explaining the exclusion from the definition of pollutant in §502(6)(B) of material pumped into a deep well incident to the production of oil or gas.
332. 554 F.2d 1310, 7 ELR 20594 (5th Cir. 1977). Despite its earlier publication in the *Federal Reporter*, *Exxon* was decided six weeks after the Seventh Circuit's decision in *U.S. Steel Corp.*, 556 F.2d 822, although that decision does not mention *U.S. Steel.*
333. *Exxon*, 554 F.2d at 1318-19.
334. EPA Gen'l Counsel Op. (Dec. 13, 1973), *quoted in Exxon*, 554 F.2d at 1320 n.21. The General Counsel reiterated this interpretation of the CWA in 1975, distinguishing *United States v. GAF Corp.*, 389 F. Supp. 1379, 5 ELR 20581 (S.D. Tex. 1975). *See Exxon*, 554 F.2d at 1320 n.21.
335. *Exxon*, 554 F.2d at 1322.
336. *Id.*
337. *Id.* at 1329.

tion that which it constantly refused to do directly."[338] Like the Texas district court in *GAF*, the Fifth Circuit left the door open to the argument that discharges to tributary groundwater were prohibited.[339]

The controversy over EPA's authority to regulate deep well injection under the CWA subsequently disappeared because Congress gave EPA direct regulatory authority over deep well injection in the Safe Drinking Water Act of 1974.[340] The controversy resurfaced tangentially when EPA ordered a steel company to take corrective remedial action with regard to its leaking hazardous waste injection wells under the Resource Conservation and Recovery Act (RCRA).[341] In *Inland Steel Co. v. EPA*,[342] the respondent raised as a defense an exemption in RCRA §6903(27) for materials in "industrial discharges which are point sources subject to permits under" CWA §402, claiming that the well was a point source adding pollutants to groundwater.[343] The Seventh Circuit's decision by Judge Richard Posner suggested that its earlier *U.S. Steel* decision might be "all wet" and held the RCRA exemption inapplicable for a variety of reasons unrelated to this discussion.[344]

The issue of whether unpermitted discharges to groundwater are prohibited by CWA §301(a) shifted from deep well injection to the discharge of pollutants from surface activities into groundwater. In this context, courts have frequently distinguished between discharges to nearby groundwater with no known or alleged connection with surface navigable water, often called isolated groundwater, and groundwater known or alleged to be tributary to nearby surface navigable water, often called tributary groundwater.[345]

A small number of decisions hold that the CWA does not regulate discharges into any groundwater, isolated or tributary.[346] The most considered

338. *Id.*
339. *Id.* at 1312, n.1.
340. Safe Drinking Water Act (SDWA), 42 U.S.C. §§300f to 300j-26, ELR Stat. SDWA §§1401-1465.
341. 42 U.S.C. §§6901-6992k, ELR Stat. RCRA §§1001-11011.
342. 901 F.2d 1419, 20 ELR 20889 (7th Cir. 1990).
343. The exclusion is in the definition of solid waste, of which hazardous waste is a subset.
344. *Inland Steel*, 901 F.2d at 1423-24.
345. Makowski, *supra* note 315, at 506, suggests a possibly more useful set of classifications of groundwater, drawn from state groundwater regulation statutes: underground streams; percolating groundwater; and subflow from surface streams.
346. *Town of Norfolk v. U.S. Army Corps of Eng'rs*, 968 F.2d 1438, 22 ELR 21337 (1st Cir. 1992), is sometimes said to be such a precedent. The decision did not consider whether a CWA permit was required for a discharge into groundwater, but whether groundwater was navigable water in the context of the Corps' consideration of alternatives in determining whether to issue a CWA §404 permit for the filling of wetlands. The First Circuit's analysis of the issue was terse, covering only 10 lines. A subsequent district court decision analyzing *Town of Norfolk* concluded that the decision did not hold "that groundwater is categorically excluded from consideration as 'waters of the United States,' but rather that such a determination requires an 'ecological judgment,' to be made according to the particular characteristics of the site." Hernandez v. Esso Standard Oil Co., 599 F. Supp. 2d 175, 179 (D.P.R. 2009). Moreover, the First Circuit was not interpreting the CWA, but instead the Corps'

of those decisions, *Umatilla Waterquality Protective Association, Inc. v. Smith Frozen Foods, Inc.*,[347] concerned discharges into allegedly tributary groundwater. The district court recited the legislative history that Congress rejected legislation covering discharges to groundwater in the CWA,[348] and the statute's repeated use of the phrase "navigable waters and ground waters,"[349] suggesting Congress recognized these to be two distinct categories of water. Finally, the decision noted that "EPA has offered no formal or consistent interpretation of the CWA that would subject discharges to groundwater to the [National Pollutant Discharge Elimination System] NPDES permitting requirement."[350] Indeed, the district court quoted from the Opinion of EPA's General Counsel referred to in the *U.S. Steel* and *Exxon* decisions that "*discharges into ground waters are not included*" in the definition of "discharge of a pollutant" in CWA §502(12).[351]

Umatilla cited two instances in which EPA restricted its regulatory authority over discharges to groundwater to conform to EPA's general counsel opinion,[352] although the court ignored the fact that the Agency had sued GAF Corp. for unpermitted deep well injection into isolated groundwater, as well as the preamble to an EPA rulemaking in which the Agency noted that the NPDES program did not apply to groundwater "unless there is a hydrological connection between the ground water and a nearby surface water body"—in other words, unless the groundwater is tributary to nearby surface water.[353] *Umatilla* criticized opinions holding that discharges to tributary groundwater were covered by the CWA's permitting program for failing to deal with EPA's disavowal of authority over discharges to groundwater.[354] It noted, however, that if EPA amended its CWA regulations to cover discharges to tributary groundwater, the court would give that interpretation *Chevron*

---

regulations, and it was not considering whether discharges to groundwater constituted discharges to navigable water for purposes of CWA jurisdiction. On close examination, *Town of Norfolk* appears to be wholly irrelevant to the inquiry here.

347. 962 F. Supp. 1312, 27 ELR 21411 (D. Or. 1997).
348. *Id.* at 1318-19.
349. *Id.* at 1318.
350. *Id.* at 1319. Subsequently, the preamble to EPA's proposed effluent guidelines for Concentrated Animal Feeding Operations (CAFOs) stated that EPA applied a direct hydrologic connection test to determine whether a discharge required a §402 permit, *see* 66 Fed. Reg. 2960 (Jan. 12, 2001), but did not reflect that statement in its final rule, *see* 40 C.F.R. pt. 412.
351. *Umatilla*, 962 F. Supp. at 1319 (emphasis added by the district court).
352. *Id. See* 38 Fed. Reg. 13528 (May 22, 1977); 44 Fed. Reg. 32854 (June 7, 1979).
353. 55 Fed. Reg. 47990, 47997 (Nov. 16, 1990).
354. Note that the court itself failed to recognize that in *GAF*, EPA asserted jurisdiction over discharges to isolated groundwater.

deference.[355] A second district court decision, *Kelley v. United States*,[356] covered much the same ground, but in lesser detail. The plaintiffs in both cases alleged that pollutants discharged to groundwater flowed into nearby surface navigable water.[357]

Other decisions have held that the CWA's jurisdiction does not extend to groundwater because the plaintiffs did not allege the discharges were to tributary groundwater[358] or alleged only "the possibility" of a hydrological connection between the groundwater and surface navigable water.[359] These decisions do not explicitly foreclose the assertion of jurisdiction over discharges to tributary groundwater.

Encouraged by a pioneering law review article,[360] a number of decisions hold that discharges of pollutants to tributary groundwater are within the jurisdiction of the CWA.[361] They all accept that the CWA's legislative history establishes Congress had no intent to protect groundwater generally from pollution or to generally regulate discharges to groundwater. They counter, however, that, in the words of an Idaho district court, "Congress's decision not to comprehensively regulate groundwater as part of the CWA, does not require the conclusion that Congress intended to exempt groundwater from

355. *Umatilla*, 962 F. Supp. at 1319 n.2. Another court refusing to interpret navigable waters to include groundwater commented that if EPA amended its regulations to include tributary groundwater, that would "pose a harder question." Village of Oconomowoc Lake v. Dayton Hudson Corp., 24 F.3d 962, 966, 24 ELR 21080 (7th Cir. 1994). *But see id.* at 966 (concurring op.).

356. 618 F. Supp. 1103, 16 ELR 20080 (W.D. Mich. 1985).

357. *Umatilla*, 962 F. Supp. at 1313-14; *Kelley*, 618 F. Supp. at 1104-05.

358. In *GAF*, for instance, EPA did not allege contaminated groundwater flowed into navigable waters. *See* United States v. GAF Corp., 389 F. Supp. 1379, 1383, 5 ELR 20581 (S.D. Tex. 1975). In *Exxon*, EPA did not allege contaminated groundwater flowed into navigable waters and the court specifically noted that it voiced no opinion on whether CWA jurisdiction would exist if the discharge was to tributary waters. Exxon Corp. v. Train, 554 F.2d 1310, 1312 n.1, 7 ELR 20594 (5th Cir. 1977).

359. *Village of Oconomowoc Lake*, 24 F.3d at 966. The "possibility" of contaminated pond water entering groundwater and thereafter migrating to *surface* navigable waters does not establish CWA jurisdiction.

360. Mary Christina Wood, *Regulating Discharges Into Groundwater: The Critical Link in Pollution Control Under the Clean Water Act*, 12 Harv. Env't L. Rev. 369-426 (1988). She has been followed by others, all arguing for the recognition of tributary groundwater as navigable. *See, e.g.*, Makowski, *supra* note 315; James W. Hayman, *Regulating Point Source Discharges to Ground Water Hydrologically Connected to Navigable Waters: An Unresolved Question of Environmental Protection Agency Authority Under the Clean Water Act*, 5 Barry L. Rev. 95 (2005); and Casey, *supra* note 315.

361. Quivira Mining Co. v. EPA, 765 F.2d 126, 15 ELR 20530 (10th Cir. 1985); Hernandez v. Esso Standard Oil Co. (Puerto Rico), 599 F. Supp. 2d 175 (D.P.R. 2009); Idaho Rural Council v. Bosma, 143 F. Supp. 2d 1169 (D. Idaho 2001), *aff'd on other grounds*, 305 F.2d 943 (9th Cir. 2002); Friends of Santa Fe Cnty. v. LAC Minerals, Inc., 892 F. Supp. 1333, 26 ELR 20135 (D.N.M. 1995); Washington Wilderness Coal. v. Hecla Mining Co., 870 F. Supp. 983, 25 ELR 20661 (E.D. Wash. 1994); Sierra Club v. Colorado Refining Co., 838 F. Supp. 1428, 24 ELR 20749 (D. Colo. 1993); McClellan Ecological Seepage Situation (MESS) v. Cheney, 763 F. Supp. 431, 20 ELR 20870 (E.D. Cal. 1989), 707 F. Supp. 1182, 19 ELR 20124 (E.D. Cal. 1988), *vacated on other grounds*, 47 F.3d 325, 25 ELR 20628 (9th Cir. 1995). *See also* New York v. United States, 620 F. Supp. 374, 381, 16 ELR 20142 (E.D.N.Y. 1985) (commenting that discharges to tributary groundwater may be subject to CWA jurisdiction, but not reaching the question).

all regulation—particularly under circumstances where the introduction of pollutants into the groundwater adversely affects the *adjoining* surface waters."[362] Although none of the decisions discuss it, the legislative history simply does not deal with tributary groundwater or pollution of surface water by groundwater.

The decisions distinguish most negative precedents as dealing with isolated rather than with tributary groundwater, and conclude that the logic of the decisions holding tributary groundwater to be protected by the statute is, in the words of a Washington district court, "compelling: since the goal of the CWA is to protect the quality of surface waters, any pollutant which enters such waters, whether directly or through groundwater, is subject to regulation."[363] They might add that allowing such pollution without a permit creates a giant loophole in the CWA; a factory adjacent to a river can evade the CWA simply by aiming its discharge pipe to dry land rather than the river, at least if its flow is low enough to percolate into the soil rather than simply running off over the surface into the river. (The same could be said of the pollution of surface waters by converting point source discharges to nonpoint source discharges.[364]) Although some of these decisions held that discharges to the tributary groundwater in question were regulated by the CWA, most were decisions on preliminary motions. When the plaintiffs prevailed on preliminary motions, the courts often commented that they would hold plaintiffs to a high standard of proof at trial that the pollutants discharged to groundwater actually reached navigable surface water.[365] Courts sometimes denied the plaintiffs' motions for summary judgment because of conflicting evidence on the facts.[366]

The plurality in *Rapanos* did not address these situations, but its repeated insistence that tributaries or wetlands must have *surface connections* with navigable water to be navigable themselves[367] suggests that those Justices would not be amenable to the argument that groundwater tributary to nearby surface water could be navigable. The plurality, however, did suggest that additions to navigable water need not be direct, implying they may be indi-

---

362. *See, e.g.*, *Idaho Rural Council*, 143 F. Supp. 2d at 1180 (emphasis added).

363. *See, e.g.*, *Washington Wilderness Coal.*, 870 F. Supp. at 990.

364. See the dissent in *United States v. Plaza Health Labs., Inc.*, 3 F.3d 643, 653-54, 23 ELR 21526 (2d Cir. 1993) (dissenting op.).

365. *See, e.g.*, *Idaho Rural Council*, 143 F. Supp. 2d at 1180; *Washington Wilderness Coal.*, 870 F. Supp. at 990.

366. *See, e.g.*, *Friends of Santa Fe Cnty.*, 892 F. Supp. at 1358; *McClellan Ecological Seepage Situation*, 763 F. Supp. at 428, 707 F. Supp. 1182, *vacated on other grounds*, 47 F.3d 325.

367. Rapanos v. United States, 547 U.S. 715, 742, 36 ELR 20116 (2006).

rect.[368] This invites the argument that the direct addition of a pollutant to groundwater flowing to nearby navigable water is an indirect addition to the navigable water into which the groundwater flows. That, however, is an interpretation of the "addition" element of the CWA offense, rather than of the "navigable waters" element. Because neither Justice Kennedy's concurring opinion nor the dissenting opinion insisted on surface connections to navigable waters, a majority of the Supreme Court might be persuaded that tributary groundwater is a jurisdictional tributary.

Discharges to groundwater are not prohibited by CWA §301(a). That conclusion is supported by the legislative history, the structure of the statute, the failure of Congress to mention groundwater in its definitions of navigable water, the failure of EPA's former definition of the waters of the United States to mention groundwater, and EPA's explicit exclusion of groundwater in its current regulation.[369] This conclusion is also supported by the plurality decision in *Rapanos*, which twice states that only wetlands "with a continuous *surface connection*" with navigable water are within the regulatory jurisdiction of §404.[370] (*Rapanos* is a §404 case and the plurality did not consider the jurisdiction of §402.) But the relevant canon of construction is to interpret a term in the same manner throughout a statute.[371] Supporting this canon, CWA §502 provides that its definitions are to be applied throughout the statute unless otherwise specifically indicated in the statute.

It might appear that groundwater could be a water of the United States because it is a tributary of navigable water, is located on a 100-year floodplain of a navigable water, or is located within the high tide line or ordinary high watermark of a navigable water.[372] But the exclusion of groundwater applies "even where [it] otherwise meet[s] the terms of paragraphs (1)(vi) through (viii) of this definition.

That seems to defeat the theoretical difference between tributary and isolated groundwater developed earlier by legal scholars, plaintiffs, and courts to

368. *Rapanos*, 547 U.S. at 743 ("The Act does not forbid 'the addition of any pollutant *directly* to navigable waters from any point source,' but rather the 'addition of any pollutant *to* navigable waters."). The argument for indirect addition by aerial spraying of pesticides near water and other discharges to air that reach water is developed in Amil Anthony, *Shotguns, Spray, and Smoke: Regulating Atmospheric Deposition of Pollutants Under the Clean Water Act*, 29 UCLA J. Envtl. L. & Pol'y 215 (2011). The argument is criticized *supra* Chapter 1, Section III.B.

369. 80 Fed. Reg. 37114, to be codified at 40 C.F.R. §122.2 waters of the United States (5)(v).

370. *Rapanos*, 547 U.S. at 742 (emphasis added). Justice Kennedy's concurring opinion rejected this conclusion, *see Rapanos*, 547 U.S. at 772-74 (Kennedy, J., concurring), as did the dissenting opinion, *see Rapanos*, 547 U.S. at 804-06 (Stevens, J., dissenting).

371. *See* Eskridge, *supra* note 14 at 324; Scalia & Garner, *supra* note 14, at 170-73.

372. 80 Fed. Reg. 37114, to be codified at 40 C.F.R. §122.2 waters of the United States (1)(vi) & (viii), as elaborated by (2)(1) & (11).

hold discharges to tributary groundwater violated the CWA. But it does not defeat the argument that pouring pollutants on land adjacent to navigable water violates the CWA in the absence of a permit. If the pollutants reach navigable water, they are indirect additions to navigable water, whether they reach navigable water through groundwater or by another mechanism. The issue simply becomes one of interpreting addition rather than interpreting navigable water.[373]

EPA should reconsider its un-nuanced exclusion of virtually all groundwater in its current definition of waters of the United States. The Agency could make a credible case for including a groundwater tributary to nearby or adjacent surface navigable water within its definition of waters of the United States. Most courts holding groundwater not to be navigable waters did so in cases where there were no allegations that the discharges were to tributary groundwater, and commented that they may have held differently if the discharges had been to tributary groundwater.[374] Such courts may well have ruled otherwise if EPA had promulgated a rule including tributary groundwater in its definition of waters of the United States.

Instead, the Agency's revised definition of waters of the United States simply excludes groundwater even if it otherwise meets the definition of waters of the United States. This exclusion is unnecessary to meet the Supreme Court's concerns: It has never considered a case in which a defendant added pollutants to groundwater that was tributary to nearby navigable water, evading the CWA's regulatory program. The majority of the Court in *Rapanos* (that is, all of the Justices other than those in the plurality) was not fixated on surface connections and might be persuaded that such groundwater could be a tributary to waters of the United States. Even if groundwater flowing into nearby navigable water is not itself navigable, adding pollutants to that groundwater could be an indirect addition of the pollutants to navigable water, a proposition with which the plurality in *Rapanos* might conceivably agree. In any event, if the pollutants are hazardous, many of the discharges to groundwater would be subject to RCRA.[375]

---

373. *See supra* Chapter 1.

374. *See, e.g.*, Village of Oconomowoc Lake v. Dayton Hudson Corp., 24 F.3d 962, 966, 24 ELR 21080 (7th Cir. 1994); Exxon Corp. v. Train, 554 F.2d 1310, 1312 n.1, 7 ELR 20594 (5th Cir. 1977); Umatilla Waterquality Prot. Ass'n, Inc. v. Smith Frozen Foods, Inc., 962 F. Supp. 1312, 1319, 27 ELR 21411 (D. Or. 1997) (groundwater at issue was tributary, but EPA rule did not cover it); United States v. GAF Corp., 389 F. Supp. 1379, 1388, 5 ELR 20581 (S.D. Tex. 1975).

375. 42 U.S.C. §6925(a).

D. EPA's Unitary Navigable Waters Theory and Water Transfer Rule[376]

A water transfer is the diversion of some or all of the water from one water body to another for intervening or subsequent human use. Water transfers are often accomplished by point sources: pumps; pipes; canals; and ditches. Assuming that the first water body is heavily polluted and the second is pristine, under traditional CWA analysis, the discharge of the polluted water into the pristine water through a point source would require a §402 permit. Under EPA's recently promulgated water transfer rule, however, the discharge is exempted from requiring a permit unless pollutants are added by human use (except for agricultural purposes) of the water between diversion and discharge.

EPA justified the rule with two theories. The first and better-articulated theory is that when viewed as a whole, the CWA strikes a grand balance between its federal program to improve and protect water quality and preexisting state programs to allocate the uses of scarce water resources, highly developed private property systems, at least in the western United States.[377] Water resource management in the arid West, where water diversions are critical to irrigated agriculture and municipal water supplies, is an important aspect of the area's economy and state legal systems. This is the Agency's "holistic theory" of the CWA. EPA's second theory supporting the water transfer rule is that all navigable waters are one, so that when polluted navigable water is discharged into a pristine navigable water, the polluted navigable water adds no pollutants to the pristine water because both waters are the same water. This is the Agency's "unitary navigable water theory."

Chapter 1, Section III.A.3., *supra*, thoroughly examines and rejects EPA's holistic theory underlying its water transfer rule. The Southern District of New York recently heard a consolidated challenge to the rule and vacated it in *Catskill Mountains Chapter of Trout Unlimited, Inc. v. U.S. Environmental Protection Agency.*[378] The Eleventh Circuit, however, in *Friends of the Ever-*

376. *See supra* Chapter 1, Section III.A.3. and III.B.3., for discussions of the water transfer rule and the unitary theory of navigable water.

377. Western water law provides that the right of the first person to divert and use water from a particular water source is superior to the rights of persons later diverting and using water from the same source. Subsequent diverters may exercise their rights only when the senior diverter has used his full allocation of water. Eastern water law, based on common law, traditionally limits the right to divert and use water to landowners adjoining a water body and requires them to share and share alike during times of water scarcity. For a detailed descriptions of both western and eastern water law, see Tarlock, *supra* note 28.

378. 8 F. Supp. 3d 500 (S.D.N.Y. 2014), 2014 WL 128544.

glades v. South Florida Water Management District,[379] had earlier accorded *Chevron* deference to the rule when it was raised as a defense in a citizen suit against an unpermitted water transfer of polluted water. *Catskill Mountains* is the most recent, and also the most comprehensive, decision on the water transfer issue. It has been appealed to the U.S. Court of Appeals for the Second Circuit.[380]

EPA enunciated its unitary navigable waters theory in its amicus brief in *South Florida Water Management District v. Miccosukee Tribe of Indians of Florida*.[381] Although its brief suggested that the Agency had always subscribed to the theory, the earlier instances it cited merely foreshadowed the theory rather than articulating it.[382] Indeed, former EPA officials including former Administrator Carol Browner filed an amicus brief indicating that EPA had not subscribed to the theory during their tenures there.[383] EPA did not evoke the theory in its Water Transfer Interpretation[384] or in its proposed Water Transfer Rule or the preamble accompanying it[385]; additionally, some of the wording used in the preamble to the proposed rule[386] and repeated more than once in the preamble to the final rule is inconsistent with the theory.[387]

Moreover, the Water Transfer Interpretation and the preamble to the proposed rule both stated that the "precise legal question addressed" was whether a water transfer was an addition; nowhere was it suggested that the definition of navigable waters was at issue. Finally, the Interpretation stated explicitly that it did "not address the meaning of . . . navigable waters."[388]

---

379. 570 F.3d 1210 (11th Cir. 2009).

380. The Second Circuit is already on record as holding that the CWA unambiguously requires permits for water transfers, *Catskill Mtns. Chapter of Trout Unlimited, Inc. v. City of New York*, 102 F.3d 1273 (2d Cir. 1996), making it likely that it will affirm the district court's decision. If it does so, then because that would create a circuit split between the Second Circuit in *Catskill Mtns.* and the Eleventh Circuit in *Friends of the Everglades*, and because the Supreme Court left the unitary navigable water theory open in *Miccosukee*, the Court may well grant certiorari.

381. 541 U.S. 95, 105-06, 34 ELR 20021 (2004).

382. *Id.* at 107-08.

383. *See* Brief for Carol Browner et al. as Amici Curiae Supporting Miccosukee Tribe of Indians of Fla., South Fla. Water Mgmt. Dist., v. Miccosukee Tribe of Indians of Fla. 541 U.S. 95, 34 ELR 20021 (2004), 2003 WL 22793539. Former EPA Administrator Browner was joined in the amicus brief by two former assistant administrators for EPA's Water Programs, and two former general counsels.

384. EPA Interpretation on Applicability of CWA §402 to Water Transfers (2005) [hereinafter EPA Interpretation], *available at* http://www.epa.gov/ogc/documents/water_transfers.pdf.

385. 71 Fed. Reg. 32887 (June 6, 2006).

386. A water transfer "conveys waters of the United States to *another* water of the United States." (Emphasis added.) *Id.* at 32891.

387. 73 Fed. Reg. 33697, 33699 (a water transfer "diverts a water of the United States to a *second* water of the United States.") (emphasis added). *Id.* at 33704.

388. EPA Interpretation, *supra* note 384, at 18, n.19.

Although all three documents reiterated that the precise[389] or principal[390] legal issue was the definition of addition, none of them enunciated the meaning of addition. Instead, all three justified the water transfer policy or rule's exclusion of water transfers from the requirement of a CWA §402 permit on the basis of a holistic interpretation of the statute favoring water transfers being regulated by authorities other than federal and state pollution control authorities.[391] EPA's frequent shifting between the elements it is interpreting and the theories it uses to interpret them in this contest is a shell game.[392]

EPA's preamble to its final rule embraces the unitary navigable waters theory as a justification for the water transfer rule, although as a subsidiary argument and in a conclusory manner:

> Thus, there is no "addition"; nothing is being added "*to*" "the waters of the United States" by virtue of the water transfer, because the pollutant at issue is already part of "the waters of the United States" to begin with . . . . When a pollutant is conveyed along with, and already subsumed entirely within, navigable waters and the water is not diverted for an intervening use, the water never loses its status as "waters of the United States," and nothing is added to those waters from the outside world.[393]

This preamble is drawn directly from EPA's brief in *Friends of the Everglades v. South Florida Water Management District*,[394] with no further authority or justification except an implicit reference to *National Wildlife Federation v. Consumers Power Co.*[395]

EPA's unitary navigable waters theory is a startling and counterintuitive notion. If Congress intended to incorporate such an unexpected theory or policy into the CWA, surely it would have explicitly articulated it in the statute or at least discussed it in the legislative history. However, Congress did not mention or even hint at the theory in either the statute or its legislative history. The only wording EPA can point to in the statute to support the notion is an inference from the presence of the word "any" before the terms "addition," "pollutant," "point source," and "person," but not before the term "navigable waters" in §§301(a) and 502(12). EPA argues that, for example, "any" before "pollutant" signifies there are many different pollutants, any

389. *See supra* Chapter 1, Section III.A.3.b.
390. *Id.*
391. *Id.*
392. Or perhaps it is evidence the documents were written by different members of a committee with insufficient editorial control to produce consistent documents.
393. 73 Fed. Reg. 33697, 33701.
394. *See id.*; Friends of the Everglades v. South Fla. Water Mgmt. Dist., 570 F.3d 1210 (11th Cir. 2009).
395. 862 F.2d 580, 19 ELR 20235 (6th Cir. 1988).

one of which could support a violation of the CWA. By the same token, the Agency argues that the absence of the word "any" before "navigable waters" signifies that there is only one aggregate navigable waters that can support a violation, not many individual navigable waters. EPA argues that this is consistent with use of the singular nouns "addition," "pollutant," and "point source," and of the plural noun "waters" in §§301(a) and 502(12). Congress did not explain why it used the plural form of navigable waters and the singular forms of the other elements.

The difference between "navigable water" and "navigable waters" is not at all clear. Indeed, in his plurality opinion in *Rapanos*, Justice Scalia used dictionary definitions to argue that the singular "water" is more extensive than the plural "waters."[396] However, Congress used the words water and waters interchangeably in the statute,[397] and elsewhere provided that unless "the context indicates otherwise," the use of a singular noun in a statute incorporates the plural of that noun, and the use of a plural noun in a statute incorporates the singular of that noun.[398] This parallels the canon of statutory construction that absent a contrary indication in a statute, "the singular includes the plural (and vice versa)."[399] Nothing in the definitions of "discharge of a pollutant" or "navigable waters" indicates differently.[400] Indeed, the CWA's water quality standards and §404 programs would be rendered meaningless by EPA's unitary navigable water theory.[401] The argument that the absence of "any" before "navigable waters" in §502(12) exempts water transfers from §402 or creates one unitary body of water from all navigable waters is inconsistent with Justice Scalia's comment that Congress does not "hide elephants in mouse holes."[402]

The CWA is premised on treating navigable waters into which pollutants are discharged as singular waters rather than as one whole navigable water. This is most evident in the CWA's water quality standards program and in the §404 permit program for filling wetlands. The CWA's water quality standards program is based on the strategy of the earlier FWPCA to regulate

396. Rapanos v. United States, 547 U.S. 715, 732, 36 ELR 20116 (2006).
397. For example, CWA §101(a) uses "waters" in (1) and "water" in (2).
398. 1 U.S.C. §1.
399. *See* Scalia & Garner, *supra* note 14, at 129-31.
400. CWA §502(7) & (12); 33 U.S.C. §1362(7) & (12).
401. It is most likely that Congress used "navigable waters" in the CWA definition of "discharge of a pollutant" only because it consciously adopted the jurisdiction and the permit program from the Refuse Act, and the Refuse Act forbade the discharge of refuse into the plural "navigable waters" without a permit from the Corps. Congress most likely used "navigable waters" in the Refuse Act because Congress used the term throughout the Rivers and Harbors Act of 1899. *See, e.g.*, 33 U.S.C. §407.
402. *See* Whitman v. American Trucking Ass'ns, Inc., 531 U.S. 457, 468, 31 ELR 20512 (2001).

the discharge of pollutants to interstate waters. The basic strategy of this program is for states to designate how they desire particular waterways to be used—for example, drinking water without treatment, body contact sports, fish propagation, non-body contact sports, or agricultural and industrial use. Next, the state designates pollutant criteria for each use, primarily concentrations of pollutants in the water that are safe for the particular use designated. Thus, different water bodies will have different uses designated and different criteria for the pollutants found in them. The states are then to reduce pollution from point and nonpoint sources in a particular water body sufficiently to meet the criteria designated for that water body.

This is inherently a regulatory strategy that requires treating each water body as a separate entity rather than as a part of one whole unitary water body. Treating navigable waters as unitary navigable water suggests one use and one set of criteria. The CWA is replete with requirements for states to regulate the water pollution of separate water bodies or portions thereof rather than of one whole body of water.[403] EPA's unitary navigable water theory is inconsistent with this strategy of water pollution control. Because Congress established the water quality standards strategy but did not mention the incompatible unitary navigable waters theory, that theory is a chimera.

If the unitary navigable water theory is inconsistent with the water quality strategy, it reads much of §404 out of the CWA. Section 404 requires a permit for adding dredged or fill material to navigable waters. The provision was enacted to regulate the disposal of dredged spoil by Corps-permitted projects to enhance waterborne transportation by maintaining or deepening navigation channels and harbors. The dredged spoil was most easily and economically disposed of in nearby water or low-lying coastal wetlands. If all navigable waters were one, no §404 permit would be needed to redeposit dredged spoil from one navigable water to another because both navigable waters would be one. Most recent §404 enforcement actions arise from unpermitted wetland clearing activities. In the typical wetland clearing case, a developer removes vegetation, levels the land, and perhaps digs a drainage ditch, redepositing the vegetative material and soil on other parts of the wetland. This redeposit is held to be the addition of material to navigable waters from a point source that requires a permit.[404] If all navigable waters are one, however, no §404 permit would be needed when material cleared from one

403. *See* CWA §302(a), 33 U.S.C. §1312(a), providing for "attainment or maintenance of . . . water quality in a *specific portion of the navigable waters* . . . ." (emphasis added). *See also supra* Chapter 1, Section III.A.3.b.

404. *Id.* at III.C.1.

part of a wetland is redeposited in another part of the wetland, for both parts of the wetland would be the same navigable water.

In the wetlands land-filling cases, another EPA theory would thwart prosecution: the Agency's theory that "addition" must be from the "outside world." Removing material from a wetland and redepositing it elsewhere in the same wetland does not add it from elsewhere.[405] Together, these two EPA theories could narrow §404 jurisdiction only to cases in which fill material from dry land is discharged into water.

EPA contends that because "Congress explicitly forbade discharges of dredged material except in compliance with" a §404 permit, the unitary navigable water theory and the outside world theory have no effect on the §404 program.[406] But its contention only tells us that dredged material is a pollutant, not whether the water into which it is deposited is navigable or whether the deposit is an addition and all elements of the CWA offense must be met before a permit is required. Perhaps the Agency is suggesting that its theory does not apply to §404, but only to §402. But EPA does not explicitly make that argument, and the argument cannot survive either the canon of statutory interpretation that one word or term must be interpreted to have the same meaning throughout a statute unless the statute provides otherwise,[407] or the statement to the same effect in the lead-in to the CWA's definition of navigable waters in §502.[408]

V. Conclusion

Congress defined navigable waters as the waters of the United States, intending that the jurisdiction of the CWA extend as far as constitutionally possible. At first, courts followed EPA's lead and legislative history to interpret the terms expansively under both §§402 and 404. The Supreme Court, however, gradually narrowed the CWA's navigable waters jurisdiction, mindful of the fact that the Court itself had developed the meaning of that term as a highway of interstate and foreign waterborne commerce under the Commerce Clause over the course of more than one-and-a-half century of its opinions. By basing the CWA's jurisdiction on a term with an already-developed meaning under the Commerce Clause, Congress invited courts to interpret the CWA's jurisdiction with some regard to its traditional connection with waterborne

405. *Id.* at III.A.3.b.
406. 73 Fed. Reg. 33697, 33703 (June 13, 2008).
407. Eskridge, *supra* note 14, at 324, Scalia & Garner, *supra* note 14, at 167-69.
408. 33 U.S.C. §1362 ("Except as otherwise specifically provided, when used in this chapter: . . . (7) the term 'navigable waters,' means the waters of the United States. . . .").

commerce. That effectively narrowed "navigable waters" under the CWA to the first of the three *Lopez* prongs of interstate and foreign commerce jurisdiction: use as a highway of commerce. The tension between congressional intent and judicial interpretation has been particularly acute under §404, where agricultural landclearing activities in wetlands sometimes appear to be moving dirt on or to dry land, rather than placing it on or in the nation's waters.

A less textual Supreme Court could reach a more expansive interpretation of navigable waters. Congress could amend the CWA to erase any mention of navigable water and explicitly state a preamble to the amendment or in the legislative history that it is using all of the jurisdiction it has under the Commerce Clause under any of the three prongs of *Lopez*, as well as its treaty powers, which it has used to protect migratory birds and their habitats. None of these changes is likely in the near future. In the meantime, EPA and the Corps had no choice but to amend their regulatory definitions of waters of the United States, acknowledging some limitations to their jurisdiction as a result of *SWANCC* and *Rapanos*. Whether their retrenchments in the new regulations are sufficient to placate the courts remains to be seen.

Chapter 4. "Point Source"

This chapter examines how the U.S. Environmental Protection Agency (EPA) and the courts have interpreted the meaning of the term "point source," the fourth jurisdictional element of the water pollution control offense. In the CWA, "point source" is an artificial construct with a statutory definition, followed by lists of examples and exclusions. Disputes over the interpretations of "point source" have produced a steady stream of reported decisions since the initial implementation of the statute. Even after four decades, many of these issues are unresolved and new issues continue to arise.

I. Congressional Action: Statutory Definition and Legislative History of Point Source

A. *Statutory Definition*

Section 502(14) of the CWA provides that:

> "[P]oint source" means any discernible, confined and discrete conveyance, including but not limited to any pipe, ditch, channel, tunnel, conduit, well, discrete fissure, container, rolling stock, concentrated animal feeding operation, or vessel or other floating craft, from which pollutants are or may be discharged.

This term does not include agricultural discharges and return flows from irrigated agriculture.[1]

The definition of point source is the most grammatically complex of the elements' statutory definitions. It has a compound structure beginning with the core definition, a "discernible, confined and discrete conveyance"; followed by an inclusive list of 11 examples; and ending with two exceptions. The comparative structural complexity of the point source definition reflects that, unlike the other elements, it is a statutory construct with no meaning apart from its statutory definition. Before the enactment of the CWA, "addition" and "pollutant" each had plain meanings and "navigable waters" had both a plain and a legal meaning. However, while both "point" and "source"

1. 33 U.S.C. §1362(14).

had plain meanings before enactment of the CWA, their combination into one term did not have a preexisting plain meaning. "Point" in its most pertinent definition means "a particular spot, place, or position in an area or on a map, object or surface"; and "source" in its most pertinent definition means "a place, person or thing from which something comes or can be obtained."[2] Significantly, these two words, alone or together, do not suggest a "discernible, confined and discrete conveyance," the statutory definition of "point source."

The other elements are on/off switches, determining whether a discharge is regulated by the CWA in one of its two permit programs. If a discharge lacks an "addition," a "pollutant," or a "navigable water," the CWA does not regulate the discharge, regardless of whether it is through a "point source." By contrast, "point source" is a switch that shuttles discharges between different CWA schemes. If a discharge meets the other elements and is through a "point source," it is regulated by one of the CWA's two permit programs. If a discharge meets the other elements but is not through a "point source," it may be regulated by the CWA's nonpoint source program. While the CWA does not mention non-additions, non-pollutants, or non-navigable waters, it does mention "nonpoint sources"[3] and provides for programs regulating them. Despite the importance of nonpoint sources to the structure of the statute and to understanding the term point source, the statute does not define nonpoint source. Nor does EPA define nonpoint source in its permit program regulations, although it does so in informal documents.[4]

Although the U.S. Congress did not define "point" or "source" generally, it defined "source" for the limited purpose of CWA §306,[5] the section requiring EPA to establish technology-based standards for new sources.[6] Section 306(a)(3) defines a source as a building or facility from which pollutants originate. This "source" is a new manufacturing facility, not a point source—that is, not a new discharge pipe leading from the facility to navigable water.[7] This makes sense, for new sources are subject to higher levels of pollution control than are existing sources. It is easier and cheaper for new sources to design and build facilities with state-of-the- art pollution control than for existing sources to retrofit existing facilities with it, particularly since new facilities can often be designed to use less water and hazardous materials,

2. Oxford English Dictionary, www.oxforddictionaries.com.
3. CWA §§105(d)(1), 201(c), 208(b)(2)(F), 319(a) & (b); 33 U.S.C. §§1255(d)(1), 1281(c), 1288(b)(2)(F), 1239(a) & (b).
4. 40 C.F.R. §122.2. *See infra* in Section II, for EPA website definitions of nonpoint source.
5. 33 U.S.C. §1316.
6. CWA §306(a)(3), 33 U.S.C. §1316(a)(3).
7. Mahelona v. Hawaiian Elec. Co., Inc., 418 F. Supp. 1328, 1334-35, 7 ELR 20031 (D. Haw. 1976).

thus generating less pollutants to treat. The same rationale would not apply if the newly constructed source was merely the pipe that carries pollutants from the facility to navigable water. Thus, "source" in both the dictionary definition and in the limited CWA definition of "new source" connotes the origin of pollutants. "Point source," on the other hand, connotes the conveyance of pollutants from their origin to their addition to navigable water. In other words, point source is the place of the addition of pollutants to navigable water rather than the origin of the pollutants.[8] When we interpret point source, we are not searching for the plain meaning of "point," "source," or "point source," but for the plain meanings of "discernible," "confined," "discrete," "conveyance," "pipe," "tunnel," or other words used in the statutory definition of the term. Indeed, the plain meaning of "source" obscures the congressional definition of "point source."

If the definition of point source was straightforward and clear, the statute's failure to define point, source, and nonpoint source would pose no problems. Unfortunately, on close reading, the definition is far from clear. First, it defines a point source as a conveyance "from which pollutants are or may be discharged." This makes the definition circular, because the CWA in turn defines "discharge" as any addition of any pollutant by any point source. Second, it defines point source as a conveyance, but lists a concentrated animal feeding operation (CAFO) as an example of a point source, although a CAFO is not a typical conveyance. Instead, a CAFO is an agricultural facility more like an industrial facility than a conveyance. Third, the two statutory exceptions contained in the definition, "agricultural stormwater discharges" and "return flows from irrigated agriculture," are not conveyances either; instead, they are liquids that are or may be conveyed. Without the exemptions, they would not be point sources. The internal inconsistencies suggest that: (1) Congress did not (and perhaps could not) clearly articulate the distinction between point source and nonpoint source; (2) Congress intended some agricultural discharges to be regulated by the CWA's permit programs and intended other agricultural discharges not to be regulated by the permit programs; and (3) determining which agricultural discharges are point sources will be a major issue in interpreting "point source."

8. South Fla. Water Mgmt. Dist. v. Miccosukee Tribe of Indians, 541 U.S. 95, 105, 34 ELR 20021 (2004). The definition of point source "makes it plain that a point source need not be the original source of the pollutant; it need only convey the pollutant to 'navigable waters.'" *Id. See also* United States v. Lucas, 516 F.3d 316, 332-33, 38 ELR 20041 (5th Cir. 2008); Dague v. City of Burlington, 935 F.2d 1343, 1354, 21 ELR 21133 (2d Cir. 1991).

The failure of Congress to define nonpoint sources[9] is particularly troublesome because its apparent meaning is anything other than a point source. But does that mean a nonconveyance or a conveyance that is indiscernible, unconfined, and indiscrete? As used in the legislative history of the CWA, and as commonly used when discussing water pollution, nonpoint sources are discharges of stormwater. The two are not synonymous, however. Stormwater may be collected and channeled by human activity, in which case it is discharged by a point source.[10] On the other hand, there are nonpoint sources that have nothing to do with stormwater.[11]

These inconsistencies help explain the difficulties that courts and EPA have encountered in construing point source and the frequent conflation by the courts, EPA, and litigants of that element with other elements of the CWA offense[12] and even with other CWA terms that are not elements of the offense.[13] Conflation also results from the fact that the elements (all nouns)

---

9. Cordiano v. Metacon Gun Club, Inc., 575 F.3d 199, 219 (2d Cir. 2009).
10. EPA's definition of "discharge of pollutants," for instance, specifically includes stormwater that is collected and channeled by man. 40 C.F.R. §122.2.
11. Examples of nonpoint sources that have nothing to do with stormwater include, for example, people and animals.
12. For conflation with "addition," see, e.g., *Rapanos v. United States*, 547 U.S. 715, 735, 36 ELR 20116 (2006) (ditches may sometimes, but not often, be both point sources and navigable waters); *National Cotton Council, v. U.S. EPA*, 553 F.3d 927, 938-39, 39 ELR 20006 (6th Cir. 2009) (EPA erroneously argued that a pesticide spray bar cannot be a point source because it did not add pollutants to water, i.e., it did not spray excess pesticides); *Peconic Baykeeper, Inc. v. Suffolk Cnty.*, 600 F.3d 180, 188, 40 ELR 20098 (2d Cir. 2010) (district court erroneously held spray bar was not a point source because it added pesticide to air rather than to water); *Dague*, 935 F.2d at 1354 (the losing party erroneously argued that a culvert could not be a point source because it did not add a pollutant to navigable water for the first time); *National Wildlife Fed'n v. Gorsuch*, 693 Fd.2d 156, 165, 13 ELR 20015 (D.C. Cir. 1982) (EPA erroneously argued that a dam was not a point source because it did not add pollutants from the outside world). For conflation with "navigable waters," see, e.g., *National Ass 'n of Home Builders v. U.S. Army Corps of Eng'rs*, 699 F. Supp. 2d 209, 215-17, 40 ELR 20104 (D.D.C. 2010) (the losing party erroneously argued that a ditch cannot be navigable water because it is a point source). For conflation with "pollutant," see, e.g., *National Cotton Council*, 553 F.3d at 938-39 (EPA erroneously argued that a pesticide spray bar was not a point source because it did not spray pollutants, but instead only a consumer product: pesticides); *United States v. West Indies Tramp., Inc.*, 127 F.3d 299, 309, 28 ELR 20202 (3d Cir. 1997) (the defendant erroneously argued that a barge could not be a point source because the barge was a pollutant when parts of it were torn off and thrown into the water). For conflation with "by any person," see, e.g., *United States v. Plaza Health Labs., Inc.*, 3 F.3d 643, 647, 23 ELR 21526 (2d Cir. 1993) (a person cannot be included in the fourth element, "point source," because person is already included in the fifth element, "by any person").
13. It is relatively common for courts, EPA, and commentators to distinguish between point source pollution and nonpoint source pollution. *See, e.g.*, League of Wilderness Defenders/Blue Mountains Biodiversity Project v. Forsgren, 309 F.3d 1181, 1184 (9th Cir. 2002) ("The issue before us is whether spraying insecticide from an aircraft . . . is point source pollution or nonpoint source pollution."); *Gorsuch*, 693 F.2d at 156 ("the point or nonpoint character of pollution is established when the pollutant first enters navigable water"). This conflates point source with pollution. Although "pollution" is defined in §502(19), 33 U.S.C. §1362(19), it is not an element of the CWA §301(a) offense. Indeed, it is quite distinguishable from "pollutant," which is an element.

are all linked by prepositions. Some courts have even treated the prepositions as elements.[14]

B. *Legislative History of Point Source*

The U.S. House of Representatives and the U.S. Senate bills both used "point source" in the definition of "discharge of a pollutant" and defined point source identically, except that the House version included "from which there is or may be a thermal discharge."[15] The omission of thermal discharges from the enacted "point source" definition is of no significance, however, because §502(6) defines "pollutant" to include "heat." Moreover, CWA §316(a)[16] provides specific regulation for thermal discharges in the §402 permit program. Nothing in the House, Senate, or Conference Reports further explains the meanings of point source, nonpoint source, the differences between the two terms, or why the permit programs are limited to point sources. Some comments made during the floor debates address these issues, but the comments are no less confusing than the statutory definition of point source.

Sen. Edmund Muskie (D-Me.), the chief sponsor of the Senate bill, on which most of the statute is based, attempted to illuminate the distinction between point source and nonpoint source:

> The discharge standard applies to point source control. Agricultural runoff is one form of agricultural activity that is a nonpoint source. It is a runoff into water that occurs perhaps miles away from the land that adjoins it. There is no effective way, as yet other than land use control, by which you can intercept that runoff and control it in the way that you do a point source. We have not yet developed technology to deal with that kind of problem. We need to find ways to deal with it, because a great quantity of pollutants is discharged by runoff, not only from agriculture but from construction sites, from streets, from parking lots, and so on, and we have to be concerned with developing controls for them.[17]

The "discharge standard" to which Senator Muskie referred is the set of uniform national technology-based standards that the CWA required EPA

14. "From" was treated as an element or close to it in *Peconic Baykeeper*, 600 F.3d at 188-90; *Gorsuch*, 693 F.2d at 174-75.
15. H.R. 11896, §502(15), H. Rep. No. 92-911, Nov. 19, 1971. *See also* S. 2770, §502(p), S. Rep. No. 92-414, Oct. 28, 1971.
16. 33 U.S.C. §1316(a).
17. 117 Cong. Rec. 38805 (Nov. 2, 1971), *reprinted in* 2 A Legislative History of the Water Pollution Control Act Amendments of 1972 [hereinafter 2 Legis. History], *available at* http://catalog.hathitrust.org/Record/001515679 (statement of Senator Muskie *also available at* http://abacus.bates.edu/muskie-archives/ajcr/1971/CWA%20Beall%20Query.shtml).

to promulgate for industrial and municipal point sources under §§301(b), 304(b)-(d), and 306.[18] His statement does not draw a crisp line between point sources and nonpoint sources and does not define nonpoint sources. Nevertheless, it makes several suggestions that may be useful in dealing with the point source/nonpoint source distinction. First, point sources typically are susceptible to pollution control by end-of-pipe technology supporting uniform national standards, while nonpoint sources typically are not.[19] Instead, nonpoint sources are more often susceptible to land use and management controls, which are usually site-specific.[20] Second, nonpoint sources are associated with stormwater runoff. Third, agricultural stormwater runoff is typically nonpoint source- related. Fourth, nonpoint sources are to a great extent from agriculture, construction sites, streets, and parking lots. Finally, when Senator Muskie stated that we have "not yet" developed technologies to control nonpoint sources and "need to," he suggested that as we gradually develop such technologies for different categories of nonpoint sources, those categories should become point sources and be regulated by the CWA's permit program.[21]

Senator Muskie's statement raises as many questions as it answers. Why cannot runoff be effectively intercepted and controlled? We routinely intercept and convey it with ditches. Moreover, we can also treat it to remove pollutants. The problem is that the great volume of stormwater often makes treatment not cost effective. His assertion that nonpoint sources are associated with runoff reminds us that flows normally are not conveyances but flow through conveyances, the exceptions in the definition of point source notwithstanding. His assertion that nonpoint sources are specifically associated with agricultural runoff is inconsistent with the definition's inclusion of CAFOs in its list of examples of point sources. What does the third sen-

18. 33 U.S.C. §§1311(b), 1314(b)-(d).

19. Pollutants discharged from nonpoint sources can be controlled by end-of-pipe technology. The polluted water flowing from many nonpoint sources can be collected and treated, such as stormwater runoff in municipal storm sewers. But such collection and treatment may be very expensive because of the large volumes involved. For the same reason, treatment of stormwaters to the same levels as treatment of municipal sewage may not be economically feasible. Management techniques may lower pollutant levels at less cost, such as street sweeping, retention ponds for parking lot runoff, green roofs, minimizing paved-over areas, and making paving less impervious.

20. Permits for point sources generally incorporate effluent limitations reflecting technology-based standards or water quality standards, rather than land use or management controls. Nonetheless, they may use best management practices to control runoff from factories, spills, and raw material storage areas. CWA §304(e). Moreover, industries must control runoff from areas where it comes into contact with industrial pollutants. CWA §402(p).

21. One court declared it clear from the legislative history that Congress "would have regulated so-called nonpoint sources if a workable method could have been derived." United States v. Earth Sciences, Inc., 599 F.2d 368, 373, 9 ELR 20542 (10th Cir. 1979).

tence mean? Stormwater runoff will not flow uncounted miles to a stream. Instead, it will percolate into the ground, evaporate into the air, or flow into the closest remote tributary of the stream. Perhaps, the Senator had in mind the suggestion by Rep. Teno Roncalio (D-Wyo.) in the context of irrigation return flows, rather than stormwater, that "[i]t is virtually impossible to trace pollutants to specific irrigation lands, making these pollutants a nonpoint source in most cases."[22] Irrigation canals often flow for miles between many irrigated farms, accumulating silt, fertilizer, pesticides, and other pollutants from all of them, so that when the much-reused irrigation water is eventually discharged to navigable water, it is difficult if not impossible to trace specific pollutants in the discharge back to particular farms. But Representative Roncalio's statement is not particularly helpful either; the definition of point source does not mention or suggest traceability.[23] Moreover, while it may be impossible to determine from a sample taken at the discharge of an irrigation ditch or canal what pollutants particular farms contributed to the discharge, it is quite possible to make that determination by sampling where the irrigation water is diverted to and discharged from each farm. It may also be possible to identify fertilizers and pesticides in a discharge by chemically fingerprinting them to fertilizers and pesticides used on particular farms.

Sen. Robert Dole (R-Kan.) reiterated the nonpoint nature of agricultural pollution. "Very simply," he stated (unknowingly ironic, as we shall see below), a nonpoint source of pollution is one that "does not confine its polluting discharge to one fairly specific location, such as a sewer pipe, a drainage ditch, or a conduit. In the area of agricultural pollution, a feedlot would generally be considered to be a nonpoint source as would pesticide and fertilizer runoff and accumulation."[24] It is not clear whether he is making a distinction based on the difference between point sources and nonpoint sources or on the difference between agricultural pollutants, such as pesticide and fertilizer runoff, and other pollutants. His observation that feedlots are nonpoint sources is demonstratively incorrect. CAFOs are included in the §502(14) list of examples of point sources. Perhaps, Senator Dole was alluding to smaller CAFOs. Senator Muskie commented that natural runoff from small CAFOs is not a point source, unless either the CAFO raises more than

22. 118 Cong. Rec. 10765 (Mar. 29, 1972), *reprinted in* 1 A Legislative History of the Water Pollution Control Act Amendments of 1972 [hereinafter 1 Legis. History], *available at* http://catalog.hathitrust.org/Record/001515679.
23. *But see* League of Wilderness Defenders/Blue Mountain Biodiversity Project v. Forsgren, 309 F.3d 1181, 1187 (9th Cir. 2002) (nonpoint sources "are identified by three characteristics," including "(2) The pollutants discharged are not traceable to any distinct or identifiable facility. . . .").
24. 2 Legis. History 1294; also included in his supplemental views published with S. Rep. No. 92-404, pp. 98-99 (1971), *reprinted in* 2 Legis. History 1513-14.

specified numbers of animals or is traversed by a stream.[25] If the CAFO is traversed by a stream, it is easy to understand how it conveys animal waste to the stream and it therefore could be a point source. If, however, a stream does not traverse the CAFO, it is more difficult to understand how the CAFO is a conveyance of the waste, as opposed to the source of the waste. In either event, the statute does not distinguish between small and large CAFOs, although EPA's regulations do.[26] So much for Senator Dole's "very simply."

Senators Dole and Muskie entered into a colloquy on the Senate floor that reiterated many of these points. Senator Dole asked "to what sort of guidance are we to look for further clarification of the terms 'point source' and 'non-point source'—especially as related to agriculture?" Senator Muskie replied that such guidance

> will be provided in regulations and guidelines of the [EPA] Administrator. The present policy with respect to the identification of agricultural point sources is generally as follows:
>
> First. If a man-made drainage ditch, flushing system or other such device is involved and if measurable waste results and is discharged into water, it is considered a "point source."
>
> Second. Natural runoff from confined livestock and poultry operations are not considered a "point source" unless the following concentrations of animals are exceeded . . .
>
> Third. Any feedlot operation which results in the direct discharge of wastes into a stream which traverses the feedlot are considered point sources without regard to the number of animals involved.[27]

C. *The CWA's Structure: Point Source/Nonpoint Source Distinction*

One court has commented that the "disparate treatment of discharges from point sources and nonpoint sources is an organizational paradigm" of the CWA.[28] Examination of the CWA's point source and nonpoint source programs may help to better frame the distinction between point sources and nonpoint sources.

25. 2 Legis. History 1299. "1,000 beef cattle; 700 dairy cows; 290,000 broiler chickens; 180,000 laying hens, etc." In its regulatory definition of a CAFO, EPA copied this "poop equivalency table." 40 C.F.R. §122.23.

26. 40 C.F.R. §122.23. CAFOs are discussed in greater detail below.

27. 1971 Cong. Rec. 38816 (Nov. 2, 1971), *reprinted in* 2 Legis. History 1298-99; *also available at* http://abacus.bates.edu/muskie-archives/ajcr/1971/CWA%20Dole%20Query.shtml.

28. Oregon Natural Desert Ass'n v. U.S. Forest Serv., 550 F.3d 778, 780, 39 ELR 20297 (9th Cir. 2008) (*quoted in* Friends of the Everglades v. South Fla. Water Mgmt. Dist. 570 F.3d 1210, 1227, 39 ELR 20118 (11th Cir. 2009).

I. Regulation of Point Sources

Point sources may not add pollutants to navigable waters without a §402 or §404 permit. EPA issues §402 permits and must approve submitted state programs for permit issuance if they are comparable to the federal program. Section 402 permits contain effluent limitations based on technology-based standards[29] and water quality standards.[30] The U.S. Army Corps of Engineers (the Corps) issues §404 permits, although EPA must approve submitted state programs comparable to the federal program for permit issuance in all but traditionally navigable waters. Section 404 permits authorize the filling of wetlands, with limitations on how much of a wetland may be filled and perhaps requiring mitigation, such as providing replacement wetlands. EPA oversees state administration of §402 programs, may veto proposed state permits if they are not in compliance with federal requirements, may enforce against violations of state-issued permits, and may withdraw approval of state §402 programs. EPA similarly oversees Corps and state administration of §404 programs. (For brevity, this chapter refers to EPA actions in developing, issuing, and enforcing permits, although such actions may be taken by the Corps or by states with EPA-approved programs.)

The most important condition in a §402 permit is that the point source cannot discharge pollutants in excess of the permit's effluent limitations (usually expressed numerically) for each pollutant it regulates. All effluent limitations must require the pollution reduction accomplished by the best technology for the particular type of facility at issue.[31] EPA regulations establish effluent guidelines on an industry-by-industry basis that permit writers can easily translate into effluent limitations for each regulated facility. The effluent guideline regulations are based on the best-performing pollution reduction technologies used or capable of being used by each category or subcategory of industry. However, if that pollution reduction is insufficient to ensure attainment or maintenance of concentrations or other criteria

---

29. CWA §§301(b)(2) & 304(b); 33 U.S.C. §§1311(b)(2) & 1314(b).
30. CWA §§303 & 304(a); 33 U.S.C. §§1313 & 1314(a).
31. CWA §301(a) specifies criteria for industrial best available control technology for toxic pollutants, best available control technology for conventional pollutants, and best available control technology for other pollutants. Section 306 also specifies criteria for best available control technology for new sources. EPA uses these criteria to promulgate regulations by industrial categories (for example, the organic chemicals and canning industries) and subcategories (such as particular fruits and vegetables in the canning industry) under §§301(b)(2) and 304(b). EPA separately established technology-based standards to be met by publicly owned treatment works (POTWs), as municipal sewage treatment plants are known among water pollution control cognoscenti, §§301(b)(1)(B) and 304(d)(1). These are performance-based standards. They do not require permittees to install particular technologies, only to remove pollutants comparable to the removal achieved by the best control technologies.

for pollutants necessary to achieve the state designated use of the receiving water, then the permit must require greater pollution reduction to achieve such water quality standards.[32] Permit holders must sample and analyze their effluents to determine whether they comply with the effluent limitations in the permits and must report the results to EPA and the state, where the reports are public information.[33] EPA, the state, and private citizens may enforce against violations.[34]

2. Regulation of Nonpoint Sources

As discussed below, EPA acknowledges that nonpoint sources have long been the leading cause of water quality problems. Sections 208 and later 319[35] encouraged states to establish programs and requirements for controlling nonpoint sources and provided initial funding for them to do so, but neither section defined nonpoint sources. Section 208 primarily required regional planning authorities to ensure that federal grant funds were spent on construction of wastewater treatment plants whose locations and capacities were efficient and reflected expected regional growth. Secondarily, §208 encouraged regional planning authorities to develop best management plans and state and local enforcement mechanisms to control pollution from the nonpoint sources for which EPA was to develop best management practices under §304(f). Section 208, however, lacked the structural mechanisms necessary for long-term effectiveness. It provided no mechanism for EPA to approve the adequacy of nonpoint programs developed by regional authorities or states. It provided no mechanism for EPA or citizens to force an uncooperative state to develop nonpoint source programs. It provided no mechanism for EPA or citizens to enforce against violations of a nonpoint source. It provided no mechanism for requiring EPA to develop a nonpoint source plan if a state does not. Finally, it did not provide for long-term funding for the state nonpoint source control programs. On top of these structural deficiencies, the regional planning organizations envisioned by §208 often were not parts of the state water pollution control authorities that the CWA expected to implement the statutes' permitting and publicly owned treatment works (POTWs) construction grant programs. Many of the state pollution control authorities viewed the regional planning organizations as potential rivals for power and fund-

32. CWA §§301(b)(1)(B) & 304(d)(1).
33. *Id.* §308.
34. *Id.* §§309 & 505.
35. 33 U.S.C. §§1288 & 1329.

ing and sought to undercut them. The combination of structural shortcomings and political feuding sounded the death knell of the program.

The 1972 statute required state water quality standards to be achieved by 1977.[36] Because the CWA included no enforceable program to control nonpoint source pollution, it is unsurprising that many segments of navigable waters had not achieved state-established water quality standards by that date or subsequently,[37] despite the fact that most industries and municipalities achieved the first required levels of their technology-based standards by that date or shortly thereafter.[38] In 1987, Congress responded to the failure to achieve water quality standards by adding §319 to the statute. The new section required states to submit reports identifying both segments of navigable waters unable to achieve water quality standards without controls on nonpoint sources and what types of nonpoint sources must be controlled to achieve water quality standards.[39] It also required states to establish programs imposing best management practices on nonpoint sources to achieve water quality standards in those segments.[40] If a state fails to submit the initial report, EPA must develop a report.[41] But if the state fails to submit the ultimate nonpoint source control program, the section provides the Agency with no authority to do so. The section does not provide either EPA or citizens with authority to force states to develop best management practices plans for nonpoint sources. While §319 goes beyond §208 in focusing on nonpoint sources, it too lacks the structural mechanism for long-term effectiveness.[42]

Section 304(f) mentions nonpoint sources in its title, but not in its text. The text directed EPA to develop best management practices for several cat-

36. CWA §301(b)(1)(C).

37. According to EPA's 2004 report to Congress on water quality:

> about 44% of assessed stream miles, 64% of assessed lake acres, and 30% of assessed bay and estuarine square miles were not clean enough to support such uses as fishing and swimming. Leading causes for impairment included pathogens, mercury, nutrients, and organic enrichment/low dissolved oxygen. Top sources of impairment included atmospheric deposition, agriculture, hydrologic modifications and unknown or unspecified sources.

The report is available at http://water.epa.gov/lawsregs/guidance/cwa/309_cl_22_305b_2004report-factsheet2004305b.pdf.

38. Of the approximately 4,000 major industrial point sources, all except approximately 860 achieved their first level of pollution control by 1979, although only one-half of the 4,000 major POTWs did so. *See* Testimony of Marvin Durning, U.S. EPA Assistant Administrator for Enforcement, Before the Subcommittee on Environmental Pollution of the Committee on Environment and Public Works, May 24, 1979, Serial No. 96-H16.

39. CWA §319(a).

40. *Id.* §319(b).

41. *Id.* §319(d).

42. Oliver A. Houck, The Clean Water Act TMDL Program: Law, Policy & Implementation (2d ed. 2000.) For the latest round of total maximum daily load (TMDL) litigation, see *American Farm Bureau Fed'n v. U.S. EPA*, No. 13-4079, 2015 WL 4069224, 45 ELR 20129 (3d Cir. July 6, 2015).

egories of activities: agricultural and silvicultural activities, including runoff from fields and forests; mining activities, including runoff from active and abandoned mines; construction activities, including runoff; deep well injection and subsurface disposal of wastes[43]; salt water intrusion[44]; and changes in water movement caused by dams, flow diversions, and other such structures.[45] Although some of these categories of activities center on nonpoint sources, others involve what could be either point sources or nonpoint sources, and some involve only point sources. Deep well injection, subsurface waste disposal, and salt water intrusion all use wells, and wells are on the statutory list of point source examples. These activities, however, usually inject pollutants into groundwater, which is generally not considered to be navigable. Salt water intrusion results from withdrawal of freshwater from wells in coastal areas and usually does not involve a discharge to surface water. Dams are usually point sources, as are the pipes, pumps, and ditches that accomplish water diversions and transfers.

This subsection does not define nonpoint sources, but reflects Congress' desire to have EPA focus on these particular pollution sources and develop solutions to them without creating serious new federal regulatory programs. Congress did not attempt in the subsection to draw distinctions between point and nonpoint sources. Indeed, the subsection overlaps to some extent with the regulatory programs Congress developed for point sources. The U.S. Supreme Court recognized in *South Florida Water Management District v. Miccosukee Tribe of Indians of Florida* that by addressing flow diversion facilities, §304(f) did not make them nonpoint sources, "if they *also* fall within the 'point source' definition."[46] In conclusion, this subsection again suggests Congress did not or could not articulate a strict line between point sources

43. Deep well injection and injection into subsurface areas are waste disposal methods regulated by the Safe Drinking Water Act (SDWA), 42 U.S.C. §§300f to 300j-26, ELR Stat. SDWA §§1401-1465, and the Resource Conservation and Recovery Act (RCRA), 42 U.S.C. §§6901-6992k, ELR Stat. RCRA §§1001-11011. Since these methods dispose of wastes through a well, which is included as an example of a point source in the CWA's definition of the term, it is odd to categorize them as nonpoint sources. Despite some initial uncertainty, they are not regulated by the CWA because they discharge to groundwater, which is not navigable water under either the statute or common understanding of that term. *See supra* Chapter 3, Section IV.C.
44. Excessive withdrawal of groundwater from aquifers near the seacoast allows salt water to replace the freshwater withdrawn, contaminating the remaining freshwater aquifer.
45. This chapter examines dams, concluding they are point sources. Flow diversions are not subject to §402 regulation because they add nothing to navigable waters, but withdraw water from navigable waters for use elsewhere. When the used water is later returned to navigable water, it may be subject to §402 regulation.
46. 541 U.S. 95, 106, 34 ELR 20021 (2004). Courts earlier had rejected arguments by industry that designation of an activity for development of nonpoint source best management practices in §304(f) exempted the activity from regulation of a point source under §402. United States v. Earth Sciences, Inc., 599 F.2d 368, 372, 23 ELR 21526 (10th Cir. 1979).

and nonpoint sources, but recognized that water quality could not be protected without moving beyond the regulation of point sources.

The CWA does not disregard nonpoint source discharges. But while its regulation of point sources is robust, uniform, and effective, its regulation of nonpoint sources is hortatory, haphazard, and of dubious effectiveness. That difference in regulatory effectiveness does not mirror the relative importance of point and nonpoint sources to the quality of the nation's water: Nonpoint sources discharge more pollutants than point sources. The CWA's failure to deal effectively with nonpoint sources is its most significant failure and the primary reason we have not achieved water quality standards 40 years after enactment of the statute. EPA could ameliorate this difficulty by amending its definition of "point source" from time to time to include former categories of nonpoint sources for which technology-based treatment standards had become available.

3. Regulation of Stormwater and Storm Sewers

After the enactment of the CWA, EPA and states with approved §402 programs were faced with the enormous task of issuing permits to more than 60,000 point source permit applicants within a fairly short period.[47] They lacked the time or resources to address stormwater and many minor point sources.[48] At the same time, EPA was making grants to cities and towns across the country to build secondary treatment plants,[49] often providing sewage treatment for the first time. The administrators of the Agency's

47. Congress enacted the CWA in October, 1972. It provided in §301(a) that the discharge of pollutants without a permit was illegal, but granted immunity until the end of 1974 to permit applicants for whom permits had not yet been issued. *See* §402(k). Thus, the statute indicates that Congress contemplated EPA would issue all permits by that date. EPA had not previously undertaken a task of this magnitude. The effluent guidelines that were to help permit writers devise effluent limitations for industrial sources were not yet promulgated; once they were, those guidelines were subjected to judicial review. Once permits were issued, many permit holders appealed their terms. EPA concentrated on issuing permits to applicants it classified as "major" point sources, that is, 15% of the applicants EPA considered to be discharging the vast majority of pollutants from point sources, based on the amounts of pollutant discharges reported in permit applications and on state evaluations of their adverse impact on local water quality. As of the end of 1975, EPA reported that of the 62,118 point sources requiring §402 permits, 9,259 were "major sources." U.S. EPA, *EPA Enforcement: A Progress Report December 1974 to December 1975* 89-90 (1976). By the end of 1975, EPA and states with approved §402 permit programs had issued permits to all but 671 of the major sources and most of the remainder were either new additions to the list of major sources, or had §402 permits issued by states with approved programs, but which those states deemed not to be final until outstanding appeals were resolved. *Id.* EPA permit writers simply did not have the resources or time to address the far more numerous stormwater point sources, few if any of which discharged as many pollutants as major sources.
48. Decker v. Northwest Envtl. Def. Ctr., 133 S. Ct. 1326, 1331-21, 43 ELR 20062 (2013).
49. Secondary treatment was the preferred treatment for municipal sewage. It constituted primary treatment by settling and secondary treatment by biological methods, usually followed by chlorination

construction grant program did not want to spend its limited resources funds making grants to deal with municipal storm sewers until all cities had secondary treatment. Thus, EPA's two largest water programs and their state counterparts were united in not wanting to address municipal storm sewers or runoff at all, or at least not until sometime in the indefinite future. As a result, EPA promulgated a regulation exempting stormwater flows and sewers from the §402 permit program, an exemption soon reversed as beyond EPA's authority, discussed in greater detail in Section III, below.

In 1987, Congress entered the fray by amending the CWA to add in §402(p) a program for gradually permitting and controlling municipal and industrial storm sewers. The provision began with a curiously worded statement in §402(p)(1) that "[p]rior to October 1, 1994, the Administrator . . . shall not require a permit under this section for discharges composed entirely of stormwater." Under §301(a), it is the CWA, not the Administrator, that requires a permit for a "discharge of any pollutant." And what is a discharge "composed entirely of stormwater"? Does that include only dischargers of stormwater carrying pollutants? If so, §301(a) already forbids it, except for the agricultural stormwater flow exemption in §502(14). If not, then the provision is meaningless, for a discharge without a permit of pollutant-free water does not violate §301(a). The provision in §402(p)(2) exempts from the temporary exemption several categories of "stormwater discharges," including discharges "associated with industrial activity" and from "municipal separate stormwater sewer systems" serving specified populations. Interpreted together, §402(p)(1) and (2) appear to require industrial stormwater dischargers and municipal separate sewer dischargers to apply for and secure permits on schedules established in §402(p)(4) and to exempt them from the requirement of having a permit in the meantime. Those subsections also appear to exempt other discharges of stormwater with pollutants from the requirement to have a permit until October 1, 1994, but not thereafter. The Court in *Decker v. Northwest Environmental Defense Center,*[50] however, seems to read "[p]rior to October 1, 1994" out of the provision, perhaps because EPA extended the exemption in the absence of a finding that the discharge must be controlled to meet water quality standards, although the decision did not mention this regulation.[51] This issue will be discussed further below.

---

or another treatment for pathogens. Such facilities normally removed 85% of settleable solids and biological oxygen demand.

50. 133 S. Ct. 1326, 43 ELR 20062 (2013).

51. 40 C.F.R. §122.6(a)(9)(i).

4. Point Source/Nonpoint Source Distinction Left Unclear

The statutory definition of point source, the structure of the statute, and the legislative history of the statute, separately or collectively, do not draw a bright line between point sources and nonpoint sources. EPA has not helped to clarify the distinction by regulation and may have muddied the waters further. We are left with a few hypotheses from the above material. Industrial and municipal waste water discharges are typically by point sources, while agricultural discharges are typically by nonpoint sources. Stormwater runoff is presumed to be discharged by nonpoint sources, but may be discharged through point sources if it is collected and channeled by man. Pollution from point sources is susceptible to end-of-pipe technology for pollution control, while pollution from nonpoint source pollution is more appropriately dealt with by land use controls or management practices. This is a rationale for subjecting point source pollution to the permit programs with nationally uniform technology-based standards and nonpoint sources to site-specific best management practices with no permit program. A corollary is that as end-of- pipe technology becomes available for particular categories of non-point sources, it would be appropriate to include those categories within the definition of point source to be regulated by the statute's permit systems.

None of these distinctions is entirely satisfactory. Stormwater runoff can be collected and be treated with end-of-pipe technology. Indeed, CWA §402(p) subjects much municipal and industrial stormwater runoff to the §402 point source permit program to do just that.[52] Moreover, agricultural runoff can be point source pollution. CAFOs, for instance, are included in the statutory definition of point sources. While it may be assumed that there is a clear divide between point source discharges and nonpoint stormwater runoff, that clarity is deceiving. When stormwater is collected and channeled by human activities, it is discharged from point sources and may be regulated by §402 permits.[53] Finally, many nonpoint sources, such as bulldozers and barges, have nothing to do with runoff.

II. Administrative Interpretations of Point Source

EPA's national pollutant discharge elimination system (NPDES) regulations in Title 40 of the *Code of Federal Regulations* define point source almost identically with the CWA's definition of the term. EPA's definition adds "landfill

52. 33 U.S.C. §342(p).
53. 40 C.F.R. §122.2 (definition of "discharge of a pollutant").

leachate collection system" at the end of the list of examples of point sources. And, in the exemption, it reverses "return flows from irrigated agriculture" and "agricultural stormwater discharges" and substitutes "runoff" for "discharges," making the exception read "return flows from irrigated agriculture and agricultural stormwater runoff." EPA probably made the last change because "point source" is an element of "discharge," making the inclusion of "discharge" as part of a listed exception to "point source" circular. Why EPA added only "landfill leachate collection system" to the list of point source examples is a mystery, for there are many other conveyances that could be added to the list. Moreover, discharges from leachate collection systems would probably be from point sources without the addition.[54] EPA's definition ends with "(See 122.3)," a phrase Congress could not have included in its definition of "point source." Section 122.3 is entitled "Exclusions" and excludes a number of large categories of discharges[55] from the requirement to secure §402 permits, even though they otherwise would constitute additions of pollutants to navigable waters from point sources. This is a curious provision.

EPA's definition of point source is augmented by the last portion of its definition of "discharge of a pollutant" in §122.2, which includes "surface runoff which is collected or channeled by man."[56] EPA's implementing regulations at 40 C.F.R. §122.26 also require treatment of industrial storm sewers carrying stormwater that comes into contact with industrial wastes and lesser, phased requirements for municipal storm sewers.

54. If a leachate collection system was discharging directly to navigable water, it would in all probability be through a pipe, already defined as a point source. If it was pumped to a tank truck for disposal elsewhere, tank trucks are held to be point sources.

55. Discharges of (a) sewage from vessels; (b) dredged and fill material regulated under CWA §404; (c) sewage and other pollutants into POTWs by indirect dischargers; (d) material in compliance with instructions of On-Scene Coordinators under 40 C.F.R. pt. 300; (e) pollutants from nonpoint source agricultural and silvicultural activities except as otherwise provided; (f) return flows from irrigated agriculture; (g) waste from privately owned treatment works, and (h) material from water transfers. This list of exclusions is vastly oversimplified, for most of the exclusions are significantly qualified. The exclusion for agricultural and silvicultural activities, for example, includes

> storm water runoff from orchards, cultivated crops, pastures, range lands, and forest lands, but not discharges from concentrated animal feeding operations as defined in §122.23, discharges from concentrated aquatic animal production facilities as defined in §122.24, discharges to aquacultural projects as defined in §122.25, and discharges from silvicultural point sources as defined in §122.27.

56. It also includes "discharges through pipes, sewers, or other conveyances owned by a State, municipality, or other person which do not lead to a treatment works; and discharges through pipes, sewers or other conveyances leading into privately owned treatment works," but not discharges by indirect dischargers into POTWs. It then defines an "indirect discharger" as a "nondomestic discharger" introducing "pollutants" to a "publicly owned treatment works." In the preamble, EPA stated its intent to incorporate the "broadest possible definition of point source consistent with the legislative intent." 55 Fed. Reg. 47990 (Nov. 16, 1990), *quoted in* Washington Wilderness Coal. v. Hecla Mining Co., 870 F. Supp. 983, 988, 25 ELR 20661 (E.D. Wash. 1994).

Early in its implementation of the §402 permit program, EPA promulgated a rule exempting several categories of point sources from the requirement of securing permits:

> all silvicultural point sources; all confined animal feeding operations below a certain size; all irrigation return flows areas less than 3,000 contiguous acres or 3,000 noncontiguous acres that use the same drainage system; all nonfeedlot, nonirrigated agricultural point sources; and separate storm sewers containing only storm water uncontaminated by any industrial or commercial activity.[57]

Environmentalists challenged EPA's regulation as beyond the Agency's statutory authority. The U.S. Court of Appeals for the District of Columbia (D.C.) Circuit agreed, holding in *Natural Resources Defense Council, Inc. v. Costle* that EPA had no authority to exempt point sources discharging pollutants to navigable waters from the permit program.[58] The court did suggest, however, that the Agency had authority to define point sources, and subsequent courts have agreed.[59] EPA followed the court's suggestion by promulgating rules defining stormwater flow point sources,[60] CAFO point sources,[61] silviculture point sources,[62] and aquacultural point sources.[63] The new definition did not cover many of the discharges exempted by the overturned rule and they therefore remain excluded from the §402 program. Courts have held, however, that the rules are valid exercises of EPA's definitional authority rather than invalid attempts to exempt point source discharges from the permit program.[64] Thus, *Costle* was a pyrrhic victory for the environmental plaintiffs.

EPA made no attempt in its permit program regulations to define nonpoint source. The Agency has made some attempt to do so elsewhere, however. In its web page on nonpoint source, EPA states that "nonpoint source pollution generally results from land runoff, precipitation, drainage, seepage or hydrologic modification. The term 'nonpoint source' is defined to mean any source of water pollution that does not meet the legal definition

57. Natural Res. Def. Council, Inc. v. Costle, 568 F.2d 1369, 1372-73, 8 ELR 20028 (D.C. Cir. 1977). The regulation is found at 40 C.F.R. §125.4 (1975).
58. 568 F.2d 1369, 8 ELR 20028 (D.C. Cir. 1977).
59. *Id.* at 1372. Subsequent courts agreeing include *United States v. Earth Sciences, Inc.*, 599 F.2d 368, 372, 23 ELR 21526 (10th Cir. 1979); *Wild Fish Conservancy v. Quilcene Nat'l Fish Hatchery*, 2009 WL 3380655 (W.D. Wash. 2009); and *U.S. Public Interest Research Grp. v. Atlantic Salmon of Maine, Inc., LLC*, 215 F. Supp. 2d 239, 249, 32 ELR 20535 (D. Me. 2002).
60. 40 C.F.R. §122.26(b)(14).
61. *Id.* §122.23(c).
62. *Id.* §122.27.
63. *Id.* §122.24 & .25.
64. *Earth Sciences, Inc.*, 599 F.2d 368; *Wild Fish Conservancy*, 2009 WL 3380655.

of 'point source' in section 502(14) of the Clean Water Act."[65] The Agency continues:

> Nonpoint source (NPS) pollution, unlike pollution from industrial and sewage treatment plants, comes from many diffuse sources. NPS pollution is caused by rainfall or snowmelt moving over and through the ground. As the runoff moves, it picks up and carries away natural and human-made pollutants, finally depositing them into lakes, rivers, wetlands, coastal waters and ground waters.[66]

III. Judicial Interpretations of Point Source

A. *Preliminary Matters*

It should be emphasized at the outset that the CWA does not authorize EPA to regulate a point source, but only to regulate the discharge of any pollutant, of which "point source" is one of several elements. The Agency made the mistake of promulgating regulations requiring CAFO point sources to apply for CWA §402 permits regardless of whether the CAFOs discharged pollutants. In *National Pork Producers Council v. U.S. EPA,* the U.S. Court of Appeals for the Fifth Circuit struck down this regulation as beyond the Agency's authority.[67] Moreover, not all point sources require §402 permits, even if they add pollutants to navigable water. For example, industries that discharge pollutants to POTWs for treatment before discharge to navigable waters are held to be point sources,[68] but EPA's regulations define them as "indirect sources" and have exempted them from the §402 permit program, subject instead to regulation under the CWA §307(b) pretreatment program.[69] Courts are also in broad agreement that EPA has substantial

65. *See* EPA's web page, http://water.epa.gov/polwaste/nps/whatis.cfm.
66. *See* EPA's web page, http://water.epa.gov/polwaste/nps/index.cfm. Also useful are several EPA guidance documents quoted or cited in *Cordiano v. Metacon Gun Club, Inc.*, 575 F.3d 199, 220-21 (2d Cir. 2009).
67. 635 F.3d 738, 749-53, 41 ELR 20115 (5th Cir. 2011). *See also* Service Oil Inc. v. EPA, 590 F.3d 545, 550, 40 ELR 20002 (8th Cir. 2009); Waterkeeper Alliance v. U.S. EPA, 399 F.3d 486, 504, 35 ELR 20049 (2d Cir. 2005).
68. RSR Corp. v. Browner, 924 F. Supp. 504, 511, 26 ELR 21353 (S.D.N.Y. 1996), *aff'd*, No. 96-6186, 1997 WL 134413 (2d Cir. Mar. 26, 1997); *affirmance vacated without explanation*, No. 96-6186 (2d Cir. Apr. 17, 1997) (defendant subject to EPA's information demand authority over point sources under CWA §308, 33 U.S.C. §1318, even though it was an indirect discharger to a POTW). *See also* U.S. EPA v. Green Forest, Ark., 921 F.2d 1394, 1398, 21 ELR 20610 (8th Cir. 1990); United States v. Velsicol Chem. Corp., 438 F. Supp. 945, 946-47 (W.D. Tenn. 1976).
69. Assuming, that is, the discharge is into a municipal system providing treatment rather than into a storm sewer. Title 40 C.F.R. §122.2 defines "discharge of a pollutant" to exclude addition of pollutants by an "indirect discharger," which it in turn defines as a nondomestic discharger introducing pollutants to a "publicly owned treatment works." *See also* 40 C.F.R. §122.3(c), which excludes indirect dischargers

authority to define point sources.[70] Perhaps, there is not as much need for a broad interpretation for point sources as for the other elements of the offense, however, because if the other elements are not met, then the discharge escapes CWA jurisdiction altogether, but if the point source element is not met, then the discharge may still be under CWA jurisdiction as a nonpoint source.

Finally, a number of courts have held that conveyances are point sources only if they are the origins of the pollutants, that is, they convey the pollutants from their point of origin. This idea was first espoused by EPA in *National Wildlife Federation v. Gorsuch,* where the Agency argued that there is an addition of a pollutant from a point source "only if the point source itself physically introduces a pollutant into water from the outside world."[71] The D.C. Circuit gave deference to EPA's interpretation. Although the interpretation in that case concerned the meaning of "addition," it could just as well have been an interpretation of "point source" as the first conveyance to introduce the pollutant into water or at least into navigable water. The Supreme Court decisively rejected this notion in *Miccosukee Tribe,* in which it held that "a point source need not be the original source of the pollutant: it need only convey the pollutant to navigable water."[72] Since *Miccosukee Tribe,* courts have usually rejected the notion, as many did earlier.

---

from the requirement to secure a §402 permit. The reason EPA excludes indirect dischargers from the permit program is that CWA §307(b) requires indirect dischargers to pretreat their wastes before discharging them to POTWs to prevent them from interfering with the POTWs or to allow the indirect sources' wastes to pass through the POTW untreated. The section also requires indirect sources to treat their wastes meeting technology-based standards to remove toxic pollutants.

Finally, CWA §402(b)(8) and (9) contemplate that POTWs will regulate indirect discharges by permits or other means to require compliance with both the interference and pass-through prohibitions and the toxic pollutant pretreatment standards. *See also* EPA's pretreatment program regulations at 40 C.F.R. pt. 403.

70. This was first recognized by *Natural Res. Def Council, Inc. v. Costle*, 568 F.2d 1369, 1377, 8 ELR 20028 (D.C. Cir. 1978), which held that EPA could not exempt point sources adding pollutants to navigable water from the §402 permit program, but had broad discretion to define point source. *See also* U.S. v. Earth Sciences, Inc., 599 F.2d 368, 372, 23 ELR 21526 (10th Cir. 1979).

71. 693 F.2d 156, 175, 13 ELR 20015 (D.C. Cir. 1982), North Carolina Shellfish Growers Ass'n v. Holly Ridge Assocs., LLC, 278 F. Supp. 2d 654, 680 (E.D.N.C. 2003) (although CWA violations cannot result from purely passive developments on a defendant's property, the active moving of land is sufficient to trigger liability under the Act). *Contra* Sierra Club v. El Paso Gold Mines, Inc., 421 F.3d 1133, 1145, 35 ELR 20175 (10th Cir. 2005) ("This is a case where if you own the leaky 'faucet,' you are responsible for its 'drips.'"); Dague v. City of Burlington, 935 F.2d 1343, 1354-55, 21 ELR 21133 (2d Cir. 1991).

72. South Fla. Water Mgmt. Dist. v. Miccosukee Tribe of Indians of Fla., 541 U.S. 95, 105, 34 ELR 20021 (2004). *See supra* Chapter 1, Section III.A.3.a.

B. *Held to Be Point Sources*

Courts have held the following to be point sources: airplanes dropping bombs or spraying pesticides[73]; aquacultural facilities[74]; barges[75]; breaks in berms[76]; bulldozers, cranes, dump trucks, loaders, plows, tractors, and other construction and earth-moving equipment[77]; canals, pumps, and levees[78]; collection pond and tank cracks, leaks, bypasses, and overflows[79]; CAFOs[80]; construction sites[81]; dams[82]; ditches[83]; entire facilities or indus-

73. Romero-Barcelo v. Brown, 478 F. Supp. 646, 664 (D.P.R. 1979), *rev'd on other grounds sub nom.* Weinberger v. Romero-Barcelo, 643 F.2d 835, 11 ELR 20391 (1st Cir. 1981), *rev'd on other grounds,* 456 U.S. 305, 12 ELR 20538 (1982); Peconic Baykeeper, Inc. v. Suffolk Cnty., 600 F.3d 180, 188-89, 40 ELR 20098 (2d Cir. 2010); League of Wilderness Defenders v. Forsgren, 309 F.3d 1181, 1183-85 (9th Cir. 2002).
74. U.S. Public Interest Research Grp. v. Atlantic Salmon of Maine, LLC, 215 F. Supp. 2d 239, 32 ELR 20535 (D. Me. 2002).
75. United States v. West Indies Transportation, Inc., 127 F.3d 299, 308-09, 28 ELR 20202 (3d Cir. 1997).
76. Reynolds v. Rick's Mushroom Serv., Inc., 246 F. Supp. 2d 449 (E.D. Pa. 2003); O'Leary v. Moyer's Landfill, Inc., 523 F. Supp. 642, 655, 12 ELR 20239 (E.D. Pa. 1981); United States v. Oxford Royal Mushroom Prods., Inc., 487 F. Supp. 852, 954, 10 ELR 20549 (E.D. Pa. 1980).
77. Parker v. Scrap Metal Processors, Inc., 386 F.3d 993, 1009, 34 ELR 20104 (11th Cir. 2004); Borden Ranch P'ship v. U.S. Army Corps of Eng'rs, 261 F.3d 810, 814, 32 ELR 20011 (9th Cir. 2001), *aff'd,* 537 U.S. 99 (2002); United States v. Pozsgai, 999 F.2d 719, 726 n.6, 23 ELR 21012 (3d Cir. 1993); Avoyelles Sportsmen's League, Inc. v. Marsh, 715 F.2d 897, 922, 13 ELR 20942 (5th Cir. 1983); United States v. Lambert, 915 F. Supp. 797, 26 ELR 21116 (S.D. W. Va. 1996); United States v. Sinclair Oil, 767 F. Supp. 200, 21 ELR 21323 (D. Mont. 1990); United States v. Weisman, 489 F. Supp. 1331, 1336-37, 10 ELR 20698 (M.D. Fla. 1980); United States v. Holland, 373 F. Supp. 665, 668, 4 ELR 20710 (M.D. Fla. 1074).
78. South Fla. Water Mgmt. Dist., v. Miccosukee Tribe of Indians of Fla., 541 U.S. 95, 34 ELR 20021 (2004); Friends of the Everglades v. South Fla. Water Mgmt. Dist., 570 F.3d 1210, 39 ELR 20118 (11th Cir. 2009).
79. *Moyer's Landfill,* 523 F. Supp. at 655.
80. National Pork Producers' Council v. U.S. EPA, 635 F.3d 738, 41 ELR 20115 (5th Cir. 2010); Concerned Area Residents for Env't v. Southview Farm, 34 F.3d 114, 123, 24 ELR 21480 (3d Cir. 1994); Community Ass'n for Restoration of Env't v. Henry Bosma Dairy, Inc., 305 F.3d 943, 955 (9th Cir. 2002); Carr v. Alta Verde Industries, Inc., 931 F.2d 1055, 21 ELR 21005 (5th Cir. 1991); Assateague Coastkeeper v. Hudson Farm, 727 F. Supp. 2d 433, 439-40, 40 ELR 20208 (D. Md. 2010).
81. North Carolina Shellfish Growers Ass'n v. Holly Ridge Assocs., 278 F. Supp. 2d 654, 679 (E.D.N.C. 2003); Molokai Chamber of Commerce v. Kukui, 891 F. Supp. 1389, 1401, 26 ELR 20303 (D. Haw. 1995).
82. S.D. Warren Co. v. Maine Bd. of Envtl. Prot., 547 U.S. 370, 36 ELR 20089 (2006); Oregon Natural Desert Ass'n v. U.S. Forest Serv., 550 F.3d 778, 787 n.4, 39 ELR 20297 (9th Cir. 2008); Greenfield Mills, Inc. v. Macklin, 361 F.3d 934, 34 ELR 20022 (7th Cir. 2004); Committee to Save Mokelumne River v. East Bay Mun. Util. Dist., 13 F.3d 305, 24 ELR 20225 (9th Cir. 1993); National Wildlife Fed'n v. Gorsuch, 693 F.2d 156, 165, 13 ELR 20015 (D.C. Cir. 1982); *North Carolina Shellfish Growers,* 278 F. Supp. 2d at 679; South Carolina Wildlife Fed'n v. Alexander, 457 F. Supp. 118, 126, 8 ELR 20757 (D.S.C. 1978); Del-Aware Unlimited v. Commissioner, Dep't of Envtl. Res., 96 Pa. Cmwlth. 361, 381-82 (Pa. 1986).
83. *Henry Bosma Diary,* 305 F.3d at 955; National Ass'n of Home Builders v. U.S. Army Corps of Eng'rs, 699 F. Supp. 2d 209, 215-16, 40 ELR 20104 (D.D.C. 2010), *vacated on other grounds,* 663 F.3d 470

trial plants[84]; facilities discharging to POTWs[85]; fields overflowing with applied manure[86]; human beings[87]; manure spreaders[88]; mine adits, shafts, pits, and tunnels[89]; mushroom-growing facilities[90]; overflowing collection and tailings ponds and lagoons[91]; piles of debris, material, and waste rock[92]; pipes and hoses[93]; railway culverts[94]; rifle ranges[95]; septic systems[96]; sluice boxes[97]; spray apparatuses attached to trucks and helicopters[98]; sump overflows[99]; systems for circulating, channeling, or draining stormwater

---

(D.C. Cir. 2011); *North Carolina Shellfish Growers,* 278 F. Supp. 2d at 679; Environmental Prot. Info. Ctr. v. Pacific Lumber Co., 469 F. Supp. 2d 803, 821, 37 ELR 20012 (N.D. Cal. 2007).

84. United States v. Earth Sciences, Inc., 599 F.2d 368, 23 ELR 21526 (10th Cir. 1979); *North Carolina Shellfish Growers,* 278 F. Supp. 2d 654; Reynolds v. Rick's Mushroom Serv., Inc., 246 F. Supp. 2d 449 (E.D. Pa. 2003); Williams Pipeline Co. v. Bayer Corp., 964 F. Supp. 1300, 1318-20 (S.D. Iowa 1997).
85. RSR Corp. v. Browner, 924 F. Supp. 504, 511, 26 ELR 21353 (S.D.N.Y. 1996).
86. Concerned Area Residents for Env't v. Southview Farm, 34 F.3d 114, 118-19, 24 ELR 21480 (3d Cir. 1994) (3d Cir. 2004); *Henry Bosma Dairy,* 305 F.3d at 955; *Rick's Mushroom Serv.,* 246 F. Supp. 2d at 456-57; United States v. Oxford Royal Mushroom Prods., Inc., 487 F. Supp. 852, 854, 10 ELR 20549 (E.D. Pa. 1980).
87. United States v. Villegas, 784 F. Supp. 6, 8-11, 22 ELR 21027 (E.D.N.Y. 1991), *rev'd sub nom.* United States v. Plaza Health Labs., Inc., 3 F.3d 643, 23 ELR 21526 (2d Cir. 1991).
88. League of Wilderness Defenders v. Forsgren, 309 F.3d at 1184; *Concerned Area Residents,* 34 F.3d 114 at 24; *Henry Bosma Dairy,* 305 F.3d at 955.
89. Sierra Club v. El Paso Gold Mines, Inc., 421 F.3d 1133, 1441, 35 ELR 20175 (10th Cir. 2005); West Virginia Highlands Conservancy v. Huffman, 588 F. Supp. 2d 678, 39 ELR 20201 (S.D. W. Va. 2009), *aff'd,* 625 F.3d 159 (4th Cir. 2010); Beartooth Alliance v. Crown Butte Mines, 904 F. Supp. 1168, 1173-74, 26 ELR 20639 (D. Mont. 1995).
90. Reynolds v. Rick's Mushroom Serv., Inc., 246 F. Supp. 2d 449 (E.D. Pa. 2003); Oxford Royal Mushroom Prods., Inc., 487 F. Supp. 852, 10 ELR 20549 (E.D. Pa. 1980).
91. Washington Wilderness Coal. v. Hecla Mining Co., 870 F. Supp. 983, 988, 25 ELR 20661 (E.D. Wash. 1994); Fishel v. Westinghouse Elec. Corp., 640 F. Supp. 442, 446, 16 ELR 20634 (M.D. Pa. 1986); O'Leary v. Moyer's Landfill, Inc., 523 F. Supp. 642, 655, 12 ELR 20239 (E.D. Pa. 1981).
92. Parker v. Scrap Metals Processors, Inc., 386 F.3d 993, 1009, 34 ELR 20104 (11th Cir. 2004); Bravos v. EPA, 324 F.3d 1166 (10th Cir. 2003); Trustees for Alaska v. EPA, 749 F.2d 549, 558, 15 ELR 20146 (9th Cir. 1984); Sierra Club v. Abston Constr. Co., Inc., 620 F.2d 41, 45, 10 ELR 20552 (5th Cir. 1980); United States v. Earth Sciences, Inc., 599 F.2d 368, 9 ELR 20542 (10th Cir. 1979); *Washington Wilderness Coal.,* 870 F. Supp. at 988.
93. United States v. Lippold, 2007 WL 3232483 at **2-3 (C.D. Ill. 2007) (uncontested).
94. Dague v. City of Burlington, 935 F.2d 1343, 1355, 21 ELR 21133 (2d Cir. 1991).
95. Stone v. Naperville Park Dist., 38 F. Supp. 2d 651, 655 (N.D. Ill. 1999); Long Island Soundkeeper Fund, Inc. v. New York Athletic Club, 1996 WL 131863 (S.D.N.Y. 1996).
96. United States v. Agosto-Vega, 617 F.3d 541, 40 ELR 20222 (1st Cir. 2010); United States v. Lucas, 516 F.3d 316, 332, 38 ELR 20041 (5th Cir. 2008); Friends of the Sakonnet v. Dutra, 738 F. Supp. 623, 629-30, 21 ELR 20055 (D.R.I. 1990).
97. *Trustees,* 749 F.2d at 558.
98. Peconic Baykeeper, Inc. v. Suffolk Cnty., 600 F.3d 180, 188-89, 40 ELR 20098 (2d Cir. 2010); National Cotton Council of Am. v. U.S. EPA, 553 F.3d 927, 39 ELR 20006 (6th Cir. 2009); League of Wilderness Defenders/Blue Mountain Biodiversity Project v. Forsgren, 309 F.3d 1181(9th Cir. 2002).
99. United States v. Earth Sciences, 599 F.2d 368, 374, 23 ELR 21526 (10th Cir. 1979); Washington Wilderness Coal. v. Hecla Mining Co., 870 F. Supp. 983, 987-89, 25 ELR 20661 (E.D. Wash. 1994).

runoff[100]; toilets connected to storm sewers[101]; tank trucks[102]; trucks with attached spreaders or sprayers[103]; and turbines.[104]

C. Held Not to Be Point Sources

Courts have held the following not to be point sources: agricultural runoff from farms and CAFO fields[105]; aquacultural facilities[106]; buildings storing trash[107]; cattle[108]; dams[109]; human beings[110]; piles of coal and rock[111]; rifle ranges[112]; seeps from a swamp into groundwater[113]; silvicultural runoff[114]; toll booth stations on bridges[115]; and waste pits with covers to minimize precipitation intrusion.[116]

100. Northwest Envtl. Def. Ctr. v. Brown, 640 F.3d 1063, 1069-73, 41 ELR 20178 (9th Cir. 2011); Greater Yellowstone Coal. v. Lewis, 638 F.3d 1143, 41 ELR 20059 (9th Cir. 2010); Carr v. Alta Verde Indust., Inc., 931 F.2d 1055, 21 ELR 21005 (5th Cir. 1991); Sierra Club v. Abston Constr. Co., Inc., 620 F.2d 41, 10 ELR 20552 (5th Cir. 1980); Sierra Club v. El Paso Gold Mines, Inc., 421 F.3d 1133, 35 ELR 20175 (10th Cir. 2005); Environmental Prot. Info. Ctr. v. Pacific Lumber Co., 469 F. Supp. 2d 803, 37 ELR 20012 (N.D. Cal. 2007); North Carolina Shellfish Growers Ass'n v. Holly Ridge Assocs., 278 F. Supp. 2d 654 (E.D.N.C. 2003).
101. United States v. Ortiz, 427 F.3d 1278, 35 ELR 20220 (10th Cir. 2005).
102. United States v. Agosto-Vega, 617 F.3d 541, 40 ELR 20222 (1st Cir. 2010) (uncontested); Concerned Area Residents for Env't v. Southview Farm, 34 F.3d 114, 24 ELR 21480 (3d Cir. 1994); United States v. Lippold, 2007 WL 3232483 at **2-3 (C.D. Ill. 2007) (uncontested).
103. League of Wilderness Defenders v. Forsgren, 309 F.3d 1181, 1184 (9th Cir. 2002).
104. S.D. Warren Co. v. Maine Bd. of Envtl. Prot., 547 U.S. 370, 36 ELR 20089 (2006); South Carolina Wildlife Fed'n v. Alexander, 457 F. Supp. 118, 126-27, 8 ELR 20757 (D.S.C. 1978).
105. National Pork Producers' Council v. U.S. EPA, 635 F.3d 738, 41 ELR 20115 (5th Cir. 2011).
106. Association to Protect Hammersly, Eld, & Totten Inlets v. Taylor Res., Inc., 299 F.3d 1007, 1018-19 (9th Cir. 2002); Wild Fish Conservancy v. Quilcene Natural Fish Hatchery, 2009 WL 3380655 (W.D. Wash. 2009).
107. Hudson Riverkeeper Fund, Inc. v. Harbor at Hastings Assocs., 917 F. Supp. 251, 257, 26 ELR 21120 (S.D.N.Y. 1996).
108. Oregon Natural Desert Ass'n v. U.S. Forest Serv., 550 F.3d 788, 39 ELR 20297 (9th Cir. 2008).
109. Froebel v. Meyer, 217 F.3d 928, 30 ELR 20746 (7th Cir. 2000); National Wildlife Fed'n v. Gorsuch, 693 F.2d 156, 13 ELR 20015 (D.C. Cir. 1982) (dams may be point sources, but they do not add pollutants to downstream flows).
110. United States v. Plaza Health Labs., Inc., 3 F.3d 643, 646, 23 ELR 21526 (2d Cir. 1993).
111. Appalachian Power Co. v. Train, 545 F.2d 1351, 6 ELR 20732 (4th Cir. 1976).
112. Cordiano v. Metacon Gun Club, Inc., 575 F.3d 199 (2d Cir. 2009) (insufficient proof to support allegations).
113. Williams Pipeline Co., v. Bayer Corp., 964 F. Supp. 1300, 1326 (S.D. Iowa 1997).
114. Decker v. Northwest Envtl. Def. Ctr., 133 U.S. 1326, 43 ELR 20062 (2013).
115. U.S. EPA ex rel. McKeown v. Port Auth. of N.Y. & N.J., 162 F. Supp. 2d 173, 188-89 (S.D.N.Y. 2001), *aff'd sub nom.* McKeown v. Delaware Bridge Auth., 23 F. App'x 81 (2d Cir. 2001).
116. Greater Yellowstone Coal. v. Lewis, 638 F.3d 1143, 1153, 41 ELR 20059 (9th Cir. 2010) (some discharges were point sources and others nonpoint sources).

D. *Easy Decisions, Difficult Decisions*

The definition of "point source" as a "discernible, confined and discrete conveyance" is a straightforward concept that is easy to apply in most factual situations. Its list of examples is long and inclusive, creating a large set of per se point sources. That set of point sources and conveyances analogous to them cover most obvious means of adding pollutants to navigable waters. Not surprisingly, more than one-half of the courts interpreting point source reached their decisions easily, based on two or fewer interpretive devices or two or fewer precedents.[117] Few decisions conduct an in-depth interpretation of point source because "any identifiable conveyance" will suffice.[118] That is not the end of the inquiry, however, since even clear point sources may be subject to a statutory or regulatory exemption, such as the exemption for "agricultural stormwater discharges."[119] The difficult cases are signaled when judicial decisions hold similar discharges to be both point sources and nonpoint sources, and when decisions use five or more interpretive devices or cite five or more precedents to decide.[120]

The difficult decisions fall into two general categories. One category involves whether human activity sufficiently collects and channels pollutants in stormwater runoff for the conveyance to be considered a point source; for example, runoff from rock and coal piles.[121] The statutory history describes nonpoint sources solely in terms of runoff. But while a large number of decisions fall into this first category, most decisions fall into the second category, examining whether a conveyance is a point source in cases where there is no runoff. In the statute broadly.

Most of the earliest opinions at both the federal district court and circuit courts of appeal levels interpreting point source were decided using a plain meaning analysis. As soon as there were reported decisions on the issue,

117. Forty of the 72 decisions interpreting point source used two or less interpretive devices, count the number of times two or less appears *infra* in Column F of Table B for Point Source. Thirty seven of the decisions use two or less precedents, *see infra* Column G of Table B for Point Source.

118. Peconic Baykeeper, Inc. v. Suffolk Cnty., 600 F.3d 180, 188, 40 ELR 20098 (2d Cir. 2010) (citing Cordiano v. Metacon Gun Club, Inc., 575 F.3d 199, 219 (2d Cir. 2009)). Less than 10% of the decisions interpreting point source use five or more interpretive devices. For the 72 decisions interpreting point source, only 5 decisions use 5 or more devices, *see infra* Column F of Table B for Point Source.

119. CWA §502(14).

120. *Compare* the categories accompanying *supra* notes 73-104, held to be point sources, with the categories accompanying notes 105-16, held not to be point sources. For instance, both include rock piles, notes 92 and 111. For decisions employing high numbers of interpretive devices or precedents, *see infra* Columns F and G in Table B for Point Source.

121. *See, e.g.*, Sierra Club v. Abston Constr. Co., Inc., 620 F.2d 41, 44-47, 10 ELR 20552 (5th Cir. 1980) (may be point sources). *But see* Appalachian Power v. Train, 545 F.2d 1351, 1372-74, 6 ELR 20732 (4th Cir. 1976) (may not be point sources).

however, precedent became the most commonly used interpretive device. The third most commonly used interpretive device was to interpret statutes broadly to effectuate the statutory purpose. In all, 18 canons were used interpreting point source.[122] More than one-half of the decisions used only one or two interpretive devices, usually precedent and plain meaning or sometimes broad interpretation. These were easy decisions. The U.S. Court of Appeals for the Tenth Circuit decision in *United States v. Earth Sciences, Inc.* claimed a particularly broad reach for the term to embrace "the broadest possible definition of any identifiable conveyance from which pollutants might enter the waters of the United States,"[123] which many decisions since have quoted.[124] And another court noted that EPA, in its promulgation of the regulations for stormwater flow, "has stated its intent 'to embrace the broadest possible definition of point source consistent with the legislative intent of the CWA.'"[125] Of course, there are bounds to broad interpretation, especially when Congress itself has articulated them, as it did in the CWA by providing for nonpoint sources.[126]

The chapter will analyze in turn the first category of difficult decisions, those involving precipitation runoff; the second category of point source decisions, those having nothing to do with runoff; and then suggest regulatory definitions of point source and nonpoint source.

E. *Stormwater Runoff Decisions*

1. Runoff

Stormwater runoff is better described as precipitation runoff because it consists of the natural flow of both rain after it hits the ground and the flow of water from snow, sleet, or ice after they melt on the ground. Senator Muskie in his above-quoted statement said that agricultural runoff "is a nonpoint

122. *See infra* Column E of Table B for Point Source, in which interpretive devices 1-8, 10, 11, 13, 16-18, and 21-24 are cited.
123. 599 F.2d 368, 373, 23 ELR 21526 (10th Cir. 1979).
124. Cordiano v. Metacon Gun Club, Inc., 575 F.3d 199, 219 (2d Cir. 2009); Dague v. City of Burlington, 935 F.2d 1343, 1354, 21 ELR 21133 (2d Cir. 1991); United States v. West Indies Transp., Inc., 127 F.3d 299, 309, 28 ELR 20202 (3d Cir. 997).
125. Beartooth Alliance v. Crown Butte Mines, 904 F. Supp. 1168, 1173, 26 ELR 20639 (D. Mont. 1995) (quoting 55 Fed. Reg. 47990, 47997 (Nov. 16, 1990)). *See also* Washington Wilderness Coal. v. Hecla Mining Co., 870 F. Supp. 983, 988, 25 ELR 20661 (E.D. Wash. 1994).
126. The definition of point source "cannot be interpreted so broadly as to read the point source requirement out of the statute," *Metacon Gun Club*, 575 F.3d at 219. For a discussion of how Congress realized that some of the goals in the CWA were unattainable and unrealistic, see *National Wildlife Fed'n v. Gorsuch*, 693 F.2d 156, 179-82, 13 ELR 20015 (D.C. Cir. 1982).

source." Courts have also called runoff "nonpoint pollution."[127] Both statements are inaccurate in terms of statutory language and can confuse the interpretive analysis. For runoff to be a "nonpoint source," stormwater would have to be a non-conveyance or an indiscernible, unconfined, and indiscrete conveyance. For runoff to be "a nonpoint pollutant," stormwater would have to be a "pollutant." The statutory definitions of point source and pollutant, however, do not include stormwater itself; rather, it is something that may be conveyed by a point source and a medium in which a pollutant may be suspended and carried. Some courts have told us that roads are the prototypical nonpoint sources:

> The most common example of nonpoint source pollution is the residue left on roadways by automobiles. Small amounts of rubber are worn off of the tires of millions of cars and deposited as a thin film on highways: minute particles of copper dust from brake linings are spread across roads and parking lots each time a driver applies the brakes; drips and drabs of oil and gas ubiquitously stain driveways and streets. When it rains, the rubber particles and copper dust and gas and oil wash off of the street and are carried along by runoff in a polluted soup, winding up in creeks, rivers, bays, and the ocean.[128]

Notice the U.S. Court of Appeals for the Ninth Circuit is discussing nonpoint source pollution. The precipitation here is natural. The road is not natural, nor are the pollutants. Is the road a conveyance? Vehicles traveling the road are conveyances of passengers and goods. Does the road in turn convey the vehicles? Not in the ordinary meaning of convey or conveyance. Does the road convey the precipitation or the pollutants? Most roads are crowned so that precipitation runs off to the side. Roads are defined, confined, and discrete as they run from one place to another, but are they defined, confined, and discrete as they convey stormwater runoff to their sides? Again, not in the ordinary meaning of convey or conveyance. Under this analysis, runoff from roads is not conveyed by the roads and the roads are nonpoint sources.

But that is not the end of the analysis. Once runoff has flowed off the roads, it often flows into roadside ditches, designed to carry stormwater runoff to the nearest surface water, often discharged from the ditches or culverts, both of which are point sources.[129] In the absence of ditches, roads might

127. Northwest Envtl. Def. Ctr. v. Brown, 640 F.3d 1063, 1076, 41 ELR 20178 (9th Cir. 2011); *Gorsuch*, 693 F.2d at 177.

128. League of Wilderness Defenders/Blue Mountains Biodiversity Project v. Forsgren, 309 F.3d 1181, 1184 (9th Cir. 2002); *see also* United States v. Earth Sciences, Inc., 599 F.2d 368, 371, 23 ELR 21526 (10th Cir. 1979); U.S. EPA ex rel. McKeown v. Port Auth. of N.Y. & N.J., 162 F. Supp. 2d 173, 189 (S.D.N.Y. 2001).

129. "Ditches" are included in the list of examples of "point sources" in CWA §502(14). Culverts have been held to be point sources, *Dague*, 935 F.2d 1354, not surprisingly since culverts are pipes that are

flood and some of the runoff would flow to surface waters, while some of it would percolate into the ground and some of it would evaporate. Courts and EPA are fond of citing roads as typical nonpoint sources,[130] but their analyses are not complete. Their reasoning

would keep roads out of the CWA §402 program, but not roadside drainage ditches and culverts. After all, municipal storm sewers carry stormwater runoff from street gutters to navigable waters and municipal storm sewers are point sources that require permits under §402(p). EPA has exempted such flows from the requirement to secure §402 permits, however, as long as they are composed "entirely of stormwater."[131]

Most courts interpret point source in terms of the general distinctions between point sources and nonpoint sources found in the legislative history and hinted at by the statute itself. At a very general level, this is the distinction between stormwater runoff that is not collected or channeled by human activity, considered to be from a nonpoint source[132]; and stormwater runoff that is collected or channeled by human activity, considered to be from a point source.[133] While this is a useful distinction, it is sometimes not dispositive because courts may believe that conveyances of stormwater flow created or channeled by minor human activity or by human activity not aimed at controlling water is insufficient to constitute a point source.[134] Moreover, many point source decisions do not involve stormwater flow at all, even decisions holding that there is no point source.

2. Getting Beyond Semantics: Dropping the Point Source/ Nonpoint Source Distinction

The problem with interpreting point source by distinguishing between point sources and nonpoint sources is that Congress did not define nonpoint sources. They could be indiscernible, unconfined, and indiscrete conveyances or they might not be conveyances at all. When the legislative history and court decisions speak of nonpoint source pollution or a nonpoint source

---

the first example of point sources listed in the statutory definition of point source, CWA §502(14). Issues related to ditches as point sources are addressed *infra* at section IIILF1.

130. *Earth Sciences*, 599 F.2d at 371; *Port Auth. of N.Y. & N.J.*, 162 F. Supp. 2d at 189; *Hecla*, 870 F. Supp. at 988 (E.D. Wash. 1994). *See also* EPA's web page at http://water.epa.gov/polwaste/nps/roadshwys.

131. 40 C.F.R. §122.26(a)(9)(i).

132. *Northwest Envtl. Def. Ctr.*, 640 F.3d at 1070-71; Greater Yellowstone Coal. v. Lewis, 628 F.3d 1143, 1153, 41 ELR 20059 (9th Cir. 2010) (seep through cover over mining pit).

133. *Northwest Envtl. Def. Ctr.*, 640 F.3d at 1070-71; *Greater Yellowstone*, 628 F.3d at 1152 (mine drainage system); Sierra Club v. El Paso Gold Mines, Inc., 421 F.3d 1133, 1142-44, 35 ELR 20175 (10th Cir. 2005); Sierra Club v. Abston Constr. Co., 620 F.2d 41, 45, 10 ELR 20552 (5th Cir. 1980).

134. Appalachian Power Co. v. Train, 545 F.2d 1351, 1372-74, 6 ELR 20732 (4th Cir. 1978).

pollutant, they are referring to pollution and pollutants, not to conveyances. The easiest way to resolve this semantic and conceptual problem is simply to eliminate the concept of nonpoint sources from the interpretation of point source. We can still use the factors that courts consider to determine whether a conveyance is a point source. The benefits to nonpoint sources of evading regulation under §402 will not be lost to them, because if there is no point source, there is no regulation under §402. Moreover, Congress has provided that some discharges from point sources are not regulated under §402, and EPA has promulgated regulations providing that other discharges from point sources are exempted from the requirement of securing a §402 permit.[135] EPA has also extended indefinitely the exemption for discharges composed entirely of stormwater from the §402 permit requirement that Congress provided only until October 1, 1994.[136] The chapter's analysis will proceed without distinguishing between point sources and nonpoint sources. It will, however, use many of the reasons courts have used to make that purported distinction. But it will use them as factors only to determine whether a particular discharge is through a point source rather than to determine whether it is through a point source or nonpoint source.

3. The Factors Test

To deal with the difficult distinction between discharges by point and nonpoint sources, courts have developed an analysis weighing specific factors under the circumstances of the cases.[137] They have not labeled their analyses as a factors-weighing test, but the factors they use are clear, although the weight courts give them is not clear or well-articulated. Although courts have used these factors to determine whether a flow is discharged by a point source or a nonpoint source, they can be applied to the narrower question of whether a flow is discharged through a point source. As discussed below, judicially developed the factors are whether: (1) human activity collects and channels stormwater flow; (2) a single source discharges pollutants or multiple, dispersed sources discharge pollutants; (3) an artificial system channels water; (4) an industrial or municipal activity produces the pollutants dis-

135. 40 C.F.R. §122.3. Query whether EPA has the authority to do so. *See generally* Natural Res. Def. Council, Inc. v. Costle, 568 F.2d 1369, 8 ELR 20028 (D.C. Cir. 1977).

136. CWA §402(p); 40 C.F.R. §122.26(a)(9)(i). Again, query whether EPA has the authority to do so. *See generally Costle*, 568 F.2d 1369.

137. EPA independently developed a more abbreviated factors test at an early date, but has not elaborated on it since. It wrote that discharges of pollutants are by nonpoint sources when the discharges are (1) "induced by natural sources"; (2) "not traceable to any discrete or identifiable facility"; and (3) "better controlled through the utilization of best management practices." 44 Fed. Reg. 24700 (June 18, 1976), *quoted in Envtl. Def. Ctr.*, 640 F.3d at 1075.

charged; and (5) the activity producing pollutants is more susceptible to end-of-pipe treatment technology and numerical effluent limitations or to best management practice. The first factor decides most cases involving stormwater flow, but is irrelevant in cases not involving stormwater runoff flow. The analysis below concludes that the fourth factor is not useful and should be disregarded. Courts have considered the remaining factors regardless of whether cases involve runoff stormwater flow. By their very nature, factors tests are not entirely satisfactory because they do not always conclusively point to one answer and, therefore, do not produce uniform results. The less than satisfactory nature of a factors test to interpret "point source," however, is mitigated by the list of examples in the definition, which point to a single answer in most cases. The factors are discussed in more detail below.

a. The Human Activity Factor

No factor has been more important or cited more often in determining whether a discharge is from a point source than whether human activity collects and channels the flow of stormwater into navigable waters.[138] If the stormwater is collected and channeled by human activity into navigable waters, it is usually from a point source. If stormwater reaches navigable water entirely through natural means, it is usually not from a point source. This reflects both legislative history and EPA's regulatory definition of "discharge of a pollutant" to include "surface runoff which is collected or channeled by man . . . ."[139] It should be remembered, however, that the statutory definition of point source says nothing about human activity; indeed, the fifth element of the §301(a) offense, "by any person," independently supplies that requirement. This factor, therefore, conflates the fourth and fifth elements of the CWA offense.

The meaning of the human activity factor is explored in greatest depth by the Fifth Circuit in *Sierra Club v. Abston Construction Co., Inc.*[140] In that case, the defendant sprayed liquid chemicals on a pile of ore to leach gold and collected the leachate to extract gold from it in a laboratory. It operated a supposedly closed loop system of sumps, ditches, hoses, pumps, and ponds to contain, collect, and recycle the leachate, thus preventing it from entering navigable water. The liquid overflowed its restraints during a storm event,

138. *Greater Yellowstone*, 628 F.3d at 1152; Cordiano v. Metacon Gun Club, 575 F.3d 199, 221 (2d Cir. 2009); Committee to Save Mokelumne River v. East Bay Mun. Util. Dist., 13 F.3d 305, 308, 24 ELR 20225 (9th Cir.1993); Environmental Prot. Info. Ctr. v. Pacific Lumber Co., 469 F. Supp. 2d 803, 822, 37 ELR 20012 (N.D. Cal. 2007).

139. 40 C.F.R. §122.2.

140. Sierra Club v. Abston Constr., Inc., 620 F.2d 41, 10 ELR 20552 (5th Cir. 1989).

either because the system was undersized, inadequately designed, poorly maintained, or simply malfunctioned. The operation also had piles of waste rock from which pollutants flowed into the stream in stormwater runoff. The court characterized the ultimate question as "whether pollutants were discharged from 'discernible, confined, and discrete conveyance(s)' either by gravitational or nongravitational means."[141] The plaintiff argued that because the original source of the pollutants was from a human activity, they were added from a point source. The defendant argued that because the pollutants were carried by rainfall and natural erosion, they were not from a point source, even if they were created as a result of human activity.

The United States, as amicus curiae, argued that because the "surface runoff [was] collected or channeled by the operator [it] constitutes a point source discharge."[142] The *Abston* court agreed with the United States. It held that:

> Gravity flow, resulting in a discharge into a navigable body of water, may be part of a point source discharge if the miner at least initially collected or channeled the water and other materials . . . [or] where miners design spoil piles from discarded overburden such that, during periods of precipitation, erosion of spoil pile walls results in discharges into a navigable body of water by means of ditches, gullies and similar conveyances . . . Nothing in the Act relieves miners from liability simply because the operators did not actually construct those conveyances, so long as they are reasonably likely to be the means by which pollutants are ultimately deposited into a navigable body of water. Conveyances of pollution formed either as a result of natural erosion or by material means, and which constitute a component of a mine drainage system, may fit the statutory definition.[143]

The court concluded that while "the point source definition 'excludes unchanneled and uncollected surface waters,' surface runoff from rainfall, when collected or channeled . . . in connection with mining activities, constitutes point source pollution."[144]

Overflows of stormwater from pollutant collection and storage systems are commonly held to be from point sources, even though the overflows are from storm events and are from facilities constructed for the purpose of collecting and channeling stormwater.[145] EPA regulations, however, may give some

141. *Id.* at 45.
142. *Id.* at 45.
143. Id.
144. *Id.* at 47.
145. Reynolds v. Rick's Mushroom Serv., Inc., 246 F. Supp. 2d 449, 456-58 (E.D. Pa. 2003); Fishel v. Westinghouse Elec. Corp., 640 F. Supp. 442, 16 ELR 20634 (M.D. Pa. 1986); O'Leary v. Moyers's Landfill, Inc., 523 F. Supp. 642, 12 ELR 20239 (E.D. Pa. 1981); United States v. Oxford Royal Mushroom Prods., Inc., 487 F. Supp. 952, 10 ELR 20549 (E.D. Pa. 1980).

relief from overflows in extraordinary storm events, such as its exemption of the zero discharge limit from CAFOs in the event of a 25-year storm event.[146]

Mining and rock piles are a subcategory in the human activity category. This factor commonly occurs in cases involving mining waste, especially with regard to runoff from rock piles of various types. The argument that the CWA generally exempted mining wastes from the §402 permit program was decisively rejected at an early date (1979) by the Tenth Circuit in *United States v. Earth Sciences, Inc.*[147] Whether piles of rock at mining and other sites are point sources, however, has been a continuing issue and one that is fact-dependent because runoff from such piles results both from natural sources, such as precipitation and gravity, and human activity, such as the placement of the piles discussed in *Abston Construction.* The earliest decision on this issue, *Appalachian Power Co. v. Train*,[148] was a challenge to EPA's promulgation of effluent guidelines applicable to steam electric generating facilities. Industry admitted that runoff from coal piles and chemical handling facilities was from point sources, but argued that runoff from construction sites was not. It is not clear from the U.S. Court of Appeals for the Fourth Circuit's opinion what the principled difference was, although the court held (at least) that the application of the guidelines to construction site runoff was sufficiently vague to vacate and remand that portion of the regulations. In doing so, the court did not reject EPA's contention that runoff collected or channeled by man was from point sources.[149] Subsequent decisions have found runoff from piles to be from point sources because the human placement of the piles dictates where natural gravity directs the runoff to navigable waters.[150]

b. The Number of Sources Factor

This factor is variously described as whether there is one identifiable, discrete source discharging pollutants or instead whether there are multiple, dispersed sources discharging pollutants. Sometimes, the factor is described as whether pollutants are traceable to a single source.[151] It is rooted in the legislative his-

146. 40 C.F.R. §122.23(a).
147. 599 F.2d 368, 371-73, 23 ELR 21526 (10th Cir. 1979); *accord* Trustees for Alaska v. EPA, 749 F.2d 549, 557-58, 15 ELR 20146 (9th Cir. 1984).
148. 545 F.2d 1351, 6 ELR 20732 (4th Cir. 1978).
149. *Id.* at 1373-75.
150. Parker v. Scrap Metal Processors, Inc., 386 F.3d 993, 1009, 34 ELR 20104 (11th Cir. 2004) (piles of scrap debris); Bravos v. EPA, 324 F.3d 1166, 1169 (10th Cir. 2003) (waste rock piles at mine); Consolidation Coal Co. v. Costle, 604 F.2d 239, 249, 9 ELR 20511 (4th Cir. 1979) (piles of coal, chemical, and construction material).
151. Cordiano v. Metacon Gun Club, Inc., 575 F.3d 199, 220 (2d Cir. 2009); League of Wilderness Defenders/Blue Mountains Biodiversity Project v. Forsgren, 309 F.3d 1181, 1184 (9th Cir. 2002)

tory distinguishing between point and nonpoint sources. A common comparison is between a factory with a pipe discharging pollutants and a road with thousands of vehicles leaving traces of rubber, lead, and other pollutants on its surface to be carried to navigable water by precipitation flow. The common assumption courts make when using this factor is that a single source is susceptible to regulation by a permit, while multiple or dispersed sources, as a practical matter, are not.[152] This assumption is sometimes true, but not always. For instance, EPA can issue general permits to multiple sources in the same area and subject to identical controls, such as oil drilling platforms in the Gulf of Mexico.[153] Congress also authorized EPA to control hundreds or thousands of municipal storm drains in one permit issued to the municipality or other jurisdiction owning them.[154] Congress gave no such authority to EPA regarding water pollution from vehicles or roads. In any event, a general permit would not help to control millions of cars, registered anywhere in the country and beyond, each leaving small bits of pollutants on a road. Because an automobile can go anywhere, it would be impossible to know what water quality standards its discharges must comply with or to know which state must certify that its operation will meet the state's standards under CWA §401. By those standards, the road itself or its drainage system, however, would be susceptible to control by CWA individual or general permits. Indeed, roads would be in a better position to capture and control such pollutants than vehicles using the roads. But, as discussed above, roads are not conveyances of pollutants to navigable waters in the sense contemplated for point sources, although their drainage ditches may be. Ditches, as point sources, are discussed below.

("nonpoint source pollution . . . is widely understood to . . . arise[ ] from many dispersed activities over large areas, and is not traceable to any single discrete source"); United States v. Plaza Health Labs., Inc., 3 F.3d 643, 652, 23 ELR 21526 (2d Cir. 1993) (Oakes, J., dissenting) (nonpoint sources are from "runoff [occurring] in diffuse patterns, over land and into navigable waters"); *Trustees*, 749 F.2d at 558; *Earth Sciences*, 599 F.2d at 371 ("oil and gas runoffs caused by rainfall on the highways, are virtually impossible to isolate to one polluter, no permit or regulatory system was established as to them"); U.S. EPA ex rel. McKeown v. Port Auth. of N.Y. & N.J., 162 F. Supp. 2d 173, 189 (S.D.N.Y. 2001); Beartooth Alliance v. Crown Butte Mines, 904 F. Supp. 1168, 1173, 26 ELR 20639 (D. Mont. 1995) (nonpoint source status "is limited to uncollected runoff water which is difficult to ascribe to a single polluted'); Washington Wilderness Coal. v. Hecla Mining Co., 870 F. Supp. 983, 988, 25 ELR 20661 (E.D. Wash. 1994) ("the touchstone for finding a point source is the ability to identify a discrete facility from which pollutants have escaped . . . non-point source designation is limited to uncollected runoff water from, for example, oil and gasoline on a highway, which is difficult to ascribe to a single polluter").

152. "Because [nonpoint source pollution] arises in such a diffuse way, it is very difficult to regulate through individual permits." League of Wilderness Defenders, 309 F.3d at 1184. *See also Earth Sciences*, 599 F.2d at 371.
153. 40 C.F.R. §122.28.
154. CWA §402(p)(3)(B)(i).

The Supreme Court skirted the issue in *Decker v. Northwest Environmental Defense Center*,[155] holding, on the basis of EPA's interpretation of its industrial stormwater regulation and its silviculture point source regulation, that collected and channeled runoff discharges from logging roads were not from point sources. The focus of the decision and the largest point of contention among the Justices was the appropriateness of the almost complete deference that lower courts had given to an agency's interpretations of its own regulations, following *Auer v. Robbins*.[156] Chief Justice John Roberts and Justice Samuel Alito concurred in *Decker*, but suggested that the Court should reconsider *Auer*. Justice Antonin Scalia dissented on this part of the decision, asserting that the Court should reconsider and overturn *Auer* and that without deference to EPA's interpretation of its own regulation, runoff from the logging roads collected and channeled by pipes, ditches, and channels to navigable waters would be discharged from point sources. Both the majority opinion[157] and Justice Scalia's dissent[158] provide detailed analyses of CWA §402(p) and EPA's industrial stormwater and silviculture regulations, 40 C.F.R. §§122.26 and 122.27. The majority's focus on EPA's interpretation of regulations applied to particular silvicultural activities diminishes the force of its decision on the general issue of whether road ditches are point sources.

155. 1 33 S. Ct. 1326, 43 ELR 20062 (2013).
156. 519 U.S. 410 (1997).
157. The majority opinion, authored by Justice Anthony Kennedy, begins with the proposition that §402(p) exempts discharges composed entirely of stormwater from requiring §402 permits, except for discharges "associated with an industrial activity," §402(p)(2)(B). The majority's analysis failed to note that the exemption expired in 1994. It stated that the CWA did not define "industrial activity," but EPA did so in its industrial stormwater regulation, 40 C.F.R. §22.26(b)(14), defining "associated with an industrial activity" to include discharges of stormwater "directly related to manufacturing, processing or raw materials storage areas in an industrial plant . . . includ[ing] . . . storm water discharges from . . . immediate access roads . . . used or traveled by carriers of raw materials, manufactured products, waste materials, or by-products used or created by the facility . . . ." EPA interpreted the phrase to include only traditional industrial operations like sawmills and other fixed and permanent operations, as opposed to logging operations that were outdoor and impermanent. The majority noted that under *Auer*, courts must defer to an agency's interpretation of its regulations if the interpretation is reasonable; it need not be the only or the best interpretation in order to warrant deference. The majority found EPA's interpretation to be reasonable and accordingly granted it deference.
158. Justice Scalia starts by casting doubt on whether the discharges are exempted by §402(p)(a) because stormwater discharges are not "'natural runoff' when they are channeled through manmade pipes and ditches, and carry with them manmade pollutants, from manmade forest roads." *Decker*, 133 S. Ct. at 1326, 1342 (Scalia, J., dissenting). Even if it were, however, §402(p)'s recapture provision for stormwater associated with industrial activities, as interpreted by EPA's regulation, 40 C.F.R. §122.27(b)(1), covers SIC Code 24, the forest products industry, including Group 2411, "Logging," for establishments primarily engaged in cutting timber, whose primary product is logs. "That," wrote Justice Scalia, "I would think, is that." *Decker*, 133 S. Ct. at 1343 (Scalia, J., dissenting). Cutting timber simply does not involve a fixed and permanent facility as argued by EPA. That addressed EPA's industrial stormwater regulation, but not its amended silvicultural point source definition consisting of four particular operations, which Justice Scalia apparently concluded, without analysis, was within EPA's authority.

Justice Scalia's dissent, however, focuses precisely on that issue, as well as on an exhaustive analysis of the regulations. His analysis on the general question is straightforward and persuasive and very much leaves open the question of whether road ditches and similar stormwater drainage systems are point sources, although they may be exempted from regulation by §402(p). While this factor is not dispositive in all cases, it is useful.

c. The Artificial System Factor

The "artificial" part of this factor duplicates and adds nothing to the first human activity factor. The "system" part of the factor, however, adds a new consideration that has been useful in some cases. A few courts have used this factor to decide that the addition of pollutants was not from a point source. None of them involve precipitation runoff, but the factor could be easily used in runoff cases. In *United States v. Plaza Health Laboratories, Inc.*,[159] the U.S. Court of Appeals for the Second Circuit held that a person is not a point source, at least not in a criminal case under the rule of lenity. Although the decision used many interpretive devices, its primary focus was on the wording of the definition of point source and the examples it includes, commenting that they "evoke images of physical structures and instrumentalities that systematically act as a means of conveying pollutants from an industrial source to navigable waters"[160] as opposed to the "myriad, random acts of human waste disposal, for example, a passerby who flings a candy wrapper into the Hudson River, or a urinating swimmer."[161] The decision as a whole is questionable, as we will examine in detail below. If the Second Circuit had made the observation that all of the examples given in the definition of point source are inanimate, while a person is animate, it would have made a pertinent observation. But to conclude that the definition suggests that point sources are only physical structures that systematically convey pollutants from an industrial source is overblown. The first and most common examples of a point source, pipes, commonly carry municipal, commercial, and agricultural waste as well as industrial waste.

In *Hudson Riverkeeper Fund, Inc. v. Harbor at Hastings Associates*,[162] the plaintiff alleged that trash falling from an abandoned and dilapidated building into an adjacent river violated CWA §301(a) because the building was a point source. The Southern District of New York cited the above-quoted

159. 3 F.3d 643, 23 ELR 21526 (2d Cir. 1993).
160. *Id.* at 646.
161. *Id.* at 647.
162. 917 F. Supp. 251, 26 ELR 21120 (S.D.N.Y. 1996).

language from *Plaza Health* and summarily held that the building was not a point source because "the discharge of material" from it "would not be deliberate or systematic."[163] If the court meant that an addition must be intentional for it to be from a point source, it is simply wrong. The Supreme Court has held that passive owners of point sources are liable for additions of pollutants to navigable waters from their point sources.[164] Moreover, civil liability under the CWA is commonly held to be strict[165] and even criminal liability may be negligent.[166] What the court meant by "systematic" is not clear. Perhaps, it meant the complained-of action must be continuing rather than a one-time event. That interpretation is wrong as well, for spills are one-time events and unpermitted spills of pollutants from point sources violate the statute.[167] Indeed, the CWA requires EPA to develop best management practices to prevent spills, for inclusion in permits to industrial dischargers.[168]

Several other decisions used the system approach to determine that a facility was a point source. In *Frobel v. Meyer,* the U.S. Court of Appeals for the Seventh Circuit, citing the Second Circuit's *Plaza Health* decision, commented that point source "connotes the terminal end of an artificial system for moving water waste, or other materials."[169] The Tenth Circuit in *Earth Sciences* answered, in part, an argument that an overflowing sump was not a point source by commenting that "[w]e have no problem finding a point source here. The undisputed facts demonstrate the combination of sumps, ditches, hoses and pumps is a circulating or drainage system to serve this mining operation."[170] The Fifth Circuit in *Abston Construction* later held that "[conveyances of pollution formed either as a result of natural erosion or by material means, and which constitute a component of a mine drainage sys-

163. *Id.* at 257.

164. South Fla. Water Mgmt. Dist. v. Miccosukee Tribe of Indians, 541 U.S. 95, 104-05, 34 ELR 20021 (2004).

165. Stoddard v. Western Carolina Regional Sewer Auth., 784 F.2d 1200, 1208, 16 ELR 20503 (4th Cir. 1986); Sierra Club v. Abston Constr. Co., Inc., 620 F.2d 41, 45, 10 ELR 20552 (5th Cir. 1980); United States v. Earth Sciences, Inc., 599 F.2d 368, 374, 23 ELR 21526 (10th Cir. 1979); West Virginia Highlands Conservancy, Inc. v. Huffman, 651 F. Supp. 2d 512, 519, 39 ELR 20201 (S.D. W. Va. 2009); Fishel v. Westinghouse Elec. Corp., 640 F. Supp. 442, 446, 16 ELR 20634 (M.D. Pa. 1986).

166. CWA §309(c)(1).

167. *See, e.g.*, United States v. Hamel, 551 F.2d 107, 7 ELR 20253 (6th Cir. 1977); Williams Pipe Line Co. v. Bayer Corp., 964 F. Supp. 1300 (S.D. Iowa 1997).

168. CWA §304(e).

169. 217 F.3d 928, 30 ELR 20746 (7th Cir. 2000). In this case the county incompletely dismantled a deteriorated dam. Predictably, the current in the stream carried sediment accumulated behind the dam, depositing it on gravel bars and native aquatic plants downstream, destroying the natural habitat of the stream. The plaintiff's theory was that the remains of the dam and the channel through which the sediment flowed were a point source. At that point the nonexistent dam was the terminal end of nothing and was not a conveyance.

170. *Earth Sciences,* 599 F.2d at 374.

tem, may fit into the statutory definition" of point source.[171] In *Williams Pipe Line Co. v. Bayer Corp.*, an Iowa district court held that an oil pipeline facility was a point source.[172] Another district court in *North Carolina Shellfish Growers Ass'n v. Holly Ridge Associates* found that an entire tract of land was a point source when it contained an "extensive network of seventeen ditches specifically designed to concentrate and accelerate the flow of stormwater from the track."[173] Finally, in *Reynolds v. Rick's Mushroom Service, Inc.*, a Pennsylvania district court found that a mushroom farmer's point source included his whole system to prevent the discharge of pollutants, consisting of "land gradations, berms . . . , sedimentation basin, the wasterwater impoundment, and the sprayer system."[174]

d. The Nature of the Polluting Activity Factor

In *Plaza Health,* the Second Circuit concluded from both the structure of the CWA[175] and its legislative history[176] that the §402 permitting program was designed to regulate industrial and municipal waste, not agricultural or random human waste. It used that conclusion to support its ruling that human beings are not point sources.[177] Later, in another context, the same circuit commented that nonpoint sources came, in part, from agricultural use of land adjacent to a river.[178] These comments are loosely in accord with the legislative history, which suggests industrial and municipal waste are regulated by the §402 permit program, but agricultural waste is not because it was nonpoint source in nature. This interpretation is overly simplified. The definition of point source includes CAFOs, which are agricultural in nature. Moreover, the exemptions for agricultural pollution are limited to "agricultural stormwater discharges" and "return flows from irrigated agriculture."[179] While these are broad-reaching exemptions, they do not include everything discharged from a farm. Moreover, there are many categories of activity that generate pollutants, beyond industrial, municipal, and agricultural. Commercial activities with point source dischargers, such as developing photographic film and prints, are regulated by the CWA permit program.[180]

171. Sierra Club v. Abston Constr. Co., Inc., 620 F.2d 41, 45, 10 ELR 20552 (5th Cir. 1980).
172. *Williams Pipe Line Co.*, 964 F. Supp. at 1318-20.
173. 278 F. Supp. 2d 654, 681 (E.D.N.C. 2003).
174. 246 F. Supp. 2d 449, 457-58 (E.D. Pa. 2003).
175. United States v. Plaza Health Labs., Inc., 3 F.3d 643, 646, 23 ELR 21526 (2d Cir. 1993).
176. *Id.* at 647.
177. The reasoning of this decision is questionable and will be examined in detail below.
178. Natural Res. Def. Council, Inc., v. Muszynski, 268 F.3d 91 (2d Cir. 2001).
179. CWA §502(14).
180. 40 C.F.R. pt. 459.

Plaza Health's conclusion, therefore, is erroneous. Indeed, some courts have rejected it altogether. The Ninth Circuit, in *Greater Yellowstone Coalition v. Lewis,* said "point and nonpoint sources are not distinguished by the kind of pollution they cause or by the activity causing the pollution, but rather by whether the pollution reaches the water through a polluting activity does not appear to be a useful one.

e. The Appropriate Control Factor

This factor also is rooted in the legislative history, which suggests that the §402 permit program was designed for point sources, for which uniform national technology-based standards for similar facilities, expressible in numbers, are appropriate. The legislative history also suggests that the §402 permit program was not designed for nonpoint sources, for which climate and geography vary so greatly that they are not susceptible to uniform standards and are more amenable to control by individual best management practices than by nationally uniform technology. While this is a useful distinction, it is somewhat vague. Is intercepting sediment at a construction site by placing bales of hay in the path of the flow of stormwater a best management practice or a use of primitive "hay bale" pollution control technology? Moreover, as we have already discussed, it is possible to intercept most runoff by trenches and ditches leading the runoff to sedimentation basins or more effective treatment technologies. The issue is largely one of cost, both in dollars and in land use. Cost is one of the factors that EPA must use in developing and promulgating effluent guidelines for industries.[181]

The Ninth Circuit has commented that the reason for the CWA's focus on point sources rather than nonpoint sources "is simply that '[differences in climate and geography make nationwide uniformity in controlling non-point pollution virtually impossible. Also, the control of non-point source pollution often depends on land use controls, which are traditionally state or local in nature.'"[182] The latter point is overstated. Most pollution control does not restrict the uses of our land, but rather tells us how we must conduct some of those uses to avoid injuring others or society as a whole. Land use control is telling a contractor what he can or cannot build on his land, not that he must place bales of hay to intercept sediment flowing in rainwater during active construction. Requiring sediment control at construction sites to prevent pollution discharges to navigable water is not inherently a state or

181. CWA §304(b)(1)(B), (b)(2)(B), & (b)(4)(B).

182. Oregon Natural Desert Ass'n v. U.S. Forest Serv., 550 F.3d 778, 785, 39 ELR 20297 (9th Cir. 2008) (quoting Marc R. Poirier, *Non-Point Source Pollution, in* Envt'l. Practice Guide §18.13 (2008)).

local interest. How and where the bales of hay are placed at a construction site to best contain sediment is entirely a local matter, dictated by the topography of the site.

At issue in *League of Wilderness Defenders/Blue Mountains Biodiversity Project v. Forsgren*[183] was whether spraying pesticides from an aircraft's spray bar on rivers in a national forest required a §402 permit. The Ninth Circuit easily concluded that the spraying apparatus was a discrete conveyance. The U.S. Forest Service, however, noted that EPA's regulations defining silvicultural point sources excluded "non-point source activities such as . . . pest and fire control . . . from which there is natural runoff'[184] and argued the exclusion included spraying pesticides for pest control. The court noted that the regulation excluded only nonpoint sources from which there is natural runoff. Since there is no natural runoff from aerial application of pesticides directly onto water, the general understanding of nonpoint sources did not apply and therefore the regulatory exclusion of point source did not apply. The court also noted that the preamble to EPA's silvicultural point source definition identified three characteristics of nonpoint sources: (1) precipitation flow; (2) number of sources; and (3) "better controlled through . . . best management practices"[185] than by end-of-pipe treatment. The best control of spraying pesticides over water is to "just say no," an end-of-pipe control.[186] The U.S. Court of Appeals for the Sixth Circuit held in *National Cotton Council v. U.S. EPA* that an aerial pesticide spray apparatus was a point source in part because it was controllable at the source.[187] Other decisions consider the silviculture point source regulations in different contexts.[188]

F. *Non-Stormwater Runoff Decisions*

Most of the decisions above concerned discharges of stormwater flow in which the distinction between point and nonpoint sources depended on the degree of human involvement in the discharge. In cases that were not clear, the courts developed the other factors to make that determination. In most point source decisions, however, stormwater flow is irrelevant. For example, when bulldozers push fill material into wetlands, the issue of whether bull-

183. 309 F.3d 1181 (9th Cir. 2002).
184. 40 C.F.R. §122.27(b)(1).
185. Wilderness Defenders, 309 F.3d at 1188.
186. Indeed, EPA is directed to promulgate technology-based effluent guidelines for many industrial sources requiring elimination of discharges altogether if "technologically and economically achievable." §301(b)(2)(B).
187. 553 F.3d 927, 940, 39 ELR 20006 (6th Cir. 2009).
188. *See* Decker v. Northwest Envtl. Def. Ctr., 133 S. Ct. 1326, 43 ELR 20062 (2013); Environmental Prot. Info. Ctr. v. Pacific Lumber Col, 469 F. Supp. 2d 803, 37 ELR 20012 (N.D. Cal. 2007).

dozers operating in that capacity are point sources has nothing to do with stormwater flow. In these decisions, courts primarily analyze whether the alleged point source is a "discernible, confined, and discrete conveyance" of pollutants and whether it is one of the examples listed in the definition or is analogous to them. Courts often use precedent and the broad interpretation canon of construction. They may also use one or more of the factors discussed above. This section discusses the cases that either fall into broad categories, have precedents on both sides of the issue, or raise particularly interesting issues.

1. The Curious Tales of Dams and Ditches

Dams and ditches are conveyances of both stormwater runoff and non-stormwater runoff. Flood control dams impound stormwater to control flooding, while other dams have nothing to do with stormwater, such as run-of-the-river dams generating electricity. Some ditches control stormwater, such as the roadside drainage systems discussed above, while other ditches have nothing to do with stormwater, such as ditches carrying water destined for irrigation. Ditches are included in the list of point source examples; dams are not. There are thousands of each, enough to present formidable challenges for the limited resources of permit writers. On the other hand, both can contribute to the degradation of water quality. Moreover, the administrative burden of issuing a multitude of permits to individual sources bay be ameliorated by issuing area permits or using similar administrative mechanisms. Each presents a unique interpretive issue.

Dams carry pollutants from upstream waters to downstream waters and they certainly are discernible, confined, and discrete. Thus, even run-of-the-river dams seem to be point sources. Power-generating dams that convey water through pipes, sluices, and tunnels through turbines and ultimately to downstream waters, even more clearly are or include discernible, confined, and discrete conveyances. At the outset of the §402 permit program, EPA made a decision that dams did not require permits. The Agency did so for policy reasons. The acknowledged reason was that the large number of dams would overwhelm permit writers, although few presented real water quality issues. The unspoken reason may have been that many dams were owned and operated by the federal government or were already subject to other federal permit programs that already involved environmental review and control.[189]

189. Environmental review was required for all major federal actions significantly affecting the environment under the National Environmental Policy Act (NEPA), 42 U.S.C. §§4321-4370f, ELR Stat. NEPA §§2-209. Section 401 of the CWA also required federal permit-issuing agencies to secure from the

The National Wildlife Federation sued EPA challenging its determination that dams were not covered by the §402 permit program; the plaintiff alleged that the Agency lacked the authority to declare that point sources adding pollutants to navigable waters did not require §402 permits. The D.C. Circuit upheld EPA's determination in *National Wildlife Federation v. Gorsuch.*[190] While it might be assumed that this decision held that dams are not point sources, it did not so hold. Instead, EPA conceded that under some circumstances, dams could be point sources,[191] and the court held that dams did not add pollutants to downstream waters. Courts today largely discount the decision because it accorded *Chevron* deference where none was due.[192] All other decisions considering the issue of whether dams are point sources have held that they are, at least under the facts of the cases, *see supra* Section III.

Ditches present particular interpretive issues because they are listed as an example of a point source but also hold bodies of water, including stormwater runoff, that could be waters of the United States, possibly meaning that they satisfy two elements of the discharge of pollutants offense. Justice Scalia, writing for the plurality in *Rapanos v. United States,* suggested that ditches could not be navigable water, in part because they were designated as point sources by the CWA.[193] He admitted that the two could overlap on occasion, but stated that they could not do so generally.[194] The problem with this analysis, however, is that ditches are man-made trenches in the earth, not bodies of water: They convey water; they are not water. This was a central point in *National Association of Home Builders v. U.S. Army Corps of Engineers,*[195] where the petitioner challenged a nationwide permit because it applied to some ditches; ditches could not be waters of the United States because they were point sources; and the two terms are mutually exclusive. The D.C. Circuit noted that the plurality in *Rapanos* did not declare that "a point source, such as a ditch, can *never* be a navigable water under the CWA."[196] It also noted that a number of courts had held that ditches were

---

state in which federal projects, or federally licensed projects, were undertaken a certification that the project would meet state water quality standards.

190. 693 F.2d 156, 13 ELR 20015 (1982).
191. *Id.* at 165.
192. *See generally* Chevron U.S.A., Inc. v. Natural Res. Def. Council, Inc., 467 U.S. 837, 14 ELR 20507 (1984).
193. 547 U.S. 715, 735, 36 ELR 20116 (2006).
194. *Rapanos*, 547 U.S. at 735-36 (The statutory definitions of "point source" and "navigable waters" suggests that they are "separate and distinct categories." It "would make little sense if the two categories were significantly overlapping." Therefore, ditches "by and large" are point sources and not navigable waters.).
195. 669 F. Supp. 2d 209, 215, 40 ELR 20104 (D.C. Cir. 2010).
196. *Id.* at 216.

navigable waters[197] and the courts that held otherwise did so on the facts rather than as a matter of law.[198] Indeed, Justice Scalia subsequently concluded that ditches could be point sources.[199] Despite the controversy, both dams and ditches convey pollutants and are or can be point sources.

EPA, in its recent rulemaking to define "waters of the United States," discussed extensively in Chapter 3, treated ditches as waters. Most importantly, it specifically provided that waters meeting the definition of tributaries of waters of the United States, are themselves, waters of the United States.[200] It then provided that ditches were tributaries if they met the definition of tributary.[201] Finally, it excluded three sets of ditches from being waters of the United States, even if they otherwise met the definition of tributaries. The three categories of excluded ditches are: (1) those with ephemeral flows and not a relocated tributary or excavated tributary; (2) those with intermittent flows and not a relocated tributary, excavated in a tributary, or draining a wetland; and (3) not flowing directly or through another water into a traditionally navigable water, interstate water or the territorial seas.[202] The preamble to the rule exhibits particular concern that tributaries that are relocated to ditches not lose their status as waters of the United States.[203]

2. Vehicles on or in Land, Water, and Air

By far, the largest number of non-runoff decisions concern whether the defendants' vehicles are point sources. All of them hold that vehicles are point sources. Typically, courts do so by reciting that point source is interpreted broadly and by citing a precedent or two, often *Avoyelles Sportsmen's League v. Marsh*.[204] They probably cite *Avoyelles* because it is the earliest appellate court decision holding bulldozers and backhoes to be point sources, not because its analysis is exhaustive.[205] Despite the absence of analytical attention to

197. Parker v. Scrap Metal Processors, Inc., 386 F.3d 993, 1009, 34 ELR 20104 (11th Cir. 2004); United States v. Deaton, 332 F.3d 698 (4th Cir. 2003); Treacy v. Newdunn, 344 F.3d 407 (4th Cir. 2003).
198. F.D. & P. Enters., Inc. v. Army Corps of Eng'rs, 239 F. Supp. 2d 509, 514 (D.N.J. 2003); Haniszewski v. Cadby, 2009 WL 316723 (W.D.N.Y. 2009).
199. "The stormwater here was discharged from logging roads through a series of pipes, ditches, and channels—all items expressly named in the definition [of 'point source']." Decker v. Northwest Envtl. Def. Ctr., 133 S. Ct. 1326, 1344, 43 ELR 20062 (2013) (Scalia, J., dissenting).
200. 80 Fed. Reg. 37114, codified at 40 C.F.R. 122.2 Waters of the United States, (1)(v).
201. *Id.* 122.2.(3)(iii).
202. 80 Fed. Reg. 37114, codified at 40 C.F.R. 122.2 Waters of the United States, (2)(iii).
203. 80 Fed. Reg. 37078.
204. 715 F.2d 897, 922, 13 ELR 20942 (5th Cir. 1983).
205. The Fifth Circuit's analysis was, in fact, cursory. "[B]ulldozers and backhoes were point sources, since they collected into windrows and piles material that may ultimately find its way back into the waters." *Avoyelles*, 715 F.2d at 922. It cited *United States v. Holland*, 373 F. Supp. 665, 668, 4 ELR 20710 (M.D. Fla. 1074), in which a Florida district court found that the point source element of

the issue, the courts are undoubtedly correct. Vehicles fit precisely within the definition as "discernible, confined and discrete conveyance[s]" The very purpose of vehicles is to convey people and material. The list of examples given in the definition includes "rolling stock" and "vessel or other floating craft," vehicles whose purposes are to convey people and materials on land and water. Although other types of terrestrial vehicles and aircraft are not listed, the nonexhaustive list includes the mentioned examples; it does not exclude others. Unlisted terrestrial vehicles and aircraft are conveyances and are analogous to rolling stock and vessels.

It was held fairly early in the §402 permit program that Navy aircraft dropping ordnance into coastal waters during bombing practice were point sources, with no analysis of the definitional issue.[206] Subsequent cases under the CWA in which aircraft (and sometimes trucks) were alleged to be point sources involve aircraft spraying pesticides into water. It is not always clear whether the aircraft (or trucks) or the spray bars attached to them were alleged to be the point sources. In either event, courts held the vehicles or the spray apparatus to be point sources, with little analysis.[207]

The problem with this easy analysis is that passenger vehicles generally, and automobiles in particular, are conveyances and therefore point sources. If so, why aren't the owners of such vehicles liable for their deposits of oil, grease, and rubber residue on roads that are ultimately discharged into navigable waters by precipitation flow or for their emissions of tailpipe materials that may fall into navigable waters below bridges? Here, a return to the defi-

---

CWA §301(a) was satisfied because defendant admitted that he had "discharged from point sources, including dump trucks, drag lines and bulldozers."

206. "It would be a strained construction of unambiguous language for the court to interpret that the release or firing of ordnance from aircraft into the navigable waters of Viequis [an island off the coast of Puerto Rico] is not 'any addition of any pollutant . . . from any point source,'" Barcelo v. Brown, 478 F. Supp. 646, 664 (D.P.R. 1979). The Supreme Court, on appeal from the First Circuit's reversal of the district court on other grounds, stated, without comment, that the district court held "the release of ordnance from aircraft or from ships into navigable waters is a discharge of pollutants." Weinberger v. Romero Barcelo, 456 U.S. 305, 309, 12 ELR 20538 (1982).

207. Peconic Baykeeper, Inc. v. Suffolk County, 600 F.3d 180, 40 ELR 20098 (2d Cir. 2010). The district court had held that the spray applicators attached to trucks and aircraft were not point sources because they discharged into air rather than into water. The Second Circuit made use of the broad interpretation canon of construction and the Ninth Circuit's reasoning in *League of Wilderness Defenders/Blue Mountains Biodiversity Project v. Forsgren*, 309 F.3d 1181, 1184 (9th Cir. 2002), to reverse. "Here, the spray apparatus was attached to trucks and helicopters, and was the source of the discharge. The pesticides were discharged 'from' the source, and not from the air . . . . The District Court's conclusion that the pesticides were not discharged from a point source was in error." *Peconic Baykeeper*, 600 F.3d at 188-89. In *National Cotton Council of America v. U.S. EPA*, 553 F.3d 927, 940, 39 ELR 20006 (6th Cir. 2009), the parties assumed that aerial pesticide spraying was from point sources, but EPA contended that the point sources did not add pollutants because they sprayed pesticides, not pollutants. The court rejected that argument. *See also Forsgren*, 309 F.3d at 1185 ("[A] n airplane fitted with tanks and mechanical spraying apparatus is a 'discrete conveyance.'").

nitions of "discharge of a pollutant" and "point source" are useful. Reordering the elements in the statute somewhat, §301(a) regulates the addition by conveyances of pollutants to navigable waters. It focuses on point sources as conveyances of pollutants rather than as the origins of pollutants. Most of the examples of point sources in the definition emphasize this. We think of the first, "pipe," as carrying pollutants, not creating them. CAFOs are the big anomaly in the list, but they are an anomaly in the definition for several reasons, which we will discuss below.

Finally, automobile-produced pollutants are not susceptible to regulation by the §402 permit program. They are classic multiple and dispersed sources, too many in number for EPA to control by permits. They do not directly convey pollutants to water. They are not susceptible to end-of-pipe treatment technology. There's not much end-of-pipe technology that an automobile owner or driver can apply to stop leaving bits of her tires on the highway. Although end-of-pipe technology may treat pollutants coming from the tailpipes of motor vehicles, those emissions are already comprehensively regulated by the Clean Air Act (CAA), both in terms of end-of-pipe and internal technology and fuel content.[208] If the technology-based standards required to be applied in permits to point sources under the CWA are not possible or appropriate, the other source of effluent limitations in the CWA, water quality standards, are not possible or appropriate either for they are developed on a local basis and we don't know what waters any particular motor vehicle's pollutants will enter. The only decision to consider this issue did so in a case alleging that tollbooths on a highway were point sources, *U.S. EPA ex rel. McKeown v. Port Authority of New York & New Jersey*.[209] The court held that the vehicles passing through the tollbooths were "the sources of the pollutants complained of" and that they were "air emissions and not water pollutants," emissions regulated by the Clean Air Act rather than the CWA.[210] These considerations do not apply to vehicles conveying pollutants to water, such as dumptrucks or bulldozers discharging fill material into wetlands.

3. People and Other Animals

In *United States v. Plaza Health Laboratories, Inc.*,[211] the Second Circuit considered the question of "whether a human being can be a point source"[212]

208. 42 U.S.C. §§7401-7671q, ELR Stat. CAA §§101-618; *see* 42 U.S.C. §§7521-7589.
209. 162 F. Supp. 2d 173, 189 (S.D.N.Y. 2001).
210. *Id.*
211. 3 F.3d 643, 23 ELR 21526 (2d Cir. 1993).
212. *Id.* at 645.

and held that the CWA "was never designed to address the random, individual polluter."[213] The court began its analysis with the statutory definition of point source, which the court noted did not include human beings in its list of examples. Moreover, the court found that the examples "evoke images of physical structures and instrumentalities that systematically act as a means of conveying pollutants from an industrial source to navigable waterways."[214] It also noted that "by any person" is already an element in a §301(a) violation and, therefore, reading point source to include a person would make "point source" and "by any person" duplicative. Such an interpretation would make the violation read, "the addition of any pollutant to navigable waters *from any person by any person* shall be unlawful,' and this simply makes no sense."[215] If the court had stopped there, its interpretation of point source would have been well-analyzed, if not unassailable.[216]

The court next analyzed the statute and its legislative history to demonstrate that the CWA was designed to regulate industrial and municipal waste, not the random polluting acts of individuals. First, the court listed a number of sections dealing with point sources,[217] implying that those sections deal exclusively with industrial and municipal discharges. They do not. They sometimes distinguish between municipal and other point sources.[218] But for the most part, they apply to all point sources, without limiting themselves to industrial point sources. The court proceeded to legislative history, which it also concluded focuses on industrial polluters.[219] For that conclusion, it cited a Senate report that distinguished between point and nonpoint sources[220] and a floor statement by Senator Dole suggesting that all agricultural discharges, including CAFO discharges, are nonpoint source pollu-

---

213. *Id.* at 646.

214. *Id.*

215. *Id.* at 647.

216. The issues with its interpretation thus far relate to its comments that the examples of point source in the definition are physical structures and structures that convey pollutants from industrial sources. Many of the examples, such as "pipe," do evoke such images. However, others, such as "channel" and "discrete fissure" do not, particularly if physical structures are equated with human-made structures, for channels and fissures occur in nature.

217. CWA §§301, 301(e), 301(g)(2), 304(b)(4)(B), 306, 308(a), 308(c), & 402(f).

218. CWA §301(b), for example, deals specifically with municipal point sources, see §301(b)(1)(B), (h), (i) & (j), while the remainder deal with point sources generally, not specifying industrial point sources. Section 306 requires EPA to promulgate standards for new sources for at least 27 categories of point sources, many of which are manufacturing facilities. These are industrial facilities. But the list includes a number of food-processing facilities. Query whether they are industrial facilities? EPA has promulgated technology-based standards for a number of commercial facilities, such as the development of photographic film and prints in 40 C.F.R. pt. 459.

219. 3 F.3d at 647.

220. *Id.*

tion.[221] These authorities do not support the Second Circuit's conclusion that Congress did not intend the CWA to include human beings as point sources. Indeed, the suggestion by Senator Dole that CAFOs are not point sources is simply wrong, undercutting whatever credibility his statement has. The court then noted the absence of case law or EPA regulations suggesting that human beings are point sources. Of course, there was no case or regulatory law that people are not point sources.

Finally, the court observed that holding a human being to be a point source would lead to the absurd result that "a passerby who flings a candy wrapper into the Hudson River, or a urinating swimmer" would violate the statute.[222] These images are easily distinguished from the facts in *Plaza Health,* where the defendant took waste from the commercial testing of blood at his laboratory in vials, which he placed below the high-water line in a bulkhead so that they would be removed by high tide. This was not a kid throwing a candy wrapper in the water, but instead the principal of a commercial activity evading the costs of proper disposal of medical waste by secretly disposing of it in navigable waters; his very secrecy is persuasive evidence of his knowing wrongdoing.

In *Oregon Natural Desert Association v. Dombeck,* the Ninth Circuit, citing *Plaza Health* as its only authority, stated that "[i]t would be strange indeed to classify as a point source something as inherently mobile as a cow" and held "the term 'point source' does not include a human being or any other animal."[223] Ten years later, the circuit court cited that decision as good authority in *Oregon Natural Desert Association v. U.S. Forest Service.*[224] Both cases were challenges to grazing licenses on federal land without CWA §401 certifications on the theory that cattle were point sources that would pollute streams and the granting of the licenses were federal actions that required §401 certification.

Humans can be conveyances. When we take clean laundry from the dryer and carry it to the bedroom, we are conveyances of clean clothes. Animals can be conveyances; horses are an obvious example. Both are discernible, confined, and discrete conveyances. As Judge Oakes pointed out in his strong dissent in *Plaza Health*, the holding that humans cannot be point sources opens a gaping door for evasion of the CWA's prescription simply by having employees form a bucket brigade from a pollutant-producing facility to the

221. *Id.*
222. *Id.*
223. 172 F.3d 1092, 28 ELR 21471 (9th Cir. 1998).
224. 550 F.3d 778, 782, 39 ELR 20297 (9th Cir. 2008).

river in place of a pipe.[225] Stripped of its erroneous arguments, the court in *Plaza Health* could have come plausibly to the opposite conclusion, as the dissent did. Unfortunately, the government poorly argued the case. If the government had argued that the vials, as containers, were the point sources, then the issue of humans as point sources would not have arisen. Indeed, the district court noted that there was an alternative to humans as point sources in the case: The fissures in the bulkhead into which the defendant inserted vials of blood to be washed away by the rising tide could be considered as point sources, thus removing the issue of whether humans could be point sources.[226]

4. CAFOs, Fish-Raising Facilities, and Other Agricultural Discharges

The definition of point source specifically provides that "agricultural stormwater discharges" and "return flows from irrigated agriculture" are not point sources and that a CAFO is a point source. Unless flowing water is a point source, however, agricultural stormwater discharges and return flows from irrigated agriculture cannot be point sources, even in the absence of the statutory exemption, because flowing water is not a conveyance; flowing water is carried by a conveyance. To give those exemptions meaning, they must be applied to additions of pollutants from agricultural stormwater and return flows from irrigation agriculture by point sources. Stormwater flow and irrigation return flow may include groundwater that originates in irrigation by flooding.[227]

Are CAFOs regulated because they are point sources, or are they exempted if they deposit their waste into navigable waters through stormwater flows or irrigation return flows? If pollutants from CAFOs enter navigable water by means other than stormwater flow or irrigation return flow, for instance in a stream flowing through the CAFO, then the exemptions do not come into play. But if pollutants from the CAFO are discharged to navigable water only by stormwater flow or irrigation return flow, then which authority governs: the definition of CAFO as a point source or the exemptions? Under the

---

225. The district court opinion, *United States v. Villegas*, 784 F. Supp. 6, 8-11, 22 ELR 21027 (E.D.N.Y. 1991), *rev'd sub nom. United States v. Plaza Health Labs., Inc.*, 3 F.3d 643, 23 ELR 21526 (2d Cir. 1993), is extremely well- analyzed, using more interpretive devices than any other decision in this chapter. *See infra* Table B, Column E.

226. *Villegas*, 784 F. Supp. at 10.

227. Fishermen Against Destruction of Env't, Inc. v. Closter Farms, Inc., 300 F.3d 1294, 1297 (11th Cir. 2002).

canon of statutory interpretation that the specific governs over the general,[228] the definition of CAFO as a point source should govern, because the exemptions of stormwater flow and irrigation return flow are general types of agricultural flow, while flow from a CAFO is a specific agricultural flow.[229] It also can be argued that unless flow from CAFOs is covered by §301(a), the inclusion of CAFOs in the definition of point source has no meaning. This is a weaker argument, however, because a CAFO would still be a point source if a stream flowed through it.

EPA has defined CAFOs in great detail in 40 C.F.R. §122.23 and courts have applied the definition as written. To be a CAFO, the operation must contain more than the number of animals specified in the regulation. EPA may also designate smaller operations as CAFOs if they significantly contribute to violations of water quality standards. Once it is a CAFO, the point source includes not just the feeding operation, but ancillary areas for raising or housing the animals and storing or disposing of manure. Thus, discharges from adjacent fields used to spread or store manure violate the statute unless performed in compliance with the regulations.[230] We have already seen that the CWA does not regulate CAFOs because they are point sources; they must also add pollutants to navigable water in order to be regulated. The CAFO must be designed not to discharge except as a result of a 25-year, 24-hour storm event and may discharge only as a result of such an event. Thus, a properly designed CAFO violates the statute if it discharges as a result of a lesser storm event.[231]

Mushroom-growing operations, like CAFOs, often dispose of their wastewater by spraying it on adjacent fields as fertilizer. When properly designed and applied, it does not escape into adjacent waterways. If it is overapplied, or applied using improperly designed or deteriorated retention facilities, it may flow into such waterways. When it does so without or in violation of a permit, it violates the statute.[232] One court remarked on the similarity of the

---

228. William N. Eskridge Jr., Dynamic Statutory Interpretation 324 (1994); Antonin Scalia & Bryan A. Garner, Reading Law 183-89 (2012).

229. Community Ass'n for Restoration of Env't v. Henry Bosma Dairy, 305 F.3d 943, 955 (9th Cir. 2002) ("because the farm itself falls within the definition of CAFO and is not subject to the agricultural exemption").

230. *Id.*

231. Carr v. Alta Verde Indus., Inc., 931 F.2d 1055, 1059-60, 21 ELR 21005 (5th Cir. 1991).

232. Reynolds v. Rick's Mushroom Serv., Inc., 246 F. Supp. 2d 449, 457 (E.D. Pa. 2003) ("point source" was the entire control system consisting of "land gradations, berms . . . , sedimentation basin, the wastewater impoundment, and the sprayer system"); United States v. Oxford Royal Mushroom Prods., Inc., 487 F. Supp. 852, 854, 10 ELR 20549 (E.D. Pa. 1980) (overapplication and escape through a break in a berm).

control system to CAFO control systems and that used to collect and contain mining waste in *Earth Sciences*.[233]

Fish-raising facilities are analogous to CAFOs and are regulated similarly. EPA defines point source to include a "concentrated aquatic animal production facility" by producing designated numbers and weights of fish raised, and provides that the Agency may designate a smaller operation as a point source if it contributes significantly to a violation of water quality standards.[234] Where a fish hatchery is below the size designated as a point source in the regulation, it is not a point source unless so designated by EPA because the operation significantly contributes to violations of water quality standards.[235] The definition includes operations raising cold water fish in "ponds, raceways, or other similar structures which discharge at least 30 days a year."[236] The Agency interprets that to include offshore net pen operations raising salmon, and its interpretation has been upheld by at least one court.[237] While CAFO discharges are usually of polluted stormwater and mushroom-growing discharges may be of polluted stormwater, fish-raising operation discharges are not of polluted stormwater.

IV. Conclusion

The statutory and regulatory definitions of point source are the most detailed and complex of the definitions of the elements of the water pollution offense. But the interpretation of point source is complicated by the statutory juxtaposition of the term with nonpoint source and the statutory failure to define nonpoint source or to draw a bright line between the two terms. The statutory history adds to the complication by suggesting that the presence of precipitation runoff is the signature of nonpoint sources, a false suggestion if the runoff is collected and channeled by human activity. Adding to the confusion is the tendency of courts and EPA to conflate point source and nonpoint source with other elements of the offense, such as by referring to a "point source pollutant." (As mentioned at the beginning of this chapter, "pollutant" is a separate element of the §301(a) offense.) Despite these complications, in most cases, the determination of whether there is a point source is easy because there is an obvious "discern-

233. *Rick'sMushroom Serv.*, 246 F. Supp. 2d at 456-577.
234. 40 C.F.R. §122.24 & app. C.
235. Wild Fish Conservancy v. Quilcene Nat'l Fish Hatchery, 2009 WL 3380655 (W.D. Wash. 2009).
236. 40 C.F.R. §122.24(a).
237. U.S. Public Interest Research Grp. v. Atlantic Salmon of Maine, LLC, 215 F.2d 239, 32 ELR 20535 (D. Me. 2002).

ible, confined and discrete conveyance" and it is often one of the examples listed in the statutory definition, such as a pipe.

At this point, the CWA robustly regulates most discharges of pollutants by point sources, but depends on state regulation of discharges of pollutants by nonpoint sources. For the most part, however, such state regulation is neither pervasive nor robust and most continuing water pollution problems can be traced primarily to nonpoint sources. The legislative history of the CWA suggested that relatively unregulated nonpoint sources should become robustly regulated point sources over time as national uniform standards became available to control water pollution from them. EPA has not expanded the definition of point source to accomplish this objective. The Agency should begin to do so. Interpreting and applying the term would be easier if EPA amended its regulatory definition of point source to incorporate what case law and the Agency's own practice have established. Distinguishing point from nonpoint sources would also be easier if EPA promulgated a regulatory definition of nonpoint source.

The following suggested definitions summarize the results of this chapter's analysis:

Nonpoint source means conveyances adding pollutants to the waters of the United States for which conveyances are either: (1) indiscernible, unconfined, and indiscrete; or (2) not the result of human activity, in whole or in part. Note: Every five years, EPA will review categories of nonpoint sources to determine if cost-effective technology has become available to treat pollutants discharged by any such category and to add to the list of point sources in the definition of point source in this section any category of nonpoint source for which such technology has become available.

Point source means any discernible, confined, discrete conveyance, resulting in whole or in part from human activity, including but not limited to any pipe, ditch, channel, tunnel, conduit, well, discrete fissure, container, rolling stock, CAFO, landfill leachate collection system, or vessel or other floating craft, from which pollutants are or may be discharged. The term includes (1) human beings deliberately discharging pollutants produced in the course of a wastegenerating activity and (2) devices collecting and channeling stormwater, but does not include devices channeling agricultural stormwater, return flow from irrigated agriculture, or runoff from roads, except as regulated under CWA §402(p). In determining whether a conveyance is a point source, the following factors should be taken into account: (1) human activity collects and channels flow to navigable waters; (2) there is a single source of pollutants, rather than multiple or dispersed sources; (3) the conveyance is systematic; and (4) the activity producing the pollutants is susceptible to end-of-pipe treatment technology with numerical effluent limitations rather than by best management practices.

Chapter 5. Lessons In Statutory Interpretation

This chapter completes this book's discussion of the statutory interpretation of the elements of the Clean Water Act (CWA)[1] offense: (1) any addition (2) of any pollutant (3) to navigable waters (4) from any point source (5) by any person, except in compliance with a CWA permit.[2] This chapter and the remainder of this book augment the first four chapters, all of which are based on the author's earlier articles in *ELR News & Analysis*.[3]

Many of the overarching observations made here are common sense. For instance, although law review editors might demand a citation for the conclusion that plain meaning and precedent are the most commonly used statutory interpretation devices, most lawyers, judges, and legal scholars would agree with that assertion from their experiences, without further authority. After all, most text means what it says and if there is precedent that particular text means what it says, so much the better. This chapter provides strong, quantified evidence to verify the commonsense proposition. The chapter's database is too narrow to provide proof positive, but there is no reason to believe that the conclusion is a function of that database. Studies of the interpretation of the CWA as a whole or of other statutes will undoubtedly reach the same conclusion and establish the proposition as a fact. Where text has no clear meaning or is ambiguous, plain meaning won't help much; and if the interpretation is a matter of first impression, precedent may not help much either. In those situations, other interpretive devices must be used.

1. 33 U.S.C. §§1251-1387, ELR Stat. FWPCA §§101-607. Section 301(a), 33 U.S.C. §1311(a), sets out the CWA offense.
2. Section §301(a) prohibits the "discharge of any pollutant" except in compliance with a CWA permit. Section 502(12), 33 U.S.C. §1362(12), defines "discharge of a pollutant" as the multi-element offense printed in the text. The statute defines "pollutant" in §502(6), "navigable waters" in §502(7), "point source" in §502(5), and "person" in §502(14). It does not define "addition." (The bookd does not examine the fifth element, "person," because it has not given rise to significant litigation or controversy.)
3. Jeffrey G. Miller, *Plain Meaning, Precedent, and Metaphysics: Interpreting the "Addition" Element of the Clean Water Act Offense*, 44 ELR 10770 (Sept. 2014); *Plain Meaning, Precedent, and Metaphysics: Interpreting the "Pollutant" Element of the Federal Water Pollution Offense*, 44 ELR 10960 (Nov. 2014); *Plain Meaning, Precedent, and Metaphysics: Interpreting the "Navigable Waters" Element of the Clean Water Act Offense*, 45 ELR 10548 (June 2015); and *Plain Meaning, Precedent, and Metaphysics: Interpreting the "Point Source" Element of the Clean Water Act Offense*, 45 ELR 11129 (Dec. 2015).

Other observations made here are similarly common sense. For instance, consider the conclusion that, for the most part, plain meaning predominates as a canon of statutory construction for the initial interpretations of a particular statutory text, but over time, precedent predominates. At the outset, there is no precedent available for most text, but hopefully there is plain meaning. Over time, as precedent becomes available, it supplements or even supplants plain meaning. The data set studied in this book demonstrates that conclusion well. It contains three data subsets in which the proposition proves true and one data subset in which it does not, and provides a logical and demonstrable reason why the outlying data subset does not conform to the general conclusion. Again, the database is too narrow to provide proof positive of the proposition, but there is no reason to believe that it is a function of the composition of the database.

Finally, some of the observations made here relate entirely to the interpretations of the particular texts studied. For instance, consider the issue of why courts have interpreted the element "navigable waters" twice as often as they have interpreted any of the other elements of the CWA offense. This chapter concludes the reason is that the ratio of decisions under CWA §402, 33 U.S.C. §1342, to decisions under CWA §404, 33 U.S.C. §1344, is strongly weighted toward §404 in decisions interpreting the navigable waters element, but not in decisions interpreting the other elements. This observation and its consequences are entirely functions of the composition of the database examined. The observation may prove useful in understanding the interpretation of this and other CWA text in which both CWA §§402 and 404 operate, but the observation is unlikely to be useful beyond interpreting the CWA. At the same time, asking analogous questions about interpretations of other statutory texts may lead to similarly illuminating observations.

The four earlier chapters examining the elements of the CWA offense consider 258 decisions—all of the reported decisions between 1974 and 2013 interpreting the first four elements[4] of the water pollution offense that the author could locate.[5] There may be others, especially in unofficial reporters. If others exist, however, they are infrequently cited and hence are not important decisions for these interpretive issues. In any event, the universe of the opinions analyzed is broad enough to support the conclusions reached in the earlier articles and to support or suggest the conclusions reached in this book. It should be noted that because many of the decisions interpret

4. The book does not examine the fifth element, "person," because it has not given rise to significant litigation or controversy.
5. See the decisions listed *infra* in Table A.

more than one element of the offense, the total number of judicial interpretations analyzed in this book is 347.[6] Some of the pertinent data from the judicial interpretations are summarized *infra* in Table B.[7] Some of that data is objective—for example, a decision in a case brought under CWA §§402 or 404. Some of the data is reasonably objective—for example, a decision uses five identified interpretive devices. Unless a decision actually announces and labels each interpretive device it uses, however, different readers may occasionally draw different conclusions on both the number and the identity of the devices a decision uses. The author is reasonably certain of his count and identification of the interpretive devices used in the decisions, but is more sure that his count and identification are consistent throughout.

The reader may be assuming that an interpretive device is a canon of statutory construction. Most are, but interpretive devices and canons of construction do not always coincide. The issue is complicated because there is no agreement on the number or wording of canons of construction. Prof. William Eskridge documents the William Rehnquist Court's use of 87 canons to interpret federal statutes.[8] Prof. Bryan Garner and the late Justice Antonin Scalia discuss 53 canons in their book on the interpretation of legal text and reject 13 ideas that others, including Eskridge, recognize or use as canons.[9] Because there is no agreement on the universe of canons, some of the interpretive devices discussed here may not appear in any existing catalogue of canons, particularly the metaphor and metamorphosis devices.

Moreover, canons of construction are directive, that is, they suggest what a particular statutory phrase means. For instance, the plain meaning canon instructs us to interpret a statute to mean exactly what it says. Some interpretive devices, however, are not directive. For instance, the metaphor device does not direct a particular meaning, but suggests a way of visualizing an abstract legal issue or proposition to make it more understandable. The earlier chapter on "Addition" observes that metaphors have been quite important in decisions interpreting that element.[10] In those decisions, the uses of metaphor

6. Count the numbers of judicial interpretations for each element listed *infra* in Table B: 63 for Addition, 70 for Pollutant, 142 for Navigable Waters, and 72 for Point Source. Add the sums to derive the total number of judicial interpretations (347) in the 258 decisions listed in *infra* Table A.

7. Table B is displayed in four subparts, one each for Addition, Pollutant, Navigable Waters, and Point Source. The data is drawn from but is slightly different than the data in the Tables B from the earlier articles cited *supra*, note 3. One difference between Table B in this book and the tables in the previously published articles is that the former use a commonly numbered list of interpretive devices set forth on the first page of Table B herein, rather than the differently numbered lists in each of the earlier tables.

8. William N. Eskridge, Dynamic Statutory Construction 323-33 (1987).

9. Antonin Scalia & Bryan A. Garner, Reading Law: The Interpretation of Legal Texts (2012). Compare with the more extensive universe of canons identified by Eskridge.

10. *See supra* Chapter 1, Section III.A.3.c., note 241.

are striking and memorable, more so than the uses of canons of construction in the decisions. This chapter considers why metaphors have been important in interpreting the "addition" element but not the other elements.

The legitimacy of the canons of construction has been subject to academic dispute for some time. Karl N. Llewellyn noted more than half a century ago that for every canon that may apply in a given situation, there is an opposite canon, leading to the question whether the canons have any value in statutory interpretation.[11] Because there is no canon instructing a court how to decide which of two competing, opposite canons governs in a particular situation, the canons may be seen as just handy devices for courts to justify the interpretive results to which they are independently inclined.

The canons of construction are sometimes described as devices to identify legislative intent when statutory language is ambiguous, raising both practical and separation-of-powers concerns.

The practical concern is whether legislative intent can exist beyond the enacted language of a statute. The usual pieces of evidence cited for legislative intent are the remarks of individual members of the U.S. Congress on the floors of the U.S. House of Representatives or the U.S. Senate, or groups of such members in committee and subcommittee reports, usually written by staff. However, remarks by individual members of Congress, by definition, do not identify the intent of Congress as a whole. Nor do reports of committees or subcommittees. The only unquestionable indication of legislative intent is the wording of the statute. The separation-of-powers concern is another side of the first concern. Gleaning congressional intent from statements by individual members of Congress, groups of members, or congressional staff allows them to usurp the Article I authority of Congress.

Legislative intent is even murkier when statutory wording is deliberately ambiguous, poorly masking political compromise. Unless a court decides that ambiguous statutory language is unconstitutionally vague, judicial interpretation of deliberately ambiguous statutory language must inevitably come down on one side or another of the congressional compromise. Do courts thereby usurp the Article I authority of Congress? This dilemma helps explains the enduring use of canons of construction, despite their recognized shortcomings. A court must decide the case before it, even when the applicable statute is ambiguous. Unless a judge throws up her hands and protests

11. Karl N. Llewellyn, *Remarks on the Theory of Appellate Decision and the Rules or Canons of How Statutes Are to Be Constructed*, 3 Vand. L. Rev. 395, 401 (1950). For a more contemporary criticism of the same nature, see Richard A. Posner, *Statutory Interpretation—In the Classroom and in the Courtroom*, 50 U. Chi. L. Rev. 800, 806.

she has no basis to interpret the statute,[12] the judge has only the means judges have used for decades, indeed for centuries, to make such resolutions: the canons of construction.[13]

The observations made in Part I of this chapter relate to statutory interpretation in general. What interpretive devices do the courts use and when do they use them? Do the interpretive devices used affect outcomes? Do the interpretive devices used change over time? If so, does the change represent evolution in jurisprudence generally or the evolution and maturation of judicial interpretations of the statutory text at issue? We might expect both. The lessons discussed in Part II of this chapter are comparative. The four elements of the water pollution offense studied in earlier chapters are very different in nature. Do those differences cause courts to approach interpreting each of the elements in singular manners? We might expect that as well.

Much of the analysis in this chapter is based on comparing numbers of decisions or judicial interpretations in the decisions calculated for different purposes. For instance, it concludes that legal analysis is deeper as a case ascends up the judicial appellate chain, based, in large part, on the numbers of devices that the U.S. Supreme Court, circuit courts of appeals, and district courts used to interpret the elements. The text states the results of these number counts, but does not reproduce the author's mathematical calculations. Footnotes explain how the author made those calculations from data *infra* in Tables A and B. The reader can verify the numbers by making the same calculations.[14] Indeed, the reader may be able to make other entirely different, but pertinent, calculations and observations from the data.

I. Conclusions Regarding Statutory Interpretation Generally

A. *Depth of Legal Analysis Increases With Level of Court*

In theory, the scrutiny of a legal issue should become more intense as a case works its way up the appellate ladder. At each step in the appeal process,

12. *See* United States v. Robison, 521 F. Supp. 2d 1247 (N.D. Ala. 2007), in which a senior judge threw up his hands and instructed the clerk to reassign the case for trial because *Rapanos v. United States*, 547 U.S. 715, 36 ELR 20116 (2006), and its progeny made it impossible for him to interpret the CWA offense element "navigable waters."
13. *See* Jonathan R. Macey & Geoffrey P. Miller, *The Canons of Statutory Construction and Judicial Preferences*, 45 VAND L. REV. 647 (1992); Lawrence C. Marshall, *The Canons of Statutory Construction and Judicial Constraints: A Response to Macey and Miller*, 45 VAND. L. REV. 673 (1992).
14. In doing so, however, the reader must be careful to observe the difference between numbers of decisions in Table A and numbers of judicial interpretations in decisions *infra* in Table B.

the number of contested legal issues usually decreases, while focus on the remaining issues usually becomes narrower and more precise. Moreover, an appellate court has the advantage of both the analyses of the legal issues by the parties before it and by the courts and parties in the courts below. When the appellate court is the Supreme Court and the federal circuits have split on the issue before the Court, these analyses could include the opinions of several federal circuits and several district courts, as well as briefs by the parties before them. And, of course, the Supreme Court will also have the advantage of analyses in amici briefs and in a brief from the Solicitor General, representing the United States as a party or as amicus.[15] The circuit courts may occasionally have the benefit of briefs by amici and the Solicitor General, but the Supreme Court almost always does.

A logical indication of the depth of judicial analysis of an issue is the number of interpretive devices a court uses in its opinion analyzing the issue. In general, a decision using multiple interpretive devices exhibits a greater depth of legal analysis than a decision using only one or two interpretive devices to resolve the issue. Not surprisingly, the numbers of devices courts used in decisions interpreting an element of the CWA offense increased as the level of the courts rendering them increased. Thus, while courts on average used 2.7 interpretive devices to analyze an element,[16] Supreme Court analyses used an average of 4.5 interpretive devices, circuit court analyses used an average of 3.0 interpretive devices, and district court analyses used an average of 2.3 interpretive devices.[17] By this objective measure, legal analysis of the contested statutory interpretation becomes deeper as the level of the interpreting court increases.

B. *Environmentally Positive Results Decrease With Level of Court*

Environmentally positive results in the context of the decisions analyzed in this book are interpretations of an element of the CWA offense finding

15. Supreme Court Rule 37.4 authorizes the Solicitor General to file an amicus brief without either the consent of the parties or a motion approved by the Court. In addition, the Court often requests the Solicitor General to file an amicus brief in cases in which the government is not a party. REBECCA MEA SALOKAR, THE SOLICITOR GENERAL: THE POLITICS OF LAW 5, 18, 142-45 (1992). For a fascinating discussion of the role of the Solicitor General generally, see LINCOLN CAPLAN, THE TENTH JUSTICE: THE SOLICITOR GENERAL AND THE RULE OF LAW (1988).

16. To calculate the average number of interpretive devices used in all judicial interpretations of the elements, add the number of interpretive devices shown *infra* in Column F of Table B for every element by all levels of court and divide by the number of interpretations for all elements by all levels of court (347), as shown *infra* in Table B.

17. To calculate the average number of interpretive devices used in judicial interpretations by a particular level of court, add the number of interpretive devices shown *infra* in Column F of Table B for every decision interpreting every element by that level of court and divide the sum by the number of decisions interpreting all of the elements by that level of court, as shown *infra* on Table B.

jurisdiction rather than denying jurisdiction—for example, interpreting an element expansively as opposed to narrowly. Finding jurisdiction authorizes greater control of water pollution, usually an environmentally positive result, while denying jurisdiction restricts control of water pollution, usually an environmentally negative result.[18] Overall, courts reached environmentally positive results in 76% of the interpretations examined here.[19] District courts reached environmentally positive results in 80% of the interpretations examined here, circuit courts in 70%, and the Supreme Court in 54%.[20]

Do not assume, from the decreasing percent of environmentally positive interpretations as the level of court increases, that appellate courts generally and the Supreme Court in particular, are hostile to water pollution control or environmental protection. The author's earlier analysis of Supreme Court opinions under the CWA determined that although the Supreme Court's CWA decisions statistically suggested an unfavorable inclination to water pollution control, they did not demonstrate that the Court had an anti-environmental bias.[21]

For instance, many of the Court's decisions were based, in whole or in part, on deference to the U.S. Environmental Protection Agency's (EPA's) interpretation of the CWA, a general doctrine of administrative law rather than one specific to the CWA or environmental protection. In such decisions, the Court asks only if the statute is ambiguous on the contested issue and, if so, whether the administering agency's interpretation of the statute on that issue is a reasonable one.[22] When courts reach environmentally negative results based on deference to EPA's interpretation of its statute, it is the Agency's interpretation of the CWA that is environmentally negative rather than the Court's interpretation; the Court is merely affirming that while EPA's interpretation may not be the only reasonable one, it is at least among the set of reasonable interpretations.[23]

---

18. While in theory there may be situations in which the reverse is true, the author does not believe any of the decisions analyzed in this book present such situations.
19. To calculate the percent of environmentally positive results for all interpretations, count the pluses (+) *infra* in Column C of Table B for all elements by all levels of court (260) and divide the sum by the total number of interpretations (347).
20. To calculate the percent of environmentally positive interpretations of the elements by a particular level of court, count the pluses (+) for interpretations by that level of court *infra* in Column C of Table B for all elements and divide that sum by the total number of interpretations by that level of court for all elements as shown *infra* on Table B.
21. Jeffrey G. Miller, *The Court's Water Pollution Jurisprudence: Is the Court All Wet?*, 24 VA. ENVTL. L.J. 125 (2005).
22. *See* Chevron, U.S.A., Inc. v. Natural Res. Def. Council, 467 U.S. 837, 14 ELR 20507 (1984); Mead Corp. v. United States, 533 U.S. 218 (2001).
23. The author's earlier article concluded, however, that while the Court may be agnostic on environmental protection, it is increasingly antagonistic toward citizen participation in administrative and enforcement processes, which in turn has led to increasingly environmentally negative results.

If the increasing percent of negative environmental interpretations of the elements observed as cases move up the judicial chain is not caused by hostility toward clean water by appellate courts, perhaps the increasingly negative environmental results are caused by the greater depth of legal analysis, which also occurs as cases move up the judicial chain. That suggestion can be tested by comparing the environmental results of lower court interpretations exhibiting greater depths of analysis with lower court interpretations generally, removing the variable of court level from the analyses. District court interpretations, of which 80% have positive environmental results, have only 67% positive environmental results when they use five or more interpretive devices.[24] Similarly, circuit court interpretations, of which 70% have environmentally positive results, have only 50% environmentally positive results when they use six or more interpretive devices.

This finding confirms that there is a correlation between increasing depth of analysis and increasing environmentally negative outcome. What does this correlation mean? One possibility is that it requires greater depth of analysis to reach an environmentally negative result. That does not make intuitive sense. The more likely possibility is that courts feel compelled to provide greater justification to the public for environmentally negative results interpreting a statute designed to protect the environment than for environmentally positive results interpreting that statute.

C. *Courts Commonly Use Only a Small Number of Interpretive Devices*

As discussed above, there is no agreed-upon catalogue of interpretive devices or canons of statutory construction. Despite a large but indeterminate number of devices available, the decisions studied in this book used only 25 devices to interpret the elements of the CWA.[25] But even that number is deceptively large, for the courts used all but a few of the 25 infrequently. Courts used the 25 devices a total of 940 times in the 347 judicial interpretations decisions

24. To calculate the percent of positive environmental results for decisions by a particular level of court using a particular number of interpretive devices, add the total number of decisions with positive results, indicated with a plus (+) *infra* in Column C on Table B for all elements by that level of court also showing the desired number of interpretive devices used *infra* in Column F on Table B, for that level of court and divide that sum by the total number of decisions by that level of court for all elements. The chapter uses six or more devices for circuit court decisions and five or more devices for district court decisions because circuit courts use more interpretive devices than district courts. Because this comparison is between average decisions and those using the largest numbers of devices, it was appropriate to use a larger cutoff for circuit court decisions than for district court decisions.

25. The devices are listed *infra* in Table B Explanation for Columns in Table B for Each Element, and are shown *infra* in Column E of Table B.

studied.[26] But in 47% of the 940 times that courts use an interpretive device, they used only two: precedent (268 times), and plain meaning (175).[27] This chapter labels these the primary devices (used more than 105 times each), at least for the purposes of these analyses.

The courts used one of only seven additional devices in another 42% of the 940 total uses: broad interpretation (104 uses); deference to agency interpretation (93); structure of the statute (70); legislative history (67); avoid absurd results (32); harmonize with other statutes (30); and avoid constitutional issues (11).[28] The chapter labels these the secondary devices (each used between 10 and 105 times). Courts used the remaining 16 interpretive devices between one and nine times.[29] The chapter labels these the tertiary or incidental devices. The courts' use of the tertiary devices is limited in another way. They used 14 of the 16 tertiary devices to interpret only one or two of the four elements.[30] These 14 narrowly used devices constitute more than one-half of the 25 total devices used and more than two-thirds of the 16 tertiary devices used.

With all of the interpretive devices available, why would courts use so few of them in these decisions? Why would they use only nine of the 25 more than 1% of the time?[31] Why would they use almost one-half of the devices to interpret only one or two of the four elements? The answers to these questions lie in the nature of interpretive devices. Most, by their very wordings, are of limited applicability. For instance, courts used the two tertiary devices "interpret exceptions narrowly" and "if a rule has an exception, in the absence of the exception, the rule applies" a total of 14 times.[32] The

26. To calculate the number of times the interpretations used interpretive devices (940), count the numbers *infra* in Column F of Table B for each element and add the four totals.

27. To calculate the number of judicial interpretations using a primary interpretive device, count the numbers of times each of interpretive devices 23 and 22 (precedent and plain meaning) is listed in Column F of Table B for all elements. To calculate the percent that figure represents of the total number of interpretations using a particular interpretive device, divide that figure by the total number of times an interpretive device was used (940).

28. To calculate the number of judicial interpretations using secondary interpretive devices, count the number of times each of interpretive devices 1, 4, 5, 7, 16, 21, and 24 is listed in Column E of Table B for all elements. To calculate the percent that figure represents of the total number of times the interpretations used an interpretive device, divide that figure by the total number of times an interpretive device was used (940).

29. To calculate the number of interpretations using tertiary interpretive devices, count the number of times each of interpretive devices 2, 3, 6, 8, 9, 10, 11, 12, 13, 14, 15, 17, 18, 19, 20, and 25 is listed *infra* in Column E of Table B for any element.

30. Count the number of interpretations for which both (1) interpretive device 1 or 2 appears *infra* in Column F of Table B for all elements; and (2) one or more of the tertiary interpretive devices is listed *infra* in Column E of Table B for any element.

31. Because none of the tertiary devices was used more than nine times, the use of any of them was less than 1% of the total of 940 uses of interpretive devices in these decisions, as calculated above.

32. Count the number of times interpretive device 11 or 12 is listed *infra* in Column E of Table B for all elements.

courts' limited use of these devices follows from the fact they come into play only to interpret statutory exceptions. Because only two of the four elements studied have exceptions, and they are narrow, courts are not likely to use these devices frequently enough to place them in the primary or secondary groups of devices.

Other tertiary interpretive devices are of similar limited applicability. For instance, courts used the rule of lenity only twice.[33] That, too, is understandable, for lenity applies only in criminal prosecutions. Although the interpretation of any of the four elements may be at issue in a criminal prosecution, the limited number of criminal prosecutions compared to other types of legal actions in the interpretations studied (31 of 347),[34] significantly lessens the likelihood of lenity being a primary or secondary interpretive device in these decisions.

Similarly, the canon of interpreting a statute to avoid constitutional issues is used only 11 times.[35] It applies only when a constitutional issue is evident, and in the cases analyzed here, the only constitutional issue apparent is with regard to federal Commerce Clause jurisdiction to regulate pollution in navigable waters—the "navigable waters" element of the CWA offense.[36] Likewise, the interpretive device of honoring federalism was used only five times.[37] As with avoiding constitutional issues, it applies only when the constitutional or traditional boundary between federal and state powers might be impinged. Again, in these decisions, federalism is most likely to be involved at the intersection of federal and state jurisdiction under the Commerce Clause. Other tertiary interpretive devices are of similarly limited applicability.[38]

Many of the other canons of construction catalogued by Eskridge, Justice Scalia, and Garner are by their nature simply not susceptible to use interpreting the elements of the water pollution offense. For instance, the seventh canon listed by Eskridge, "'[m]ay' is usually precatory, whereas 'shall' is usually mandatory,"[39] is irrelevant to interpretation of the elements of the CWA

33. Count the number of times interpretive device 17 is listed *infra* in Column E of Table B for any element.
34. Count the number of times "Crim." is listed *infra* in Column D of Table B for any element.
35. Count the number of times interpretive device 5 is listed *infra* in Column E of Table B for any element.
36. Interpretive device 5 is listed *infra* in Column E in Table B 10 times for navigable waters, and only once for any other element.
37. Interpretive device 15 is listed *infra* in Column E of Table B only for navigable waters.
38. For instance, the canon of interpreting waivers applies only when a statutory provision includes a waiver. There are no waivers in any of the elements of the CWA. In the judicial interpretation decisions analyzed in this book, the issue arises in the context of parties waiving their rights to litigate a legal issue by entering a consent decree. United States v. Rueth Dev. Co., 335 F.3d 598 (7th Cir. 2003). The canon of interpreting a statute to harmonize with other statutes applies only when two statutes overlap or possibly collide.
39. *See* Eskridge, *supra* note 8, at 323.

offense because none of the elements or their statutory definitions use either "may" or "shall." Although the author makes no systematic attempt here to narrow the canons identified by those authorities to canons that may be susceptible to interpreting the elements of the CWA offense, many other canons they identify are irrelevant to those elements.[40] In conclusion, only a handful of the interpretive devices are generally useful; many canons are useful only in limited situations not at issue when interpreting the elements of the CWA offense.

D. *Uses of Interpretive Devices Change Over Time*

Neither the identity nor the order of the frequency of use of the primary interpretive devices changed between the earliest and latest sets of interpretations.[41] In the earliest set of interpretations, the two most used devices and the numbers of their uses were precedent (42) and plain meaning (37).[42] In

---

40. For instance, see *id.* at 325 (the "rule against congressional interference with the president's authority over foreign affairs and national security"); *id.* (the "rule against congressional curtailment of the judiciary's 'inherent powers' or its equity powers "); *id.* at 326 (the "presumption against application of federal statutes to state and local political processes"); *id.* at 327 (the "[r]ule against state taxation of Indian tribes and reservation activities"); and *id.* at 328 (the "rule against extraterritorial application of U.S. law").

41. The earliest and latest sets of interpretations used in this analysis are the 20% of the judicial interpretations that are the first in time and the 20% of the judicial interpretations that are the latest in time interpreting each element. The earliest and latest sets of interpretations cannot be reduced to periods of years for all elements because they are calculated by percentages of interpretations of each element rather than by years. Even for a particular element, the uniform percentages of early and late decisions sometimes does not conform to particular years, in which case district court decisions are dropped until the requisite percentage is reached. For ease of identification, these judicial interpretations are highlighted *infra* in Table B. The highlighted interpretations are visually easy to distinguish as the earliest and latest for each court, except for the Supreme Court's interpretations of "pollutant," all of which are highlighted. The first Supreme Court interpretation of pollutant is in the earliest set of interpretations and the second and third are in the latest set of interpretations. For addition, there are 12 early and 12 late interpretations. The earliest are 100, 107, 108, 110, 111, 116, 123, 125, 238, 245, 247, and 249; and the latest judicial interpretations are 5, 12, 18, 24, 26, 27, 37 136, 145, 152, 154, and 160. For pollutant, there are 14 early and 14 late judicial interpretations. The earliest are 10, 11, 105, 110, 117, 120, 122, 124, 235, 236, 238, 241, 246, and 247; and the latest are 3, 27, 40, 53, 130, 133, 146, 152, 157, 159, 166, 167, 169, and 170. For navigable waters, there are 27 early and 27 late judicial interpretations. The earliest are 108, 109, 112, 114, 115, 116, 118, 120, 121, 126, 127, 235, 236, 237, 239, 241, 243, 248, 249, 250, 252, 253, 254, 255, 256, 257, and 258; and the latest interpretations for navigable waters are 1, 15, 16, 19, 21, 22, 23, 24, 25, 26, 29, 30, 128, 129, 130, 131, 132, 134, 135, 137, 139, 142, 142, 143, 144, 1471, 15, 16, 19, 21, 22, 23, 24, 25, 26, 29, 30, 128, 129, 130, 131, 132, 134, 135, 137, 139, 142, 142, 143, 144, 147, and 153. For point sources, there are 14 early and 14 late judicial interpretations. The earliest interpretations are 108, 110, 111, 113, 116, 119, 123, 240, 241, 242, 244, 247, 251, and 259; and the latest judicial interpretations for point source are 2, 12, 13, 14, 17, 19, 20, 22, 27, 138, 140, 141, 148, and 149.

42. Count the number of times interpretive devices 23, 22, and 16 (precedent, plain meaning, and legislative history) are listed *infra* in Column E of Table B for all elements for the earliest set of interpretations identified in note 41.

the latest set of interpretations, the two most used devices and their numbers of uses were precedent (58) and plain meaning (38).[43]

During the same period, however, the order and the frequency of use of the secondary interpretive devices changed considerably. In the earliest set of interpretations, the numbers of uses for the secondary interpretive devices were legislative history (26), broad interpretation (25), structure of the statute (17), deference (10), harmonize with other statutes (6), and avoid absurd results (2).[44] In the latest set of interpretations, the numbers of uses for the secondary interpretive devices were deference (24 uses), broad interpretation (17), structure (15), avoid absurd results (6), harmonize with other statutes (8), and legislative history (5).[45] Although the secondary interpretive devices include "avoid constitutional issues," the decisions falling within the earliest and latest sets of decisions do not use it. Between the earliest and latest sets of interpretations, legislative history moved from the most used of the secondary devices to the least used and deference moved from the fourth most used to the most used. Although broad interpretation was the second most used device in both periods, its use declined from 24 in the earliest period to 17 in the latest period.

The most remarkable change over time is the drastic decline in the use of legislative history to interpret the elements, coupled with a less substantial decline in the use of the broad interpretation canon. In the earliest set of decisions, courts used legislative history in 26 interpretations, while in the latter period, they used legislative history in only five interpretations. Use of the broad interpretation canon declined from 25 to 17 interpretations between the same periods. The two are causally linked because legislative history emphasizes that "navigable waters" should be interpreted broadly.[46] Without reading the legislative history, courts will not know that Congress intended them to interpret the navigable waters term broadly. Thus, uses of the legislative history and broad reading devices go hand-in-hand. While the courts' use of both of these devices declined, the use of precedent between

43. Count the number of times interpretive devices 23 and 22 (precedent and plain meaning) are listed *infra* in Column E of Table B for all elements for the latest set of interpretations identified in note 41.

44. Count the number of times interpretive devices 16, 4, 24, 6, 21, and 1 (legislative history, broad interpretation, structure of the statute, deference, harmonize with other statutes, and avoid absurd results) are listed *infra* in Column E of Table B for all elements in the earliest set of interpretations identified in note 41.

45. Count the number of times interpretive devices 7, 4, 24, 1, 21, and 16 (deference, broad interpretation, structure of the statute, avoid absurd results, harmonize with other statutes, and legislative history) are listed *infra* in Column E of Table B for all elements in the latest set of interpretations identified in note 41.

46. *See supra* discussion of legislative history of "navigable waters" in Chapter 3, Section II.B.

the two periods increased from 42 to 58, the use of plain meaning increased from 37 to 38, and the use of deference increased from 10 to 24. Although the increases in the uses of precedent and deference may in part reflect the substitution of these devices for the recently disfavored legislative history and broad interpretation devices, the increases also represent the growth over time of the number of precedents available for citation and the number of EPA interpretations to which deference is accorded.

These changes in the uses of interpretive methods may have caused or at least influenced the outcomes of decisions. In the earliest set of interpretations, courts reached environmentally positive results in 53 of the 67 interpretations (80%), while during the most recent set of interpretations, they reached environmentally positive results in 48 of the 67 interpretations (72%).[47] The drop was most remarkable in interpretations of "navigable waters," where environmentally positive results dropped from 25 (92%) to 19 (70%) between the same periods.[48] In these interpretations of navigable waters, the use of legislative history dropped from 16 in the earliest interpretations to one in the latest ones and the use of broad interpretation dropped from 17 in the earliest interpretations to eight in the latest ones.[49] As noted above, congressional admonition to interpret an element broadly is more prominent in the legislative history of navigable waters than in the legislative history of the other elements.[50] The precipitous decline in judicial cognizance of these congressional voices for the liberal construction of the CWA offense's elements generally, and of navigable waters in particular, appears to have caused the decline in the courts' environmentally positive interpretations of the elements between the earlier and later decisions.

The decline in the use of legislative history to interpret the elements generally and navigable water particularly is attributable directly to attacks on the legitimacy of legislative history as an interpretive device by the new textualists, led by Justice Scalia.[51] Their main argument is that the intent of Congress is embodied in the statute itself rather than in committee reports or

---

47. Calculate the number of times (+) is listed *infra* in Column C of Table B for all elements for the earliest and latest sets of interpretations identified in note 41 and divide by the total number of the earliest and latest sets of interpretations.
48. Calculate the number of times (+) is listed in Column C of Table B for the Navigable Waters element for the earliest and latest sets of interpretations identified in note 41. Divide the sums by 27, the number of interpretations of Navigable Waters in each period.
49. Count the number of times interpretive device 4 is used *infra* in Column E of Table B for all elements, for the earliest and latest sets of interpretations identified in note 41.
50. Compare the discussions of legislative history regarding navigable waters in *supra,* Chapter 3, Section II.B., with the discussions of legislative history regarding addition, pollutant and point source *supra,* Chapter 1, Section I., Chapter 2, Section I.B., and Chapter 4, Section I.B.
51. *See* Eskridge, *supra* note 8, at 207-38; Scalia & Garner, *supra* note 9, at 391-96.

remarks of individual legislators.[52] Even if statements by individual members or groups of members can be taken at face value,[53] they represent only the intent of individual members or members of committees (or of the majorities of their members), not of Congress itself. If the intent of Congress is represented only in the statute it enacted, then the plain meaning of its text is the primary and most important interpretive device.[54]

While this criticism of legislative history suggests caution against overreliance on it, the criticism does not warrant abandoning it altogether. The criticism is not the only or necessarily the best analysis of the subject.[55] And when applied to the CWA, it has an unduly negative impact, for it ignores the overwhelming legislative support for the statute and its broad interpretation.[56] Of course, Justice Scalia is averse to the broad interpretation canon as well.[57]

The other conspicuous changes in the numbers of uses of the interpretive devices between the early and late periods are the relative uses of plain meaning and precedent. In the earlier period, courts used precedent in 42 interpretations and plain meaning in 37. In the later period, courts used precedent in 58 interpretations, an increase of 16, and plain meaning in 38 interpretations, an increase of one.[58] The much greater increase in the use of precedent than in the use of plain meaning reflects the growing body of precedent available to support interpretation of the elements. At the outset of CWA litigation, almost by definition, there was no precedent (except for navigable waters) to draw on and close examination of the plain meaning, legislative history, and structure

52. *See* Scalia & Garner, *supra* note 9, at 369-90. *See also* Bank One Chicago, N.A. v. Midwest Bank & Trust Co., 516 U.S. 264, 279-80 (1996) (Scalia, J., concurring in part and concurring in the judgment); United States v. Estate of Romani, 523 U.S. 517, 535-37 (1998) (Scalia, J., concurring); Blanchard v. Bergeron, 489 U.S. 87, 97-99 (1989) (Scalia, J., concurring); Thompson v. Thompson, 484 U.S. 174, 191-92 (1998) (Scalia, J., concurring).

53. Query whether individual members of Congress voting to enact the CWA in 1972 intended a particular meaning for addition or for the other elements and, if they did, whether we can trust the members to have disclosed their true intent in their floor remarks or otherwise. Would they have voted against the 200-page-long CWA if they did not agree with its definition of one of the elements? If not, was their intent to have another definition? What interpretive value is their intent?

54. Justice Scalia tells us plain meaning is the basic and most important canon. Scalia & Garner, *supra* note 9, at 56-58, 69-77. Although not a textualist, Eskridge agrees. Eskridge, *supra* note 8, at 223. Others contend plain meaning is the starting place for any interpretation of a statute and hence is not a canon of construction; the canons come into play only if there is no plain meaning. *See* Abner J. Mikva & Eric Lane, Legislative Process 689 (3d ed. 2009).

55. Stephen Breyer, *On the Uses of Legislative History in Interpreting Statutes*, 65 S. Cal. L. Rev. 845 (1992).

56. To enact the CWA, Congress voted by a supermajority to override President Richard Nixon's veto. *See* Rapanos v. United States, 547 U.S. 715, 787, 36 ELR 20116 (Stevens, J., dissenting).

57. Scalia & Garner, *supra* note 9, 364-66 ("The false notion that remedial statutes should be liberally construed" is one of their "Thirteen Falsities Exposed.").

58. Compare the number of times interpretive devices 22 and 23 (plain meaning and precedent) are listed *infra* in Column E of Table B for all elements, for the earliest and latest sets of interpretations identified in note 41.

of the statute were the primary means available to courts to interpret it. Indeed, in 1980 (eight years after the enactment of the CWA), there were only 30 precedents interpreting a few issues under the four elements, and 20 of them were precedents for interpreting navigable waters.[59]

After several decades of CWA litigation, however, there are now more than 347 interpretations of the elements[60] illuminating multiple aspects of the elements and enabling courts in many cases to decide issues based on precedent alone. Indeed, the number of decisions using only one interpretive device grew from 16 in the earliest set of interpretations to 24 in the latest set, and 15 of the 24 later interpretations use precedent as the sole device, while only four of the earlier 16 interpretations did so.[61] The scant increase in the use of plain meaning over time does not reflect jurisprudential disfavor of that device—indeed, it is the preferred analytical tool of the same new textualists who advocate the abandonment of legislative history. Rather, it indicates a steady use of plain meaning in contrast to the great increase in the use of precedent as the number of precedents available for use increases.

E. *Interpretive Devices Are Not All Outcome-Neutral*

One apparent question arising from this study is whether use of particular interpretive devices leads to predominantly positive or negative environmental results. Because the devices were not designed only to interpret environmental statutes, however, the question is a broader one. Are particular interpretive devices designed to lead to predominately expansive or restrictive interpretations of statutes? Comparing the percentage of all interpretations studied reaching environmentally positive results with the percentage of interpretations reaching environmentally positive results using particular interpretive devices may shed some light on this issue. Overall, the decisions studied in this book reached environmentally positive results or expansive interpretations in 76% of their interpretations of the four elements.

Predictably, environmentally positive results or expansive interpretations were similar for decisions using the precedent and plain meaning devices: 76% for precedent and 73% for plain meaning.[62] Because

59. Count the number of judicial interpretations dated 1980 or before *infra* in Column B of Table B for all elements.

60. While the book examines 258 reported decisions with 347 judicial interpretations of the elements through 2013, the number of precedents has continued to grow after that date.

61. Compare the number of times 1 is listed *infra* in Column F of Table B for all elements in the earliest interpretations and the latest interpretations identified above, and compare the number of times in those interpretations that the sole device listed *infra* in Column E is 23 (precedent).

62. To calculate the percent of environmentally positive results using interpretive device 22 or 23 (plain meaning and precedent), count the total number of times that 22 or 23 is listed *infra* in Column E

these two primary interpretive devices were used in most of the decisions studied, their positive environmental results or expansive interpretations could not differ markedly from the overall environmental results of all the decisions. In that respect, their use interpreting the elements is environmentally neutral. The only frequently used device that reached environmentally positive results or expansive interpretations significantly more often than the average is broad interpretation, used in decisions reaching environmentally positive results in 83% of the times used.[63] That should be no surprise, for the canon explicitly calls for expansive interpretations of remedial statutes.

Most of the other devices reached positive environmental results or expansive interpretations in less than 75% of their uses. That is consistent with the earlier conclusion that courts used more interpretive devices to reach environmentally negative decisions or restrictive statutory interpretation than to reach environmentally positive decisions. Devices used to achieve environmentally positive results or restrictive interpretations 50% or less of the time, however, may depart sufficiently from the norm to be examined as potential environmentally negative or restrictive devices. They are: rule of the last antecedent, used four times, 50% environmentally positive results; avoid constitutional issues, used 11 times, 45% environmentally positive results; *ejusdem generis*, used five times, 40% environmentally positive results; equity, used once, 0% environmentally positive results; *expressio unius*, used three times, 33% environmentally positive results; honor federalism, used five times, 40% environmentally positive results; rule of lenity, used twice, 50% environmentally positive results; *noscitur a sociis*, used twice, 50% environmentally positive results; and harmonize with other statutes, used 31 times, 42% environmentally positive results.[64]

Courts used most of these devices so infrequently that there is insufficient data to analyze. Two of the devices, however, are secondary interpretive devices: avoid constitutional issues and harmonize with other statutes. These two devices are similar in that they invite or justify abandoning plain meaning, precedent, or even the integrity of the text being interpreted, here the

of Table B for all elements, count the number of times those interpretations have a plus (+) *infra* in column C of Table B for all elements and divide the second sum by the first sum.

63. Although the decisions using the devices of metaphor and interpret waivers narrowly reached environmentally positive results 100% of the time, courts used metaphor only five times and the waiver device only twice, too few to signal that they may be positively oriented interpretive devices. Calculate the percent of environmentally positive results using interpretive device 4 (broad interpretation) in the same manner as for interpretive devices 22 or 23 (plain meaning and precedent), but substituting interpretive device 4 for interpretive devices 22 or 23.

64. Calculate the percent of environmentally positive results using any of these interpretive devices (3, 5, 6, 8, 9, 14, 15, 17, 20, 21) in the same manner as for interpretive device 22 or 23.

CWA, in favor of the effectiveness of another text, such as the U.S. Constitution or another statute. As applied in a case arising under the CWA, they are conceptually environmentally negative devices and the results of their use in fact yield environmentally negative decisions in a significant percentage of cases. This analysis applies as well to some of the tertiary interpretive devices, for instance, honor federalism. That canon too invites abandoning the plain meaning of a statute in favor of an interpretation more favorable to states' rights.

II. Comparative Statutory Interpretation: How and Why the Methods Used to Interpret Each Element Differ

A. How the Natures of the Elements Differ

The compositions of the elements of the CWA offense are very different. "Addition" is abstract and, because it is the action element, is arguably the most important element. Yet, the CWA does not define addition. The statute defines "pollutant" as an exclusive list of 18 materials or categories of materials, with two exceptions.[65] Some of the listed materials are not commonly thought of as pollutants, for example, rock and sand; and many of the categories of materials are very broad, for example, "biological material" and "industrial waste." "Navigable waters" has a long history as a term of art developed by the Supreme Court to establish federal jurisdiction under the Commerce Clause[66] over waters used in the past for navigation, used presently for navigation, or usable in the future for navigation, with reasonable improvements.[67] The statute, however, defines navigable waters with the short phrase "waters of the United States,"[68] having no apparent meaning and no apparent connection with navigability. "Point source" has no common meaning, but is a legislative construct in the CWA. The statute defines point source as a "discernible, confined and discrete conveyance," followed by an inclusive list of 11 examples, and providing two exclusions.[69] Two of the definitions include lists, but they use lists differently. The definition of pollutant limits pollutants to listed materials and categories of materials, while the definition of point source is a "conveyance" and includes a non-limiting list of examples.

---

65. CWA §502(6), 33 U.S.C. §1362(6).
66. U.S. Const. art. III, cl. 8.
67. *See supra* Chapter 3, Section I.
68. CWA §502(7), 33 U.S.C. §1362(7).
69. CWA §502(14), 33 U.S.C. §1362(14).

The compositions of these elements are so diverse that courts naturally use different approaches to interpret them. This part of the chapter will explore those differences.

B. *How the Methods Used to Interpret Each Element Differ*

1. Addition

Addition is the action element in the offense. Because offenses are actions, addition is the central element of the offense. Because addition is the only element without a statutory definition, a regulatory definition, or even legislative history bearing directly on its meaning, it is likely to be the most difficult of the elements to interpret. This likelihood is increased because addition is an abstract idea rather than a concrete thing, like sand, a conveyance, or water (words used in the statutory definitions of the other elements).

Indeed, the author found addition the most difficult of the elements to interpret in this book. One tangible sign of the difficulty was that the author was able to summarize in two paragraphs interpretations with environmentally positive and environmentally negative results in the chapters examining pollutant or point source,[70] but not in the chapters examining addition or navigable waters.[71] The chapters on addition and navigable waters are comparable in length, and both are appreciably longer than the chapters on the other two elements. Because the chapter on navigable waters analyzes twice as many judicial interpretations as the chapter on addition, however, their comparable lengths suggests that addition is the more difficult of the two terms to interpret.[72] Of course, there is no agreed-upon scale of difficulty for interpreting different statutory texts, as there is for performing different dives in diving competitions. However, several observations made from the data set studied in this book lend objective weight to the suggestion that addition is the most difficult element of the CWA offense to interpret.

First, when an element is difficult to interpret, courts can be expected to use more interpretive tools than when an element is easy to interpret. The relative difficulty in interpreting the elements, therefore, is reflected by the average number of devices courts have used to interpret each of them: 3.0 devices for interpreting addition, 2.8 for point source, 2.6 for navigable

70. *See supra* Chapter 2, Section II.A., and Chapter 4, Section III.B. & C.
71. *See supra* Chapters 1 & 3.
72. While it could be argued that the greater number of decisions interpreting navigable waters suggests that that element is the most difficult to interpret, the greater number of navigable waters decisions is more likely explained by another factor, *see infra* Section II.B.3.

waters, and 2.5 for pollutant.[73] Alternatively, the greater number of devices the courts used to interpret addition could reflect the relatively greater percent of environmentally negative decisions interpreting addition than interpreting the other elements and the courts' apparent need to provide greater justification for environmentally negative than for environmentally positive decisions.[74] Although these numbers are not hugely different, together with the other observations made here, they suggest that addition is the most difficult element to interpret.

Next, there are fewer judicial interpretations of addition than interpretations of any of the other elements. There are 63 judicial interpretations of addition; 70 of pollutant; 142 of navigable waters; and 72 of point source.[75] Why does the lower number of addition interpretations suggest relative difficulty of interpretation? Because addition is not defined and is abstract, parties have a hard time framing legal issues based on it, harder than framing issues based on better-defined and less-abstract elements. With fewer parties making an issue of the addition element, fewer court decisions interpret it. The differences between the numbers of interpretations of addition, pollutant, and point source are not huge, but, again, when supported by other factors, they are suggestive that addition is the most difficult element to interpret.

Next, addition is the only element for which courts used the interpretive devices of metaphor and metamorphosis.[76] Neither are canons of construction[77]: They do not instruct us how to interpret an element; rather, they provide visual images to help understand abstract ideas. This does not help us decide which of several interpretations of addition is correct, as some canons of construction, such as legislative history, may do directly. But it does help to understand possible interpretations of addition. Metamorphosis is the phenomenon by which one thing becomes another, as a caterpillar slowly morphs into a butterfly in the cocoon. Often, metamorphosis is fanciful, as a man becoming a beetle in Kafka's novel, *Metamorphosis*, or a man becoming a rhinoceros in Ionesco's play, *The Rhinoceros.*

73. To calculate the average number of devices used to interpret a particular element, add the number of interpretive devices used in Column F of Table B for the interpretations of that element and divide that sum by the number of interpretations for that element listed *infra* on Table B.

74. Only 68% of the decisions interpreting addition were environmentally positive, while 77% of the decisions interpreting pollutant, 77% of the decisions interpreting navigable waters, and 75% of the decisions interpreting point source were environmentally positive. To calculate the percent of environmentally positive interpretations of an element, count the number of interpretations of that element with a (+) *infra* in Column C of Table B for all levels of court and divide the sum by the total number of interpretations of that element.

75. *See supra* note 6.

76. *See infra* Column E of Table B for interpretive devices 18 and 19 (metaphor and metamorphosis).

77. Neither Eskridge nor Justice Scalia and Garner identify them as canons of construction in their books on the canons. *See* Eskridge, *supra* note 8; Scalia & Garner, *supra* note 9.

Thus, in *United States v. Deaton*, the court wrote "once that material was excavated from the wetland, its redeposit in that same wetland *added* a pollutant where none had been before."[78] Several decisions, following arguments by the government, posit that sediment dredged from a wetland suddenly morphs into a pollutant when it is replaced in the same wetland,[79] a fanciful metamorphosis enabling there to be an addition of a pollutant to the wetland and therefore a violation of the CWA. The use of a fanciful metamorphosis in these interpretations, although elegant, suggests that the argument that a violation has occurred may be equally fanciful.

Another interpretive fiction suggested by EPA is the notion that all navigable waters are one, the so-called unitary navigable waters theory. The theory initially sounds plausible, alluding to the scientific notions of the hydrological cycle and the hydrosphere, in which all waters[80] are connected. But the totality of navigable waters is far from either the hydrological cycle or the hydrosphere. Both the hydrological cycle and the hydrosphere include water vapor in the atmosphere, clouds, precipitation, non-navigable surface water, groundwater, and water beyond the control of the United States, while EPA does not claim that any of them are navigable water under CWA §502(7) or any other CWA text. While EPA's unified navigable water fantasy appears to relate to the definition of "navigable waters," its sole purpose is to establish that when a person withdraws polluted water from one body of navigable water and discharges it into a less polluted body of navigable water, he does not add pollutants to the second body of water because both bodies are the same navigable water. Indeed, EPA claims the notion flows from the addition element.[81]

Whenever EPA's interpretations depart so far from the wording of the CWA, they are suspect as pure fantasy. This is illustrated by the metaphors associated with EPA's arguments. For instance, to illustrate the abstract idea

78. 209 F.3d 331, 335-36, 30 ELR 20508 (4th Cir. 2000). The whole passage reads:

> The idea that there could be an addition of a pollutant without an addition of material seems to us entirely unremarkable, at least when an activity transforms some material from a nonpollutant into a pollutant, as occurred here. In the course of digging a ditch across the Deaton property, the contractor removed earth and vegetable matter from the wetland. Once it was removed that material became "dredged spoil," a statutory pollutant and a *type* of material that up until then was not present on the Deaton property. It is of no consequence that what is now dredged spoil was previously present on the same property in the less threatening form of dirt and vegetation in an undisturbed state. What is important is that once that material was excavated from the wetland, its redeposit in that same wetland *added* a pollutant where none had been before.

The court did not explain why earth and vegetable matter were less threatening before they were removed from the ditch bottom than when they were replaced there.

79. *See supra* Chapter 1, Section III.C.1.

80. Or almost all. Some pockets of groundwater have been isolated from other waters for millions of years. But it was once surface water, as perhaps it could be again in the far-off future.

81. *See supra* Chapter 1, Section III.A.3.b., and Chapter 5, Section D.

of addition (or non-addition), the Supreme Court used the metaphor of taking "a ladle of soup from a pot, lift[ing] it above the pot, and pour[ing] it back into the pot" to suggest how a discharger could remove polluted water from a water body and pour it back into the same water body without adding anything to the water body.[82] The better metaphor in this situation would be two soup pots on the same stove, one with split pea soup and another with chicken noodle soup. Lifting a ladle of split pea soup and placing it into the chicken noodle soup adds pollutants to the chicken noodle soup (imagine green chicken noodle soup), even though both vessels are soup pots, both are filled with soup, and both are on the same stove. Several other decisions used metaphors in their analysis of addition.[83]

2. Pollutant

The definition of pollutant is an exclusive list of specific substances, such as rock and sand, and broad categories of substances, such as "biological materials." The exclusivity of the list, the specific nature of the materials listed, and the broad nature of the listed categories of materials suggest that plain meaning will be the predominant device used to interpret the element, the interpretation of the pollutant element will be relatively easy, and the results will be largely environmentally positive.

Courts used precedent and plain meaning more than any other devices to interpret all of the elements. But while they generally used precedent more than plain meaning, courts used plain meaning in 49 of the 70 interpretations of pollutant, and used precedent in only 36 of them. While courts also used plain meaning more than precedent to interpret addition, they did so only by a very slight margin.[84] Moreover, courts used only one interpretive device in 25 of their 70 interpretations of pollutant, and 16 of those 25 interpretations used plain meaning.[85]

---

82. South Fla. Water Mgmt. Dist. v. Miccosukee Tribe of Indians, 541 U.S. 95, 34 ELR 20021 (2004) (quoting Catskill Mountains Chapter of Trout Unlimited, Inc. v. New York, 273 F.3d 481, 492 (2d Cir. 2001)). Both decisions rejected the concept of unitary navigable waters. Both decisions questioned or rejected the concept that discharging a polluted navigable water into a pristine navigable water did not add a pollutant to the pristine water. *See supra* Chapter 1, Section III.A.3.c., note 241.
83. *See supra* Chapter 1, Section III.C.3.
84. Courts used plain meaning 52 times and precedent 51 times to interpret addition. They used plain meaning 22 times and precedent 131 times to interpret navigable waters. And they used plain meaning 42 times and precedent 63 times to interpret point source. To calculate how many decisions interpreting an element used a particular interpretive device, see *supra* note 27.
85. Count the number of decisions with only interpretive device 22 *infra* in Column E of Table B for pollutant.

The predominance of plain meaning in the interpretation of pollutant is a strong indication that the element is relatively easy to interpret. This is confirmed by the relatively smaller number of devices the courts used to interpret pollutant than to interpret the other elements. As noted earlier, the courts used 25 interpretive devices to interpret the four elements. But courts did not use all 25 of the devices to interpret any one of the elements; they used 12 devices to interpret pollutant, 13 to interpret addition, 18 to interpret point source, and 20 to interpret navigable waters.[86]

As discussed above, decisions interpreting more difficult text use more interpretive devices than decisions interpreting less difficult text. The average number of devices used to interpret pollutant is less than the average number of devices used to interpret the other elements: 2.5 for pollutant, 2.8 for point source, 2.6 for navigable waters, and 3.0 for addition. Moreover, courts used only one device to interpret pollutant in 25 of the 70 interpretations of the term.[87] Indeed, the sole device in 16 of the 25 interpretations of pollutant was plain meaning, while precedent was the dominant sole device used to interpret the other elements.[88]

The positive environmental results of decisions interpreting the elements of the CWA offense are 78% for pollutant, 75% for point source, 77% for navigable waters, and 68% for addition.[89] As explained above, courts use more interpretive devices to decide difficult cases and to decide cases with environmentally negative results. Many of the environmentally negative results interpreting pollutant were in decisions in which the pollutant was pervasively regulated by another statute.[90]

The differences in none of these comparisons are great enough by themselves to prove the point. But together they are persuasive that pollutant is relatively easy to interpret and that plain meaning drives its interpretation.

86. To calculate the number of different devices used to interpret an element, count the number of different devices (numbers 1 to 25) *infra* in Column E of Table B for that element.

87. To calculate the percent of interpretations in which the courts used only one device to interpret a particular element, count the number of decisions with a 1 *infra* in Column F of Table B for that element and divide the sum by the total of decisions interpreting that element. Courts used only one device in 24% of the 62 interpretations of addition, 17% of the 71 interpretations of point source, and 28% of the 136 interpretations of navigable waters.

88. Count the number of decisions with a 1 *infra* in Column F of Table B for addition and divide the sum by the total of decisions interpreting addition (62).

89. To calculate the percent of positive environmental results of decisions interpreting any element, count the number of interpretations of that element with a plus (+) *infra* in Column C of Table B for that element and divide the sum by the number of interpretations of that element.

90. *See supra* Chapter 2, Section II.G., notes 49-57.

3. Navigable Waters

Because navigable waters is a term of art the Supreme Court developed in a dozen decisions over a century and a half to define the extent of federal Commerce Clause jurisdiction over waterways, precedent has driven the interpretation of that element from the outset. Courts used precedent to decide whether waters were navigable under the CWA from the first decisions addressing the issue.[91] There was no comparable body of earlier precedent to help interpret the other elements. Courts have used precedent far more often than plain meaning to interpret navigable waters: 131 times as opposed to 22 times. But courts used plain meaning somewhat more than precedent to interpret the other elements: 153 times as opposed to 144 times.[92] This may change over time as more precedent becomes available to interpret the other elements.

Because the interpretation of navigable waters is so precedent-driven, the author organized much of the analysis in the chapter on navigable waters by comparing the analyses and outcomes of lower court interpretations in groups, starting with lower court interpretations occurring before any Supreme Court ruling on the subject, proceeding with lower court interpretations between successive Supreme Court rulings, and ending with lower court interpretations after the latest Supreme Court ruling. There were noticeable differences between the analyses and outcomes of the later groups of interpretations, reflecting the influence of the latest Supreme Court precedent.[93]

For the same reasons, courts have used constitutionally based devices to interpret the phrase. Indeed, courts used the canons "construe statutes narrowly to avoid constitutional issues" and "honor federalism" to interpret navigable waters and did not use them to interpret other elements of the CWA offense.[94]

Despite the fact that the Supreme Court already had defined navigable waters in some detail, Congress defined the term in the CWA as the "waters of the United States," a term having no obvious or traditional meaning. Why would Congress have done so? The relevant committee reports and floor statements of the drafters of the statute are quite clear on this: Congress

---

91. *See infra* Column E of Table B for Navigable Waters and note the earliest court of appeals and district court decisions, using interpretive device 23 (precedent).
92. Count the number of times interpretive devices 22 or 23 (plain meaning and precedent) are used *infra* in Column E of Table B for each of the elements addition, pollutant, and point source and add the three sums.
93. *See supra* Chapter 3, Section IV.A.3.-6.
94. Compare the number of times interpretive devices 5 and 15 (interpret statutes to avoid constitutional issues and honor federalism) are used *infra* in Column E of Table B for navigable waters and for addition, pollutant, and point source.

defined the term as the waters of the United States to cast aside the strait-jacket of the traditional meaning of navigable waters and replace it with a broader term reaching the full extent of Congress' power over waters under the Constitution.[95] An interpreter of navigable waters as it appears in the CWA would be aware of this if she has read the relevant legislative history, but would be puzzled by it if she has not. At the same time, the legislative history does not explain why Congress failed to jettison the term navigable waters altogether and directly use the term waters of the United States as the jurisdictional element of the CWA offense.

It is noted above that the use of legislative history to interpret the elements of the CWA offense declined precipitously between the earliest and latest sets of interpretations because courts questioned its efficacy.[96] That decline is most aggravated for interpretations of navigable waters. In the earliest set of interpretations, courts used legislative history in 16 interpretations of navigable waters and in 11 interpretations of the other elements. In the latest set of interpretations, courts used legislative history in one interpretation of navigable waters and in three interpretations of the other elements.[97]

There was a parallel decline in the courts' use of the canon of broad interpretation in decisions interpreting navigable waters. In the earliest set of interpretations, courts used broad interpretation in 17 decisions interpreting navigable waters, while in the latest set of interpretations, they used it in eight interpretations of navigable waters. In decisions interpreting the other elements, however, courts used broad interpretation steadily in both sets of interpretations, employing it in seven interpretations during the earliest period and in eight interpretations in the latest set.[98] The decline in the use of broad interpretation for navigable water can be directly attributed to the decline in the use of legislative history generally, because the legislative history of navigable waters explicitly and repeatedly urges that the phrase be broadly interpreted and earliest decisions cited that history as a reason to broadly interpret the term.

The most striking difference between interpretations of navigable waters and interpretations of the other elements, however, is between the numbers of interpretations of each element: 63 for addition, 70 for pollutant, 142 for navigable waters, and 72 for point source. The numbers of interpretations of pollutant and point source, the two elements with list-based definitions,

95. *See supra* Chapter 3, Section II.B.

96. *See supra* note 41.

97. Count the number of decisions listing interpretive device 16 *infra* in Column E of Table B for each element in the earliest and latest sets of decisions.

98. Count the number of decisions listing interpretive device 4 *infra* in Column E of Table B for each element in the earliest and latest sets of decisions.

are very close, while the number of interpretations of addition is slightly less. But the number of interpretations of navigable waters is approximately twice the number of interpretations of any of the other elements. Navigable waters is the numerical leader of the elements in other respects. The number of Supreme Court interpretations of navigable waters in the CWA doubles the numbers of the Court's interpretations of any of the other elements: two for addition, three for pollutant, six for navigable waters, and two for point source.[99] However, this is probably a function of the underlying differences in the total numbers of interpretations of the different elements.

The courts also employed more canons of construction to interpret navigable waters (20 canons used) than to interpret addition (13), pollutant (12), or point source (18). While this partly reflects the much larger number of interpretations of navigable water than of the other elements, it also reflects the difficulty courts have had in reconciling the congressional desire that courts interpret navigable waters broadly, as suggested by the congressional definition of navigable waters as the waters of the United States, with the continued congressional use of the more restrictive navigable waters as the element of the CWA offense.

What, then, accounts for the far greater number of interpretations of navigable waters than interpretations of any of the other elements? The answer is not evident from the nature of the term, its definition, or its legislative history. Instead, the answer lies in the predominance of interpretations of navigable waters in cases arising under §404, compared to its lesser importance in interpretations of the other elements. Of the 142 interpretations of navigable waters, 81 or 57% were in cases arising under §404, while of the 208 interpretations of the other elements, 64 or 32% were in cases arising under §404.[100] Indeed, when the number of interpretations in cases brought under §404 are subtracted from the total number of interpretations of each element, the numbers of the remaining interpretations, almost all in cases brought under §402,[101] are far more similar: 42 for addition, 40 for pollutant, 61 for navigable waters, and 59 for point source.[102]

---

99. See the numbers of Supreme Court decisions, for each element *infra* in Table A or B.

100. Count the number of §404 interpretations listed for each element *infra* in Column H of Table B. To calculate the relevant percentages, divide the number of §404 interpretations for each element by the total number of interpretations for each element.

101. Some decisions are under the CWA §401, 33 U.S.C. §1341, the Oil Pollution Act (OPA), 33 U.S.C. §§2701-2761, ELR Stat. OPA §§1001-7001, and the National Environmental Policy Act (NEPA), 42 U.S.C. §§4321-4370f, ELR Stat. NEPA §§2-209.

102. Subtract the number of §404 decisions listed for each element from the total number of decisions listed for each element.

Significantly, cases brought under §404 do not raise serious issues under some of the other elements. For instance, the courts are unanimous that earthmoving equipment used in filling wetlands are point sources[103] and that fill material is a "pollutant."[104] The addition element has not been a fruitful target for challenge in decisions in cases brought under §404.[105] This leaves §404 defendants little to argue, except that the periodically dry wetlands they are developing are not navigable waters. From the outset, it was counterintuitive to many that wetlands, particularly those that are dry much of the year, are navigable waters.[106] Hence, the majority of decisions in cases brought under §404 focused on navigable waters, 61% or 81 of 142, while a minority of interpretations in cases brought under §404 focused on one of the other elements in the CWA offense, 31% or 64 of 205.[107]

Another partial explanation for the greater number of interpretations of navigable waters than of the other elements results from EPA's regulatory treatment of the element. For the other elements, EPA either did not promulgate a regulatory definition (addition) or promulgated definitions that differed only slightly from the statutory definitions (pollutant and point source).[108] EPA did not promulgate a regulatory definition of navigable waters, but instead promulgated a definition of waters of the United States, the congressional definition of navigable waters. By doing so, EPA attempted to redirect attention from the historical use and meaning of the term navigable waters to the apparently broader statutory waters of the United States definition. Moreover, EPA's definition of waters of the United States is very long and detailed, extending far beyond the traditional understanding of navigable waters.[109] This signaled the government's intention to implement the CWA's permit programs, particularly the §404 program, beyond the traditional understanding of navigable waters, inviting additional litigation.

Many of the factors discussed above suggest restrictive rather than expansive interpretations of navigable waters: the use of the canons of "honor federalism" and "interpret statutes narrowly to avoid constitutional issues"

103. *See supra* Chapter 4, Section III.F.2.
104. *See supra* Chapter 2, Section II.F.2.
105. *See supra* Chapter 1, Section III.C.
106. Indeed, neither the wording nor the legislative history of §404 as it was enacted in 1972 mention wetlands nor hint that the section was intended to protect wetlands from filling. *See supra* Chapter 3, Section II.B.2.
107. Compare the number of §404 interpretations *infra* in Column H of Table B for navigable waters with the numbers of §404 decisions for addition, pollutant, and point source. To calculate the percent of §404 decisions under an element, divide the total of §404 interpretations decisions for that element by the total number of interpretations for that element.
108. 40 C.F.R. §122.2. *See infra*, Appendix I.
109. *Id.*

exclusively to interpret navigable waters; virtually abandoning use of legislative history to interpret the element; and curtailing the use of the canon of "broad construction" while the legislative history of navigable waters urged use of that canon. The cumulative impact of these factors is the decline in environmentally positive results interpreting navigable waters from 25 or 97% in the earliest set of interpretations to 19 or 70% in the latest set of interpretations,[110] while the percent of environmentally positive interpretations of all of the elements between those periods remained virtually the same.[111]

4. Point Source

Because the term point source has no inherent meaning, the plain meaning canon of construction is not of much use interpreting the phrase. On the other hand, the CWA's definition of point source is the most grammatically complex and detailed of its definitions of the elements. The statute defines it as a "discernible, confined and discrete conveyance," followed by a non-exclusive list of 11 examples and ending with two exclusions. Plain meaning can be used to interpret the statutory definition, as can interpretive devices centering on lists and exclusions. The complexity of the definition reflects the fact that the term point source is a legislative construct with no previous meaning. That complexity suggests relatively more interpretive devices may be susceptible to use in understanding it that in understanding the other elements.

Because the statutory definition of point source includes a long list of examples, cases in which discharges are through a listed example are usually easy to decide. Indeed, more than one-half the decisions interpreting point source use two or fewer interpretive devices and two or fewer precedents.[112] Where a case does not involve a discharge through a listed conveyance, however, the term point source may be particularly difficult to interpret both because of the complexity of the definition and because it is the only element in the offense to have an opposite in the CWA, a "nonpoint source." The CWA provides very different programs to control discharges from point

---

110. To calculate the percent of positive environmental decisions for the earliest and latest sets of interpretations of navigable waters, *see supra* note 48. To calculate the percent of positive environmental decisions for the earliest and latest sets of interpretations of other elements, repeat the same procedure using the number of positive results *infra* in Column C of Table B for each element and dividing by the total number of interpretations in that set of decisions shown *supra* in note 41.

111. To calculate the percent of positive environmental decisions for the earliest or latest set of interpretations of all elements, count the number of positive results marked by (+) *infra* in Column C of Table B for that set of interpretations and divide by 67 (the number of the earliest interpretations for all elements).

112. Count the number of times that 1 or 2 is listed *infra* in Column E of Table B for point source (41 of the 72 interpretations listed for point source in that column).

sources and nonpoint sources,[113] but it provides no definition of nonpoint source and draws no distinct line between the two. Although by its construction, nonpoint source is the opposite of a point source, that does not tell us whether a nonpoint source is something other than a conveyance or is a conveyance that is not discernible, confined, and discrete, or both.

Both the statute and its legislative history suggest that Congress could not draw a clear line between point and nonpoint sources.[114] Both suggest that precipitation runoff and agricultural waste are usually discharged by nonpoint sources, but not always.[115] Although the statute and its legislative history suggest only these two differences between point and nonpoint sources, most of the discharges of pollutants that courts have held to be from nonpoint sources had nothing to do with either precipitation runoff or agricultural wastes.[116]

The complexity of the statutory definition of point source lends itself to use of a wide variety of devices to interpret it. For instance, interpretive devices bearing on lists and exceptions, including *ejusdem generis, expression unius*, and *noscitur a sociis*, are susceptible for use interpreting only the elements that have lists and exceptions. But the list in the definition of point source is exclusive, while the list in the definition of pollutant is inclusive. The existence of statutory programs to address pollution from both point sources and nonpoint sources suggest the structure of the statute should be commonly used to interpret the point source definition to direct the pollutant discharges at issue into one program or the other. Courts used the structure of the statute interpretive device in 18 interpretations of point source.[117]

Because of the grammatical complexity of the statutory definition of point source, and because Congress and EPA failed to draw a distinct line between point and nonpoint sources, cases in which there is no discharge through a listed example may be unusually difficult to decide. Indeed, courts used 18 of the 25 interpretive devices to construe point source. They used only 13 devices to interpret addition and 12 to interpret pollutant. Although courts used 20 devices to interpret navigable water and 18 to interpret point source, it took them 142 interpretations to do so for navigable water, but only 72 to do so for point source, using an average of 2.8 devices to interpret point source and an average of 2.6 devices to interpret navigable waters.

113. *See supra* Chapter 4, Section III.B. & F.
114. *Id.*
115. *Id.*
116. *Id.*
117. Count the number of times (24) that interpretive device appears *infra* in Column E for Navigable Waters in Table B.

The failure of Congress or EPA to draw a discernible line between point sources and nonpoint sources also explains why courts are only a few steps away from adopting a factors test to help them distinguish between the two in situations involving precipitation flow or agricultural wastewater.[118] The same factors test may be just as useful in many other situations.[119]

III. Conclusions

Studying the interpretations of the four elements of the CWA offense informs us of their meanings. Assuming that those elements are reasonably representative of other statutory text, this study also provides lessons on statutory interpretation itself. Courts used a total of only 25 interpretive devices in the course of 342 interpretations of the elements, using only nine of them in 10 or more interpretations. On average, courts used 2.6 devices in each interpretation; slightly more than 25% of the interpretations used only a single device.

Most of the time courts used a single device, they used plain meaning or precedent. Indeed, plain meaning and precedent were the predominant devices courts used throughout the interpretations, with plain meaning generally dominating at the outset of interpretations, when few precedents were available, and precedent generally dominating later, as the number of precedents available increased. This was not true for the interpretation of navigable waters, for which precedents were available from the outset and which dominated its interpretation throughout.

Other changes in the devices used by courts occurred over time—for instance, the use of legislative history decreased precipitously over time as developing jurisprudence questioned its legitimacy. Courts used a maximum of nine devices in any interpretation, and used five or more devices in only slightly more that 15% of the interpretations. The number of devices used in an interpretation increased as the level of court and the depth of legal analysis increased. The Supreme Court used twice as many devices in its average opinion as district courts used in their average opinions. The Supreme Court rarely used a single interpretive device, while district courts used a single device in nearly one-third of their interpretations.

A few of the interpretive devices inherently encourage expansive interpretations of statutory text (for example, interpret remedial statutes broadly to achieve their goals), while others inherently encourage restrictive inter-

118. *See supra* Chapter 4, Section III.E.
119. *See supra* Chapter 4, Section III.F.

pretations (for example, interpret statutes to avoid constitutional issues or to harmonize with other statutes). In the context of environmental statutes, expansive and restrictive interpretations translate into environmentally positive or environmentally negative decisions. It seems likely that these patterns would be observed interpreting a great deal of other statutory text.

Statutory texts may be different in nature; even the elements of the offense studied in this book are different in nature. Those differences result in courts using different analyses and devices to interpret the elements. A simple example is that statutory text with a list of covered items invites questions whether the list is inclusive or exclusive and, if it is inclusive, whether the item at issue in the case is on the list or, if not, whether the item is similar to or different from the items listed. These questions and their associated canons of construction simply do not apply to statutory text lacking a list of covered items.

Differences in the natures of texts analyzed also lead to the ease or difficulty of interpreting the texts, as is evident from the differences in the natures of the CWA elements studied in this analysis. For instance, pollutant, with its exclusive list of specific materials and broad categories of materials, is relatively easy to interpret by determining whether the material at issue in a case is on the list or within a category of material on the list. On the other hand, addition, which is undefined and abstract, is relatively difficult to interpret. Although there is no scale of difficulty for interpreting statutory texts, several objective factors comparing the judicial interpretations of the addition and pollutant elements of the CWA suggest that addition is more difficult to interpret than pollutant.

More sophisticated analyses comparing the interpretations of different statutory texts or different parts of a statutory text may help understand the text or its interpretation. For instance, the chapter on navigable waters suggests that one of the reasons there are so many more decisions interpreting that element than any of the other elements is that courts were comfortable interpreting it from the outset of CWA litigation: Courts had been interpreting the term for a century and a half in other contexts, whereas they had no prior familiarity with the other elements of the CWA offense.[120] This suggestion seemed plausible and the Court's historic development of the meaning of navigable waters certainly explained a great deal about judicial interpretation of that phrase.

On closer examination, however, the real reason there are so many more decisions interpreting navigable waters than interpreting any of the other ele-

120. *See* Miller, 45 ELR 10548, *supra* note 3.

ments of the CWA offense is that there are twice as many decisions in cases brought under CWA §404 questioning the existence of navigable waters as there are questioning the existence of any other element. That, in turn, probably results from the comparative strength of legal challenges to the meaning of navigable waters in §404 cases, compared to legal challenges to the meanings of the other elements. This factor is not apparent from analyzing navigable waters decisions alone, but becomes discernible only from understanding all four elements and the decisions analyzing them. The author recognized the importance of §404 in interpreting navigable waters when he wrote the third article in the *ELR News & Analysis* series, but did not recognize its full importance until he had finished analyzing the meanings of all four elements for this book and wrote this comparative analysis.

Table A: Decisions Interpreting the Elements of the Water Pollution Offense

Reference Number and Decision		Element Interpreted[a]
U.S. Supreme Court Decisions		
1.	Los Angeles Cnty. Flood Control Dist. v. Natural Res. Def. Council, 133 S. Ct. 710, 43 ELR 20004 (2013)	3
2.	Decker v. Northwest Envtl. Def. Ctr., 133 S. Ct. 1326, 43 ELR 20062 (2013)	4
3.	Coeur Alaska, Inc. v. Southeast Alaska Conservation Council, 557 U.S. 261, 39 ELR 20133 (2009)	2
4.	Rapanos v. United States, 547 U.S. 715, 36 ELR 20116 (2006)	3
5.	S.D. Warren Co. v. Maine Bd. of Envtl. Prot., 547 U.S. 370, 36 ELR 20089 (2006)	1
6.	South Fla. Water Mgmt. Dist. v. Miccosukee Tribe of Indians, 541 U.S. 95, 34 ELR 20021 (2004)	1, 3, 4
7.	Solid Waste Agency of N. Cook Cnty. v. U.S. Army Corps of Eng'rs, 531 U.S. 159, 31 ELR 20382 (2001)	3
8.	International Paper Co. v. Ouellette, 479 U.S. 481, 17 ELR 20327 (1987)	3
9.	United States v. Riverside Bayview Homes, Inc. 474 U.S. 121, 16 ELR 20086 (1985)	3
10.	Weinberger v. Barcelo-Romero, 456 U.S. 305, 12 ELR 20538 (1982)	2
11.	Train v. Colorado Pub. Interest Research Grp., 426 U.S. 1, 6 ELR 20549 (1976)	2
U.S. Circuit Court of Appeals Decisions		
12.	Natural Res. Def. Council v. County of Los Angeles, 673 F.3d 880, 41 ELR 20109 (9th Cir. 2011), *rev'd on other grounds*, 133 S. Ct. 710, 43 ELR 20180 (2013)	1, 4
13.	Northwest Envtl. Def. Ctr. v. Brown, 640 F.3d 1063, 41 ELR 20178 (9th Cir. 2011)	4
14.	National Pork Producers Council v. U.S. EPA, 635 F.3d 738, 41 ELR 20115 (5th Cir. 2011)	4
15.	United States v. Donovan, 2011 WL 5120605, 42 ELR 20328 (3d Cir. 2011)	3
16.	Precon Dev. Corp. v. U.S. Army Corps of Eng'rs, 633 F.3d 278, 41 ELR 20071 (4th Cir. 2011)	3
17.	Greater Yellowstone Coalition v. Lewis, 628 F.3d 1143, 41 ELR 20059 (9th Cir. 2010)	4

a. 1 = Addition; 2 = Pollutant; 3 = Navigable Waters; and 4 = Point Source.

Reference Number and Decision		Element Interpreted[a]
18.	West Virginia Highlands Conservancy, Inc. v. Huffman, 625 F.3d 159, 40 ELR 20014 (4th Cir. 2010)	1
19.	United States v. Agosto-Vega, 617 F.3d 541, 40 ELR 20222 (1st Cir. 2010)	3, 4
20.	Peconic Baykeeper, Inc. v. Suffolk Cnty., 600 F.3d 180, 40 ELR 20098 (2d Cir. 2010)	4
21.	United States v. Milner, 583 F.3d 1174, 39 ELR 20232 (9th Cir. 2009)	3
22.	Cordiano v. Metacon Gun Club, Inc., 575 F.3d 199 (2d Cir. 2009)	3, 4
23.	United States v. Bailey, 571 F.3d 791, 39 ELR 20148 (8th Cir. 2009)	3
24.	Friends of Everglades v. South Fla. Water Mgmt. Dist., 570 F.3d 1210, 39 ELR 20118 (11th Cir. 2009)	1, 3
25.	Ohio Valley Envtl. Coal. v. Aracoma Coal Co., 556 F.3d 1717 (4th Cir. 2009)	3
26.	United States v. Cundiff, 555 F.3d 200, 39 ELR 20025 (6th Cir. 2009)	1, 3
27.	National Cotton Council of Am. v. U.S. EPA, 553 F.3d 927, 39 ELR 20006 (6th Cir. 2009)	1, 2, 4
28.	Oregon Natural Desert Ass'n v. U.S. Forest Serv., 550 F.3d 778, 39 ELR 20297 (9th Cir. 2008)	4
29.	United States v. Lucas, 516 F.3d 316, 38 ELR 20041 (5th Cir. 2008)	3, 4
30.	United States v. Robinson, 505 F.3d 1208, 37 ELR 20265 (11th Cir. 2007)	3
31.	Northern Cal. River Watch v. City of Healdsburg, 496 F.3d 993, 37 ELR 20202 (9th Cir. 2007)	3
32.	United States v. Moses, 496 F.3d 984, 37 ELR 20206 (9th Cir. 2007)	3
33.	San Francisco Baykeeper v. Cargill Salt Div., 481 F.3d 700 (9th Cir. 2007)	3
34.	United States v. Johnson, 467 F.3d 56, 36 ELR 20218 (1st Cir. 2006)	3
35.	United States v. Gerke Excavation, Inc., 464 F.3d 723, 36 ELR 20200 (7th Cir. 2006)	3, 4
36.	Northern Cal. River Watch v. City of Healdsburg, 457 F.3d 1023, 36 ELR 20163 (9th Cir. 2006)	3
37.	Catskill Mountains Chapter of Trout Unlimited, Inc. v. City of New York, 451 F.3d 77, 36 ELR 20111 (2d Cir. 2006)	1, 3

Reference Number and Decision		Element Interpreted[a]
38.	United States v. Hubenka, 438 F.3d 1026, 36 ELR 20043 (10th Cir. 2006)	3
39.	Baccarat Freemont Dev., LLC v. U.S. Army Corps of Eng'rs, 425 F.3d 1150, 35 ELR 20212 (9th Cir. 2005)	3
40.	Fairhurst v. Hagener, 422 F.3d 1146 (9th Cir. 2005)	2
41.	Sierra Club v. El Paso Gold Mines, Inc., 421 F.3d 1133, 35 ELR 20175 (10th Cir. 2005)	1, 4
42.	United States v. Gerke Constr., Inc., 412 F.3d 804 (5th Cir. 2005), *judgment vacated on other grounds*, 548 U.S. 901 (2006)	4
43.	Waterkeeper Alliance, Inc. v. U.S. EPA, 399 F.3d 486, 35 ELR 20049 (2d Cir. 2005)	1
44.	Parker v. Scrap Metal Processors, Inc., 386 F.3d 993, 34 ELR 20104 (11th Cir. 2004)	3, 4
45.	United States v. Phillips, 367 F.3d 846 (9th Cir. 2004)	3
46.	Greenfield Mills, Inc. v. Macklin, 361 F.3d 934, 34 ELR 20022 (7th Cir. 2004)	1
47.	In re Needham, 354 F.3d 340 (5th Cir. 2003)	3
48.	No Spray Coal., Inc. v. City of New York, 351 F.3d 602, 34 ELR 20007 (2d Cir. 2003)	1
49.	Treacy v. Newdunn Assocs., LLP, 344 F.3d 407 (4th Cir. 2003)	3
50.	United States v. Rueth Dev. Co., 335 F.3d 598 (7th Cir. 2003)	3
51.	United States v. Deaton, 332 F.3d 698 (4th Cir. 2003)	3
52.	Alabama Rivers Alliance v. Federal Energy Regulatory Comm'n, 325 F.3d 290 (D.C. Cir. 2003)	1
53.	Northern Plains Res. Council v. Fidelity Exploration & Dev. Co., 325 F.3d 1155 (9th Cir. 2003)	2
54.	Kentuckians for Commonwealth, Inc. v. Rivenburgh, 317 F.3d 425 (4th Cir. 2003)	2
55.	League of Wilderness Defenders v. Forsgren, 309 F.3d 1181 (9th Cir. 2002)	4
56.	Altman v. Town of Amherst, N.Y., 47 F. App'x 62 (2d Cir. 2002)	2
57.	Community Ass'n for Restoration of Env't v. Henry Bosma Dairy, 305 F.3d 943 (9th Cir. 2002)	3, 4
58.	United States v. Krillich, 303 F.3d 784 (7th Cir. 2002)	3
59.	Fishermen Against Destruction of Env't v. Closter Farms, Inc., 300 F.3d 1294 (11th Cir. 2002)	4
60.	Association to Protect Hammersley, Eld, & Totten Inlets v. Taylor Res., Inc., 299 F.3d 1007 (9th Cir. 2002)	2, 4
61.	Miccosukee Tribe of Indians of Fla. v. South Fla. Water Mgmt. Dist., 280 F.3d 1364, 32 ELR 20475 (11th Cir. 2002)	1

Reference Number and Decision		Element Interpreted[a]
62.	Catskill Mountains Chapter of Trout Unlimited, Inc. v. City of New York, 273 F.3d 481 (2d Cir. 2001)	1, 3
63.	Borden Ranch P'ship v. U.S. Army Corps of Eng'rs, 261 F.3d 810, 32 ELR 20011 (9th Cir. 2001)	1, 2, 3, 4
64.	Rice v. Harken Exploration Co., 250 F.3d 264, 31 ELR 20599 (5th Cir. 2001)	3
65.	Headwaters, Inc. v. Talent Irrigation Dist., 243 F.3d 526, 31 ELR 20535 (9th Cir. 2001)	2, 3
66.	Froebel v. Meyer, 217 F.3d 928, 30 ELR 20746 (7th Cir. 2000)	4
67.	United States v. Deaton, 209 F.3d 331, 30 ELR 20508 (4th Cir. 2000)	1, 2
68.	Driscoll v. Adams, 181 F.3d 1285, 29 ELR 21387 (11th Cir. 1999)	2, 3
69.	United States v. TGR Corp., 171 F.3d 762, 29 ELR 21059 (2d Cir. 1999)	3
70.	Resource Investments, Inc. v. U.S. Army Corps of Eng'rs, 151 F.3d 148, 28 ELR 21407 (10th Cir. 1998)	2
71.	National Mining Ass'n v. U.S. Army Corps of Eng'rs, 145 F.3d 1399, 28 ELR 21318 (D.C. Cir. 1998)	1
72.	Waste Action Project v. Dawn Mining Corp., 137 F.3d 1426, 28 ELR 21035 (9th Cir. 1998)	2
73.	United States v. Wilson, 133 F.3d 251. 28 ELR 20299 (4th Cir. 1997)	1, 2, 3
74.	United States v. West Indies Transp., Inc., 127 F.3d 299, 28 ELR 20202 (3d Cir. 1997)	4
75.	North Carolina v. Federal Energy Regulatory Comm'n, 112 F.3d 1175, 27 ELR 20929 (D.C. Cir. 1997)	1
76.	Chemical Weapons Working Grp., Inc., v. U.S. Dep't of Army, 111 F.3d 1485, 27 ELR 21130 (10th Cir. 1997)	1, 2
77.	United States v. Eidson, 108 F.3d 1336, 27 ELR 20853 (11th Cir. 1997)	2, 3
78.	Dubois v. U.S. Dep't of Agric., 102 F.3d 1273, 27 ELR 20622 (1st Cir. 1996)	1, 3
79.	Hughey v. JMS Dev. Corp., 78 F.3d 1523, 26 ELR 20924 (11th Cir. 1996)	2
80.	Sierra Club, Lone Star Chapter v. Cedar Point Oil Co., Inc., 73 F.3d 546, 26 ELR 20522 (5th Cir. 1996)	2
81.	Leslie Salt Co. v. United States, 55 F.3d 1388, 25 ELR 21406 (9th Cir. 1995)	2

Reference Number and Decision		Element Interpreted[a]
82.	Concerned Area Residents for Env't v. Southview Farm, 34 F.3d 114, 24 ELR 21480 (3d Cir. 1994)	2, 4
83.	Village of Oconomowoc Lake v. Dayton Hudson Corp., 24 F.3d 962, 24 ELR 21080 (7th Cir. 1993)	3
84.	Committee to Save Mokelumne River v. East Bay Mun. Util. Dist., 13 F.3d 305, 24 ELR 20225 (9th Cir. 1993)	1, 2, 4
85.	United States v. Plaza Health Labs., Inc., 3 F.3d 643, 23 ELR 21526 (2d Cir. 1993)	2, 4
86.	United States v. Pozgsai, 999 F.2d 719, 23 ELR 21012 (3d Cir. 1993)	2, 3
87.	Hoffman Homes, Inc. v. Administrator, U.S. EPA, 999 F.2d 256, 23 ELR 21139 (7th Cir. 1993)	3
88.	United States v. Schallom, 998 F.2d 196 (4th Cir. 1993)	2
89.	United States v. Law, 979 F.2d 977, 23 ELR 20466 (4th Cir. 1992)	1
90.	Save Our Community v. U.S. EPA, 971 F.2d 1155 (5th Cir. 1992)	1
91.	Town of Norfolk v. U.S. Army Corps of Eng'rs, 968 F.2d 1438, 22 ELR 21337 (1st Cir. 1992)	2, 3
92.	American Mining Cong. v. U.S. EPA, 965 F.2d 759, 22 ELR 21135 (9th Cir. 1992)	1
93.	Dague v. City of Burlington, 935 F.2d 1343, 21 ELR 21133 (2d Cir. 1991)	4
94.	Carr v. Alta Verde Indust., Inc., 931 F.2d 1055, 21 ELR 21005 (5th Cir. 1991)	4
95.	Rybachek v. U.S. EPA, 904 F.2d 1276, 20 ELR 20973 (9th Cir. 1990)	1, 2
96.	Inland Steel Co. v. U.S. EPA, 901 F.2d 1491, 20 ELR 20889 (7th Cir. 1990)	3
97.	Natural Res. Def. Council v. U.S. EPA, 863 F.2d 1420, 19 ELR 20225 (9th Cir. 1988)	3
98.	National Wildlife Fed'n v. Consumers' Power Co., 862 F.2d 580, 19 ELR 20235 (6th Cir. 1988)	1, 2, 3
99.	Bersani v. U.S. Army Corps of Eng'rs, 850 F.2d 36, 18 ELR 20874 (2d Cir. 1988)	2
100.	United States v. M.C.C. of Fla, Inc., 772 F.2d 1501, 15 ELR 21091 (11th Cir. 1985)	1, 2
101.	Conant v. United States, 786 F.2d 1008, 16 ELR 20453 (11th Cir. 1986)	3

Reference Number and Decision		Element Interpreted[a]
102.	Quivira Mining Co. v. U.S. EPA, 765 F.2d 126, 15 ELR 20530 (10th Cir. 1985)	3
103.	Trustees for Alaska v. U.S. EPA, 749 F.2d 549, 15 ELR 20146 (9th Cir. 1984)	4
104.	United States v. City of Fort Pierre, 747 F.2d 464, 15 ELR 20177 (8th Cir. 1984)	3
105.	Orleans Audubon Soc'y v. Lee, 742 F.2d 901, 15 ELR 20030 (5th Cir. 1984)	2, 3
106.	Utah v. Marsh, 740 F.2d 799, 14 ELR 20683 (10th Cir. 1984)	3
107.	Tennessee Valley Auth. v. Tennessee Water Quality Control Bd., 717 F.2d 992, 14 ELR 20598 (6th Cir. 1983)	1
108.	Avoyelles Sportsmen's League v. Marsh, 715 F.2d 897, 13 ELR 20942 (5th Cir. 1983)	1, 3, 4
109.	United States v. Tilton, 705 F.2d 429, 13 ELR 20583 (11th Cir. 1983)	3
110.	National Wildlife Fed'n v. Gorsuch, 693 F.2d 156, 13 ELR 20015 (D.C. Cir. 1982)	1, 2, 4
111.	Missouri ex rel. Ashcroft v. U.S. Dep't of Army, 672 F.2d 1297, 12 ELR 20368 (8th Cir. 1982)	1, 4
112.	Deltona Corp. v. United States, 657 F.2d 1184, 11 ELR 20905 (Ct. Cl. 1981)	3
113.	Sierra Club v. Abston Constr. Co., 620 F.2d 41, 10 ELR 20552 (5th Cir. 1980)	4
114.	United States v. Texas Pipe Line Co., 611 F.2d 345, 10 ELR 20184 (5th Cir. 1980)	3
115.	United States v. Byrd, 609 F.2d 1204, 9 ELR 20757 (7th Cir. 1979)	3
116.	United States v. Earth Sci., Inc., 599 F.2d 368, 9 ELR 20542 (10th Cir. 1979)	1, 3, 4
117.	Minnehaha Creek Watershed Dist. v. Hoffman, 597 F.2d 617, 9 ELR 20334 (8th Cir. 1979)	2
118.	Leslie Salt Co. v. Froehilke, 578 F.2d 742, 8 ELR 20480 (9th Cir. 1978)	3
119.	Natural Res. Def. Council v. Costle, 568 F.2d 1369 (D.C. Cir. 1977)	4
120.	U.S. Steel Corp. v. Train, 556 F.2d 822, 7 ELR 20419 (7th Cir. 1977)	2, 3
121.	Exxon Corp. v. Train, 554 F.2d 1310, 7 ELR 20594 (5th Cir. 1977)	3

Reference Number and Decision		Element Interpreted[a]
122.	United States v. Hamel, 551 F.2d 107, 7 ELR 20253 (6th Cir. 1977)	2
123.	Appalachian Power Co. v. Train, 545 F.2d 1351, 6 ELR 20732 (4th Cir. 1976)	1, 4
124.	FMC Corp. v. Train, 539 F.2d 973, 6 ELR 20382 (4th Cir. 1976)	2
125.	American Iron & Steel Inst. v. U.S. EPA, 526 F.2d 1027, 6 ELR 20068 (3d Cir. 1975)	1
126.	California v. U.S. EPA, 511 F.2d 963 (9th Cir. 1975)	3
127.	United States v. Ashland Oil & Transp. Co., 504 F.2d 1317, 4 ELR 20784 (6th Cir. 1974)	3
U.S. District Court Decisions		
128.	Stillwater of Crown Point Homeowners' Ass'n v. Kovich, 2011 WL 4818511 (N.D. Ind. 2011)	3
129.	Deerfield Plantation Phase II v. U.S. Army Corps of Eng'rs, 2011 WL 2746232 (D.S.C. 2011)	3
130.	United States v. Brink, 795 F. Supp. 2d 565 (S.D. Tex. 2011)	2, 3
131.	San Francisco Baykeeper v. West Bay Sanitary Dist., 791 F. Supp. 2d 719 (N.D. Cal. 2011)	3
132.	United States v. Freedman Farms, Inc., 786 F. Supp. 2d 1016 (E.D.N.C. 2011)	3
133.	Gouger v. U.S. Army Corps of Eng'rs, 779 F. Supp. 2d 588 (S.D. Tex. 2011)	2
134.	Ohio Valley Envtl. Coal., Inc. v. Coal-Mac, Inc. v, 775 F. Supp. 2d 900, 41 ELR 20141 (S.D. W. Va. 2011)	3
135.	United States v. Vierstra, 2011 WL 1064526 (D. Idaho 2011)	3
136.	Concerned Area Residents for Env't v. Nelson Faria Dairy, Inc., 2011 WL 61882 (E.D. Wash. 2011)	1
137.	Gulf Restoration Network v. Hancock Cnty. Dev., LLC, 772 F. Supp. 2d 761 (S.D. Miss. 2011)	3
138.	Assateague Coastkeeper v. Alan & Kristin Hudson Farm, 727 F. Supp. 2d 433, 40 ELR 20208 (D. Md. 2010)	4
139.	United States v. Donovan, 2010 WL 3000058 (D. Del. 2010)	3
140.	United States v. Righter, 2010 WL 2640189 (M.D. Pa. 2010)	4
141.	National Ass'n of Home Builders v. U.S. Army Corps of Eng'rs, 699 F. Supp. 2d 209, 40 ELR 20104 (D.D.C. 2010), *vacated on other grounds*, 663 F.3d 470 (D.C. Cir. 2011)	3, 4
142.	Benjamin v. Douglas Ridge Rifle Club, 673 F. Supp. 2d 1210, 40 ELR 20283 (D. Or. 2009)	3
143.	Stephens v. Koch Foods, LLC, 667 F. Supp. 2d 768 (E.D. Tenn. 2009)	3

Reference Number and Decision		Element Interpreted[a]
144.	Precon Dev. Corp. v. U.S. Army Corps of Eng'rs, 658 F. Supp. 2d 752 (E.D. Va. 2009)	3
145.	Catskill Mountains Chapter of Trout Unlimited, Inc. v. U.S. EPA, 630 F. Supp. 2d 295 (S.D.N.Y. 2009)	1
146.	Alabama Rivers Alliance, Inc. v. U.S. Army Corps of Eng'rs, 697 F. Supp. 2d 1251 (N.D. Ala. 2009)	2
147.	Hernandez v. Esso Standard Oil Co. (Puerto Rico), 599 F. Supp. 2d 175 (D.P.R. 2009)	3
148.	Ogeechee-Canooche Riverkeeper, Inc. v. T.C. Logging, Inc., 2009 WL 2390851 (S.D. Ga. 2009)	4
149.	West Virginia Highlands Conservancy v. Huffman, 2009 WL 2705854 (S.D. W. Va. 2009), *aff'd*, 625 F.3d 159, 40 ELR 20014 (4th Cir. 2010)	4
150.	Wild Fish Conservancy v. Quilcene National Fish Hatchery, 2009 WL 3380655 (W.D. Wash. 2009)	4
151.	United States v. Acquest Transp., LLC, 2009 WL 2157005 (W.D.N.Y. 2009)	4
152.	Peconic Baykeeper, Inc. v. Suffolk Cnty., 585 F. Supp. 2d 377 (E.D.N.Y. 2008), *rev'd in part*, 600 F.3d 180, 40 ELR 20098 (2d Cir. 2010)	1, 2, 4
153.	American Petroleum Inst. v. Johnson, 541 F. Supp. 2d 165, 38 ELR 20081 (D.D.C. 2008)	3
154.	United States v. Fabian, 522 F. Supp. 2d 1078, 37 ELR 20083 (N.D. Ind. 2007)	1, 3
155.	United States v. Robinson, 521 F. Supp. 2d 1247 (N.D. Ala. 2007)	3
156.	United States v. Bailey, 516 F. Supp. 2d 990 (D. Minn. 2007)	3
157.	D'Olive Bay Restoration & Preservation Comm., Inc. v. U.S. Army Corps of Eng'rs, 513 F. Supp. 2d 1261 (S.D. Ala. 2007)	2
158.	United States v. Cundiff, 480 F. Supp. 2d 940, 37 ELR 20082 (W.D. Ky. 2007)	3
159.	Simbury-Avon Preservation Soc'y, LLP v. Metacon Gun Club, Inc., 472 F. Supp. 219 (D. Conn. 2007)	2, 3
160.	Environmental Prot. Info. Ctr. v. Pacific Lumber Co., 469 F. Supp. 2d 803, 37 ELR 20012 (N.D. Cal. 2007)	1, 3, 4
161.	National Ass'n of Home Builders v. U.S. Army Corps of Eng'rs, 2007 WL 259944, 37 ELR 20028 (D.D.C. 2007)	1
162.	United States v. Lippold, 2007 WL 3232483 (C.D. Ill. 2007)	3, 4
163.	Friends of Everglades v. South Fla. Water Mgmt. Dist., 2006 WL 3635465 (S.D. Fla. 2006)	1

Reference Number and Decision		Element Interpreted[a]
164.	United States v. Marion L. Kincaid Trust, 3 F. Supp. 2d 680 (E.D. Mich. 2006)	3
165.	United States v. Chevron Pipe Line Co., 437 F. Supp. 2d 605, 36 ELR 20131 (N.D. Tex. 2006)	3
166.	Tungett v. Papierski, 2006 WL 51148 (E.D. Tenn. 2006)	2
167.	Florida Wildlife Fed'n v. U.S. Army Corps of Eng'rs, 401 F. Supp. 1298 (S.D. Fla. 2005)	2
168.	United States v. Adams Bros. Farming, Inc., 369 F. Supp. 2d 1166 (C.D. Cal. 2003)	3
169.	City of Southacres v. Waterworth, 322 F. Supp. 992 (S.D. Tex. 2004)	2
170.	National Wildlife Fed'n v. Norton, 332 F. Supp. 2d 170 (D.D.C. 2004)	2
171.	North Carolina Shellfish Growers Ass'n v. Holly Ridge Assocs., LLC, 278 F. Supp. 2d 654 (E.D.N.C. 2003)	1, 2, 3, 4
172.	Reynolds v. Rick's Mushroom Servs., Inc., 246 F. Supp. 2d 449 (E.D. Pa. 2003)	4
173.	FD&P Enters., Inc. v. U.S. Army Corps of Eng'rs, 239 F. Supp. 2d 509 (D.N.J. 2003)	3
174.	United States v. Hummel, 2003 WL 1845365 (N.D. Ill. 2003)	1
175.	United States v. RGM Corp., 222 F. Supp. 2d 780 (E.D. Va. 2002)	3
176.	U.S. Pub. Interest Research Grp. v. Atlantic Salmon of Maine, LLC, 215 F. Supp. 2d 239 (D. Me. 2002)	1, 2, 4
177.	California Sportfishing Protecting Alliance v. Diablo Grande, Inc., 209 F. Supp. 2d 1059 (E.D. Cal. 2002)	3
178.	Kentuckians for Commonwealth, Inc. v. Rivenburgh, 204 F. Supp. 2d 927 (S.D. W. Va. 2002), *rev'd on other grounds*, 317 F.3d 425 (4th Cir. 2003)	2
179.	United States v. Newdunn Assocs., 195 F. Supp. 2d 751 (E.D. Va. 7002)	3
180.	Greenfield Mills, Inc. v. O'Bannon, 189 F. Supp. 2d 893 (N.D. Ind. 2002), *aff'd in part, rev'd in part*, 361 F.3d 934 (7th Cir. 2004)	1, 2
181.	United States v. Rueth Dev. Co., 189 F. Supp. 2d 874 (N.D. Ind. 2002)	3
182.	Sierra Club & Mineral Pol'y Ctr. v. El Paso Gold Mines, Inc., 2002 WL 33932715 (D. Colo. 2002)	1, 3, 4
183.	Altman v. Town of Amherst, N.Y., 190 F. Supp. 2d 467 (W.D.N.Y. 2001), *rev'd*, 47 F. App'x 62 (2d Cir. 2002)	2

Reference Number and Decision		Element Interpreted[a]
184.	United States ex rel. McKeown v. Port Auth. of N.Y & N.J., 162 F. Supp. 2d 173 (S.D.N.Y. 2001)	1, 2, 4
185.	United States v. Interstates Gen'l Co., 152 F. Supp. 2d 843 (D. Md. 2001)	3
186.	Idaho Rural Council v. Bosma, 143 F. Supp. 2d 1169 (D. Idaho 2001)	3
187.	United States v. Buday, 138 F. Supp. 2d 1282 (D. Mont. 2001)	3
188.	Aiello v. Town of Brookhaven, 136 F. Supp. 2d 81 (E.D.N.Y. 2001)	3
189.	American Mining Cong. v. U.S. Army Corps of Eng'rs, 120 F. Supp. 2d 23 (D.D.C. 2000)	1
190.	Pronsolino v. Marcus, 91 F. Supp. 2d 1337, 30 ELR 20460 (N.D. Cal. 2000)	2
191.	No Spray Coal., Inc. v. City of New York, 2000 WL 1401458 (S.D.N.Y. 2000)	1
192.	Stone v. Naperville Park Dist., 38 F. Supp. 2d 651 (N.D. Ill. 1999)	4
193.	United States v. Bay-Houston Towing Co., 33 F. Supp. 2d 596, 29 ELR 21011 (E.D. Mich. 1999)	1
194.	United States v. Hallmark Constr. Co., 30 F. Supp. 2d 1033, 28 ELR 21438 (N.D. Ill. 1998)	1, 3
195.	Frobel v. Meyer, 13 F. Supp. 2d 843 (E.D. Wis. 1998)	1, 4
196.	United States v. United Homes, Inc., 1999 WL 117701 (N.D. Ill. 1999)	2
197.	United States v. Gulf Park Water Co., Inc., 972 F. Supp. 1056 (S.D. Miss. 1997)	2, 3
198.	Williams Pipe Line Co. v. Bayer Corp., 964 F. Supp. 1300 (S.D. Iowa 1997)	3, 4
199.	Umatilla Waterquality Protective Ass'n, Inc. v. Smith Frozen Foods, Inc., 962 F. Supp. 1312, 27 ELR 21411 (D. Or. 1997)	3
200.	American Mining Cong. v. Corps of Eng'rs, 951 F. Supp. 267, 27 ELR 20589 (D.D.C. 1979)	1
201.	Georgia v. City of East Ridge, Tenn., 949 F. Supp. 1571, 27 ELR 20782 (N.D. Ga. 1996)	3
202.	RSR Corp. v. Browner, 924 F. Supp. 504, 26 ELR 21353 (S.D.N.Y. 1996)	4
203.	United States v. Lambert, 915 F. Supp. 797, 26 ELR 21116 (S.D. W. Va. 1996)	4
204.	Hudson Riverkeeper Fund, Inc. v. Harbor at Hastings Assocs., 917 F. Supp. 251, 26 ELR 21120 (S.D.N.Y. 1996)	4

Reference Number and Decision		Element Interpreted[a]
205.	Long Island Soundkeeper Fund, Inc. v. N.Y. Athletic Club, 1996 WL 131863 (S.D.N.Y. 1996)	2, 4
206.	Beartooth Alliance v. Crown Butte Mines, 904 F. Supp. 1168, 26 ELR 20639 (D. Mont. 1995)	1, 2, 3, 4
207.	Friends of Santa Fe Cnty. v. LAC Minerals, Inc., 892 F. Supp. 1333, 26 ELR 20135 (D.N.M. 1995)	3
208.	Molokai Chamber of Commerce v. Kukui, 891 F. Supp. 1389, 26 ELR 20303 (D. Haw. 1995)	4
209.	United States v. Banks, 873 F. Supp. 650, 25 ELR 20776 (S.D. Fla. 1995)	3
210.	Washington Wilderness Coal. v. Hecla Mining Co., 870 F. Supp. 983, 25 ELR 20661 (E.D. Wash. 1994)	3, 4
211.	Hudson River Fishermen's Ass'n v. Arcuri, 862 F. Supp. 73, 25 ELR 20460 (S.D.N.Y. 1994)	2
212.	Sierra Club v. Colorado Refining Co., 838 F. Supp. 1428, 24 ELR 20749 (D. Colo. 1993)	3
213.	Salt Pond Assocs. v. U.S. Army Corps of Eng'rs, 815 F. Supp. 766, 23 ELR 21026 (D. Del. 1993)	1
214.	Golden Gate Audubon Soc'y, Inc. v. U.S. Army Corps of Eng'rs, 796 F. Supp. 1306, 23 ELR 20267 (N.D. Cal. 1992)	3
215.	United States v. Villegas, 784 F. Supp. 6, 22 ELR 21027 (S.D.N.Y. 1991), *rev'd sub nom.* United States v. Plaza Health Labs., Inc., 3 F.3d 643, 23 ELR 21526 (2d Cir. 1993)	4
216.	United States v. Zanger, 767 F. Supp. 1030, 22 ELR 20231 (N.D. Cal. 1991)	3
217.	United States v. Sinclair Oil Co., 767 F. Supp. 2000, 21 ELR 21323 (D. Mont. 1990)	1, 4
218.	McClellan Ecological Seepage Situation v. Weinberger, 763 F. Supp. 431, 20 ELR 20870 (E.D. Cal. 1989)	3
219.	Hudson River Fishermen's Ass.n v. City of New York, 751 F. Supp. 1088, 21 ELR 20647 (S.D.N.Y. 1990), *aff'd without opinion*, 940 F.2d 649, 21 ELR 21226 (2d Cir. 1991)	2
220.	Friends of Sakonmet v. Dutra, 738 F. Supp. 623, 21 ELR 20055 (D.R.I. 1990)	4
221.	Dague v. City of Burlington, 732 F. Supp. 458, 20 ELR 21001 (D. Vt. 1989, *aff'd on other grounds*, 935 F.2d 1088 (2d Cir. 1990), *rev'd on other grounds*, 505 U.S. 557, 22 ELR 21099 (1992)	2
222.	West Virginia Coal Ass'n v. Reilly, 728 F. Supp. 1301 (D.N.J. 1989)	3
223.	United States v. Malibu Beach, Inc., 711 F. Supp. 1301, 19 ELR 21247 (D.N.J. 1989)	3

Reference Number and Decision		Element Interpreted[a]
224.	Hanson v. United States, 710 F. Supp. 1105, 19 ELR 21074 (E.D. Tex. 1989)	2, 3
225.	McClellan Ecological Seepage Situation v. Weinberger, 707 F. Supp. 1182, 19 ELR 20124 (E.D. Cal. 1988)	3
226.	United States v. Larkins, 657 F. Supp. 76, 17 ELR 20783 (W.D. Ky. 1987)	3
227.	National Wildlife Fed'n v. Consumers Power Co., 657 F. Supp. 989, 17 ELR 20801 (W.D. Mich. 1987)	1, 2, 4
228.	Fishel v. Westinghouse Elec. Corp., 640 F. Supp. 442, 16 ELR 20634 (M.D. Pa. 1986)	4
229.	Kelley v. United States, 618 F. Supp. 1103, 16 ELR 20080 (W.D. Mich. 1985)	3
230.	Track 12, Inc. v. District Eng'r, 618 F. Supp. 448, 16 ELR 20163 (D. Minn. 1985)	3
231.	Higbee v. Starr, 598 F. Supp. 323 (W.D. Ark. 1984)	2
232.	United States v. Saint Bernard Parish, 589 F. Supp. 617, 14 ELR 20794 (E.D. La. 1984)	3
233.	United States v. Lambert, 589 F. Supp. 366, 14 ELR 20588 (M.D. Fla. 1984)	3
234.	United States v. Champitti, 583 F. Supp. 483 (D.N.J. 1984)	3
235.	United States v. Robinson, 570 F. Supp. 1157, 14 ELR 20056 (M.D. Fla. 1983)	2, 3
236.	United States v. Bradshaw, 541 F. Supp. 880, 12 ELR 20629 (D. Md. 1981)	2, 3
237.	Bayou Des Familles Dev. Corp. v. U.S. Army Corps of Eng'rs, 541 F. Supp. 1025, 13 ELR 20055 (E.D. La. 1982)	3
238.	National Wildlife Fed'n v. Gorsuch, 530 F. Supp. 1291, 12 ELR 20268 (D.D.C. 1982)	1, 2
239.	United States v. Lee Wood Contractors, Inc., 529 F. Supp. 119, 12 ELR 20421 (C.D. Mich. 1981)	3
240.	O'Leary v. Moyer's Landfill, Inc., 523 F. Supp. 642, 12 ELR 20239 (E.D. Pa. 1981)	4
241.	United States v. Weisman, 489 F. Supp. 1331, 10 ELR 20698 (M.D. Fla. 1980)	2, 3, 4
242.	United States v. Oxford Royal Mushroom Prods., Inc., 487 F. Supp. 952, 10 ELR 20549 (E.D. Pa. 1980)	4
243.	American Dredging Co. v. Dutchyshyn, 480 F. Supp. 957 (E.D. Pa. 1979)	3

Reference Number and Decision		Element Interpreted[a]
244.	Romero-Barcelo v. Brown, 478 F. Supp. 646 (D.P.R. 1979), *rev'd on other grounds sub nom.* Weinberger v. Barcelo-Romero, 643 F.2d 835, 11 ELR 20391 (1st Cir. 1981), *rev'd on other grounds*, 456 U.S. 305, 12 ELR 20538 (1982)	4
245.	Avoyelles Sportsmen's League v. Alexander, 473 F. Supp. 525, 11 ELR 20315 (W.D. La. 1979)	1
246.	United States v. Frezzo Brothers, Inc., 461 F. Supp. 266, 9 ELR 20139 (E.D. Pa. 1978), *aff'd*, 602 F.2d 1123, 9 ELR 20556 (3d Cir. 1979)	2
247.	South Carolina Wildlife Fed'n v. Alexander, 457 F. Supp. 118, 8 ELR 20757 (D.S.C. 1978)	1, 2, 4
248.	Commonwealth of Puerto Rico v. Alexander, 438 F. Supp. 90, 7 ELR 20751 (D.P.R. 1977)	3
249.	United States v. Velsicol Chem. Corp., 438 F. Supp. 945 (W.D. Tenn. 1976)	1, 3
250.	Wyoming v. Hoffman, 437 F. Supp. 114, 8 ELR 20001 (D. Wyo. 1977)	3
251.	Mahelona v. Hawaiian Elec. Co., Inc., 418 F. Supp. 1328, 7 ELR 20031 (D. Haw. 1976)	4
252.	Conservation Council of N.C. v. Costanzo, 398 F. Supp. 653, 5 ELR 20666 (E.D.N.C. 1975)	3
253.	Sun Enters., Inc. v. Train, 394 F. Supp. 211 (S.D.N.Y. 1975)	3
254.	P.F.Z. Props., Inc. v. Train, 393 F. Supp. 1370 (D.D.C. 1975)	3
255.	Natural Res. Def. Council v. Callaway, 392 F. Supp. 685, 5 ELR 20285 (D.D.C. 1975)	3
256.	United States v. Phelps Dodge Corp., 391 F. Supp. 1181, 5 ELR 20308 (D. Ariz. 1975)	3
257.	United States v. GAF Corp., 389 F. Supp. 1379, 5 ELR 20581 (S.D. Tex. 1975)	3
258.	United States v. Holland, 373 F. Supp. 665, 4 ELR 20710 (M.D. Fla. 1974)	3, 4

Table B: Judicial Interpretations of Each Element

Table B is displayed in four subparts, one each for Addition, Pollutant, Navigable Waters, and Point Source.

Explanation of Columns in Table B for Each Element

Column A. Number of decision from Table A.

Column B. Year of decision. The latest and earliest sets of decisions are highlighted (see Chapter 5, note 41).

Column C. Positive or negative environmental results for the element. Positive is an expansive interpretation of the element; negative is a restrictive interpretation. The absence of a (+) or (–) in the column means that the decision discussed the element but did not hold what it meant. The presence of both a (+) and a (–) in the column means that there were mixed results; for example, the interpretation was positive, but the facts did not support a positive result.

Column D. Type of Decision. Cit. S. = citizen suit. Crim. = criminal prosecution. Enf. = civil enforcement by the government. Jud. Rev. = judicial review.

Column E. Interpretive devices used for the element, from the list below.

Column F. Number of interpretive devices used for the element.

Column G. Number of precedents used to interpret the element.

Column H. Statutory section for which the element is relevant: CWA §§401, 402, 404, National Environmental Policy Act (NEPA), or Oil Pollution Act (OPA).
Column I (Navigable Waters only). Type of Water: G=groundwater. I=isolated water. O=other water. T=tributary. U=unitary navigable water. W=wetland.

Interpretive Devices

1. Absurd results, avoid
2. Administrative issues, avoid
3. Antecedent, rule of last
4. Broad interpretation
5. Constitutional issues, avoid
6. Deference to agency interpretation of statute
7. Deference to agency interpretation of agency rule
8. *Ejusdem generis*, general term interpreted based on specific terms accompanying it
9. Equity
10. Every word, give meaning to
11. Exception, interpret narrowly
12. Exception proves the rule
13. Exclusive/inclusive lists
14. *Expressio unius*, expression of one thing suggests exclusion of others
15. Federalism, honor
16. Legislative history
17. Lenity, rule of
18. Metaphor
19. Metamorphosis
20. *Noscitur a sociis*, term interpreted in context of other terms accompanying it
21. Other statutes, harmonize with
22. Plain meaning
23. Precedent
24. Structure of statute
25. Waiver, interpret narrowly

Table B: Addition

A	B	C	D	E	F	G	H
U.S. Supreme Court Decisions							
5.	2006	+	Jud. Rev.	7, 16, 20, 22, 23, 24	6	2	§401
6.	2004	+	Cit. S.	1, 4, 7, 22, 23, 24	6	2	§402
U.S. Circuit Court of Appeals Decisions							
12.	2011	+	Cit. S.	13, 22, 23	3	3	§402
18.	2010	+	Cit. S.	1, 4, 7, 22, 23	5	3	§402
24.	2009	–	Cit. S.	4, 7, 16, 22, 23	5	6	§402
26.	2009	+	Enf.	7, 22, 23, 24	4	4	§404
27.	2009	+	Jud. Rev.	1, 4, 7, 13, 19	5	–	§402
37.	2006	+	Cit. S.	4, 7, 21, 22, 23, 24	6	5	§402
41.	2005	+	Cit. S.	7, 18, 22, 23, 24	5	5	§402
43.	2005	+	Cit. S.	4, 7, 22, 23, 24	5	1	§402
46.	2004	+	Cit. S.	4, 7, 22, 23, 24	5	4	§404
48.	2003	+	Jud. Rev.	21, 22	2	–	§401
52.	2002	+	Cit. S.	22, 23	2	1	§402
61.	2001	+	Cit. S.	4, 7, 16, 18, 22, 23	6	3	§402
62.	2001	+	Jud. Rev.	4, 19, 22, 23	4	5	§404
63.	2001	+	Cit. S.	21, 22, 23	3	3	§402
67.	2000	+	Enf.	4, 19, 22, 23	4	5	§404
71.	1998	–	Jud. Rev.	1, 19, 21, 22, 23, 24	6	5	§404
73.	1997	–	Crim.	1, 22, 23	3	1	§404
75.	1997	–	Jud. Rev.	20, 22, 23, 24	4	2	§401
76.	1997	–	Cit. S.	1, 4, 21, 23	2	1	§402
78.	1996	+	Cit. S.	1, 18, 22, 23	4	3	§402
84.	1993	+	Cit. S.	22, 23	2	2	§402
89.	1992	+	Crim.	22, 23	2	3	§402
90.	1992	–	Cit. S.	7, 22, 23	3	4	§404
92.	1992	+	Jud. Rev.	4, 7, 16, 22	4	–	§402
95.	1990	+	Jud. Rev.	22, 23	2	2	§402
98.	1988	–	Jud. Rev.	1, 7, 22, 23, 24	5	1	§402
100.	1985	+	Enf.	4, 23	2	1	§404
107.	1983	–	Cit. S.	23	1	2	§402
108.	1983	+	Cit. S.	3, 16, 22, 23	4	2	§404
110.	1982	–	Jud. Rev.	4, 7, 16, 22, 23, 24	6	4	§402
111.	1982	–	Cit. S.	22	1	–	§402
116.	1979	+	Cit. S.	22	1	–	§402

A	B	C	D	E	F	G	H
123.	1976	–	Jud. Rev.	22	1	–	§401
125.	1975	–	Jud. Rev.	22	1	–	§402
U.S. District Court Decisions							
136.	2011	–	Cit. S.	22, 23	2	1	§402
145.	2009	+	Jud. Rev.	7, 22, 23	3	4	§402
152.	2008	+	Cit. S.	21, 22, 23	3	7	§§402/404
154.	2007	+	Enf.	23	1	4	§404
160.	2007	+	Cit. S.	23	1	3	§402
161.	2007	–	Jud. Rev.	23	1	1	§404
163.	2006	+	Cit. S.	4, 7, 16, 18, 22, 23, 24	7	10	§402
171.	2003	+	Cit. S.	23	1	2	§402
174.	2003	+	Enf.	22, 23	2	6	§404
176.	2002	+	Cit. S.	22, 23	2	1	§404
180.	2002	–	Cit. S.	22, 23, 24	3	5	§402
182.	2002	+	Cit. S.	23	1	6	§404
184.	2001	–	Enf.	22	1	–	§402
189.	2000	+	Jud. Rev.	22, 23	2	1	§404
191.	2000	–	Cit. S.	21, 22, 23	3	5	§402
193.	1999	+	Enf.	23	1	6	§404
194.	1988	+	Enf.	23	1	2	§404
195.	1988	–	Cit. S.	22, 23	2	6	§§402/404
200.	1997	–	Jud. Rev.	4, 7, 16, 21, 22, 23, 24	7	6	§404
206.	1995	+	Cit. S.	22	1	–	§402
213.	1993	–	Enf.	7, 16, 22	3	–	§402
217.	1990	+	Enf.	22, 23, 24	3	3	§404
227.	1987	+	Cit. S.	22, 23	2	5	§402
238.	1982	+	Cit. S.	4, 22, 23	3	1	§402
245.	1979	+	Cit. S.	4, 16, 22	3	–	§402
247.	1978	+	Cit. S.	22, 23	2	1	§402
249.	1976	+	Enf.	22	1	–	§402

Table B: Pollutant

A	B	C	D	E	F	G	H
U.S. Supreme Court Decisions							
3.	2009	–	Jud. Rev.	1, 7, 22, 24	4	–	§404
10.	1982	+	Cit. S.	22	1	–	§402
11.	1976	–	Cit. S.	12, 16, 21, 24	4	–	§402

A	B	C	D	E	F	G	H
U.S. Circuit Court of Appeal Decisions							
27.	2009	+	Jud. Rev.	7, 21, 22, 23	4	6	§402
40.	2005	–	Cit. S.	1, 7, 21 22, 23	5	3	§402
53.	2003	+	Cit. S.	7, 12, 22, 23, 24	5	6	§402
54.	2003	–	Jud. Rev.	7, 21, 24	3	–	§404
56.	2002	+	Cit. S.	7, 23	2	5	§402
60.	2002	–	Cit. S.	1, 8, 22, 23, 24	5	3	§402
63.	2001	+	Enf.	22, 23	2	2	§404
65.	2001	+	Cit. S.	1, 7, 21, 22, 23	5	1	§402
67.	2000	+	Enf.	22, 23	2	1	§404
68.	1999	+	Cit. S.	22, 23	2	1	§402
70.	1998	+	Jud Rev.	16, 22, 24	3	–	§404
72.	1998	–	Cit. S.	7, 16, 21, 22, 23	5	1	§402
73.	1997	+	Crim.	22	1	–	§404
76.	1997	–	Cit. S.	1, 21	2	–	§402
77.	1997	+	Crim.	22	1	–	§402
79.	1996	+	Cit. S.	23	1	2	§402
80.	1996	+	Cit. S.	4, 7, 12, 13, 16, 22, 23	7	8	§402
81.	1995	+	Enf.	22	1	–	§402
82.	1994	+	Cit. S.	22	1	1	§402
84.	1993	+	Cit. S.	22	1	–	§402
85.	1993	+	Crim.	22	1	–	§404
86.	1993	+	Enf.	22	1	–	§404
88.	1993	+	Crim.	22	1	–	§402
91.	1992	–	Jud. Rev.	22	1	–	§404
95.	1990	+	Jud. Rev.	22	1	–	§402
98.	1988	+	Cit. S.	22, 23	2	1	§402
99.	1988	+	Jud. Rev.	22	1	–	§404
100.	1985	+	Enf.	16	1	–	§404
105.	1984	–	Cit. S.	22	1	–	§404
110.	1982	–	Cit. S.	7, 13, 16, 24	4	–	§402
117.	1979	+	Jud. Rev.	4, 7, 12, 16, 22, 24	6	–	§404
120.	1977	+	Jud. Rev.	12, 14, 16, 22, 24	5	–	§402
122.	1977	+	Crim.	16, 21, 22, 23, 24	5	3	§402
124.	1976	+	Jud. Rev.	22, 24	2	–	§402
U.S. District Court Decisions							
130.	2011	+	Enf.	22	1	–	§404
133.	2011	+	Jud. Rev.	22	1	–	§404

A	B	C	D	E	F	G	H
146.	2009	+	Jud. Rev.	22	1	–	§404
152.	2008	–	Cit. S.	7, 21, 22, 23	4	10	§402
157.	2007		Jud. Rev.	22	1	–	§404
159.	2007	+	Cit. S.	7, 22, 23	3	2	§402
166.	2006	+	Cit. S.	4, 22, 23	3	5	§404
167.	2005	+	Jud. Rev.	22	1	–	§404
169.	2004	+	Jud. Rev.	22	1	–	§404
170.	2004	+	Jud. Rev.	22	1	–	§404
171.	2003	+	Cit. S.	22, 23, 24	3	6	§§402/404
176.	2002	+	Cit. S.	22, 23	2	10	§402
178.	2002	+	Cit S.	4, 7, 16, 22, 24	5	–	§404
180.	2002	–	Cit. S.	22, 24	2	–	§§402/404
183.	2001	–	Cit. S.	7, 21, 22, 23	4	3	§402
184.	2001	–	Cit. S.	21	1	–	§402
190.	2000	+	Jud. Rev.	4, 7, 16, 23, 24	5	5	§402
196.	1999	–	Enf.	16, 22, 23, 24	4	6	§§402/404
197.	1997	+	Enf.	22, 23	2	1	§402
205.	1996	+	Cit. S.	7, 22, 23	3	4	§402
206.	1995	+	Cit. S.	4, 22, 23	3	2	§402
211.	1994	+	Cit. S.	22	1	–	§402
219.	1990	+	Cit. S.	4, 7, 22, 23, 24	5	3	§402
221.	1989	+	Cit. S.	22, 24	2	–	§402
224.	1989	+	Enf.	23	1	1	§404
227.	1987	+	Cit. S.	7, 22, 23	3	1	§401
231.	1984	+	Cit. S.	22	1	1	§402
235.	1983	+	Enf.	22	1	–	§404
236.	1981	+	Enf.	22, 23	2	3	§404
238.	1982	+	Cit. S.	1, 4, 22, 23, 24	5	–	§402
241.	1980	+	Enf.	7, 22, 24	3	–	§404
246.	1978	+	Crim.	22	1	–	§402
247.	1978	+	Cit. S.	1, 22	2	–	§404

Table B: Navigable Waters

A	B	C	D	E	F	G	H	I
U.S. Supreme Court Decisions								
1.	2013	+	Jud. Rev.	23	1	2	§402	T
4.	2006	–	Enf.	1, 5, 7, 10, 15, 16, 22, 23, 24	9	19	§404	T, W

A	B	C	D	E	F	G	H	I
6.	2004	+	Cit. S.	7, 15, 22, 23, 24	5	5	§402	O, U
7.	2001	–	Jud. Rev.	1, 4, 5, 12, 15, 16, 23, 24	8	2	§404	I
8.	1987		Cit. S.	16	1	1	§402	O
9.	1985	+	Enf.	1, 4, 5, 7, 16, 22, 24	7	–	§404	T, W
U.S. Circuit Court of Appeals Decisions								
15.	2011	+	Enf.	23	1	5	§404	T, W
16.	2011	–	Jud. Rev.	7, 23	2	6	§404	W
19.	2010	+	Crim.	4, 23	2	4	§§402/404	T
21.	2009	–	Enf.	4, 21, 22, 23	4	5	§404	W
22.	2009	+	Cit. S.	4, 7, 23	3	5	§402	W
23.	2009	+	Enf.	23	1	6	§404	W
24.	2009	–	Cit. S.	1, 4, 7, 10, 15, 21, 22, 23, 24	9	10	§402	O, U
25.	2009	–	Cit. S.	7, 11, 21, 23	4	2	§402	T, U
26.	2009	+	Enf.	23	1	8	§404	T, W
29.	2008	+	Crim.	23	1	3	§404	W
30.	2007	–	Crim.	23	1	6	§402	T
31.	2007	+	Cit. S.	4, 11, 23	3	7	§402	T, W
32.	2007	+	Crim.	7, 23	2	9	§§402/404	T
33.	2007	–	Cit. S.	7, 13, 14, 23	4	9	§402	W
34.	2006	+	Enf.	23	1	3	§404	W
35.	2006	–	Enf.	23	1	2	§404	W
36.	2006	+	Cit. S.	11, 23	2	6	§402	W
37.	2006	+	Cit. S.	1, 7, 15, 23, 24	5	4	§402	U
38.	2006	+	Crim.	4, 5, 7, 10, 23	5	17	§404	T
39.	2005	+	Jud. Rev.	23	1	5	§404	W
44.	2004	+	Cit. S.	16, 23	2	3	§402	T
45.	2004	+	Crim.	7, 16, 17, 23	4	7	§402	T
47.	2003	+	Bankr.	5, 7, 22, 23	4	7	OPA	T, W
49.	2003	+	Enf.	4, 23	2	5	§404	T, W
50.	2003	+	Enf.	23, 25	2	6	§404	W
51.	2003	+	Enf.	5, 7, 23	3	13	§404	T, W
57.	2002	+	Cit. S.	23	1	3	§402	T
59.	2002	+	Enf.	23, 25	2	4	§404	W
62.	2001	+	Cit. S.	1, 7, 23	3	2	§402	O, U
63.	2001	+/–	Enf.	23	1	7	§404	W
64.	2001	–	Cit. S.	16, 21, 22, 23	4	8	OPA	G
65.	2001	+	Cit. S.	22, 23	2	8	§402	T

A	B	C	D	E	F	G	H	I
68.	1999	+	Cit. S.	23	1	2	§402	T
69.	1999	+	Crim.	4, 11, 23	3	5	§402	T
73.	1997	–	Crim.	5, 23	2	1	§404	T, W
77.	1997	+	Crim.	4, 23	2	9	§402	T
78.	1996	+	Cit. S.	1, 4, 23, 24	4	4	§402	O, U
83.	1994	–	Cit. S.	4, 7, 14, 16, 23	5	7	§402	G, I
86.	1993	+	Enf.	4, 7, 16, 22, 23, 24	6	12	§404	T, W
87.	1993	–	Jud. Rev.	7	1	–	§404	W
91.	1992	–	Jud. Rev.	23	1	4	§404	W
96.	1990	–	Enf.	1, 7, 11, 22, 23	5	5	§402	G
97.	1988	+	Jud. Rev.	16, 21, 22, 23	4	3	§402	O, W
98.	1988	–	Cit. S.	7, 23	2	3	§402	U
101.	1986	+	Enf.	4, 11, 23	3	1	§404	W
102.	1985	+	Jud. Rev.	4, 7, 23	3	3	§402	G, T
104.	1984	–	Enf.	1, 4, 9	3	–	§401	W
105.	1984	–	Jud. Rev.	6	1	–	§404	W
106.	1984	+	Decl. Jud.	4, 23	2	12	§404	T
108.	1983	+	Cit. S.	4, 7, 16, 23	4	8	§404	W
109.	1983	+	Enf.	16, 22, 23	3	1	§404	W
112.	1981	+	Taking	4, 23	2	7	§404	W
114.	1980	+	Enf.	23	1	1	§402	T
115.	1979	+	Enf.	4, 16, 23	3	4	§404	W
116.	1979	+	Enf.	4, 16, 23	3	2	§402	T
118.	1978	+	Enf.	1, 4, 16, 21, 23	5	2	§404	W
120.	1977	+	Jud. Rev.	3, 4, 7, 10, 13	5	–	§402	G
121.	1977	–	Jud. Rev.	7, 12, 16, 21, 22, 23, 24	7	1	§402	G
126.	1975	+	Jud. Rev.	16, 23	2	2	§402	T
127.	1974	+	Enf.	4, 16, 23	3	4	§402	T
U.S. District Court Decisions								
128.	2011	+	Cit. S.	23	1	4	§404	T, W
129.	2011	–	Jud. Rev.	7, 23	2	3	§404	I, W
130.	2011	+	Enf.	23	1	7	§404	T
131.	2011	+	Cit. S.	23	1	5	§402	T
132.	2011	–	Crim.	23	1	7	§404	W
134.	2011	+	Cit. S.	23, 24	2	3	§402	T
135.	2011	+	Crim.	4, 22, 23, 24	4	10	§402	T
137.	2011	+	Cit. S.	23	1	5	§404	T, W

A	B	C	D	E	F	G	H	I
139.	2010	+	Enf.	23	1	8	§404	T, W
141.	2010	+	Jud. Rev.	13, 23, 24	3	3	§404	T
142.	2009	+	Cit. S.	4, 23	2	7	§402	T, W
143.	2009	+	Cit. S.	4, 23	2	2	§402	T
144.	2009	+	Jud. Rev.	7, 23	2	7	§404	W
147.	2009	+	Cit. S.	4, 7, 16, 23	4	14	§402	G
153.	2008	–	Jud. Rev.	23	1	5	§404	O
154.	2007	+	Enf.	7, 23	2	9	§404	T, W
155.	2007	+	Crim.	23	1	6	§404	W
156.	2007	+	Enf.	7, 23	2	8	§404	W
158.	2007	+	Enf.	23	1	4	§404	W
159.	2007	–	Cit. S.	23	1	1	§404	W
160.	2007	–	Cit. S.	23	1	4	§402	T
162.	2007	+	Crim.	23	1	7	§404	T
164.	2006	+	Enf.	23	1	6	§404	W
165.	2006	–	Enf.	4, 5, 22, 23	4	8	§402	T
168.	2003	+	Enf.	7, 23	2	13	§404	T, W
169.	2003	+	Enf.	4, 23	2	6	§404	T, W
171.	2003	+	Cit. S.	23	1	10	§§402/404	T, W
173.	2003	–	Jud. Rev.	4, 7, 23	3	12	§404	W
175.	2002	–	Enf.	1, 5, 7, 16, 23	5	2	§404	W
177.	2002	+	Cit. S.	23	1	6	§402	G,T
179.	2002	–	Enf.	5, 7, 16, 22, 23	5	2	§404	T, W
181.	2002	+	Enf.	23, 25	2	3	§404	W
182.	2002	+	Cit. S.	23	1	3	§402	T
185.	2002	+	Enf.	23, 25	2	1	§402	W
186.	2001	+	Cit. S.	4, 16, 23	3	9	§404	G, T
187.	2001	+	Crim.	4, 16, 23	3	8	§404	T, W
188.	2001	+	Cit. S.	22, 23	2	6	§402	T
194.	1998	–	Enf.	23	1	2	§404	W
197.	1997	+	Enf.	4, 23	2	1	§402	T
198.	1997	+	Cit. S.	4, 23	2	4	§402	G, W
199.	1997	–	Cit. S.	2, 4, 7, 16, 22, 23, 24	7	11	§402	G
201.	1996	+	Cit. S.	4, 23	2	4	§402	T
206.	1995	+	Cit. S.	23	1	3	§402	T
207.	1995	+/–	Cit. S.	4, 22, 23	3	13	§402	G
209.	1995	+	Enf.	23	1	3	§404	W

A	B	C	D	E	F	G	H	I
210.	1984	+	Cit. S.	4, 7, 16, 23	4	7	§402	G
212.	1993	+	Cit. S.	4, 23	2	9	§402	G
214.	1992	+	Cit. S.	6, 23	2	1	§404	W
216.	1991	+	Enf.	23	1	3	§404	T
218.	1989	+	Cit. S.	4	1	–	§402	G
222.	1989	+	Jud. Rev.	7, 24	2	–	§404	T
223.	1989	+	Enf.	23	1	2	§404	W
224.	1989	+	Enf.	23	1	1	§404	T, W
225.	1988	+	Cit. S.	4, 7, 16, 22, 23, 24	6	7	§402	G
226.	1987	+	Enf.	7, 23	2	1	§404	W
229.	1985	–	Cit. S.	4, 7, 16, 23, 24	5	4	§402	G
230.	1985	+	Jud. Rev.	22, 23	2	2	§404	W
232.	1984	+	Enf.	4, 16, 23	3	7	§404	T, W
233.	1984	+	Enf.	7, 23	2	3	§404	W
234.	1984	+	Enf.	4, 16, 23	3	3	§404	W
235.	1983	+	Enf.	4, 23	2	3	§404	W
236.	1981	+	Enf.	23	1	1	§404	W
237.	1982	+	Jud. Rev.	7, 16, 21, 23	4	3	§404	W
239.	1981	+	Enf.	4, 23	2	2	§404	W
241.	1980	+	Enf.	4, 16, 23	3	5	§404	W
243.	1979	+	Jud. Rev.	23	1	10	§404	W
248.	1977	+	Jud. Rev.	4, 16, 21, 23	4	2	§404	T
249.	1976	+	Enf.	4, 16	2	–	§402	T
250.	1977	+	Jud Rev.	4, 16, 23	3	8	§404	W
252.	1975	+	Cit. S.	7, 23	2	5	§404	W
253.	1975	+	Cit. S.	4, 23	2	2	§404	W
254.	1975	+	Jud. Rev.	16, 23	2	1	§404	W
255.	1975	+	Jud. Rev.	10	1	–	§§402/404	W
256.	1975	+	Crim.	4, 22, 23	3	1	§402	T
257.	1975	–	Enf.	4, 16, 22, 24	4	2	§402	G
258.	1974	+	Enf.	4, 16, 21, 23, 24	5	5	§404	W

Table B: Point Source

A	B	C	D	E	F	G	H
U.S. Supreme Court Decisions							
2.	2013	–	Cit. S.	3, 6, 22, 24	4	–	§402
6.	2004	+	Cit. S.	4, 8, 22	3	–	§402

A	B	C	D	E	F	G	H
U.S. Circuit Court of Appeals Decisions							
12.	2011	+	Cit. S.	22, 23, 24	3	4	§402
13.	2011	–	Cit. S.	16, 22, 23, 24	4	4	§402
14.	2011	–	Jud. Rev.	22, 23	2	4	§402
17.	2010	+	Cit. S.	22, 23	2	3	§402
19.	2010	+	Crim.	22	1	–	§402
20.	2010	+	Cit. S.	4, 22, 23	3	3	§402
22.	2009	–	Cit. S.	4, 7, 22, 23, 24	5	7	§402
27.	2009	+	Jud. Rev.	4, 22, 23, 24	4	4	§402
28.	2008	–	Cit. S.	23, 24	2	3	§402
29.	2008	+	Crim.	16, 22, 23	3	6	§402
35.	2005	+	Crim.	23	1	4	§404
41.	2005	+	Cit. S.	7, 22, 23, 24	4	5	§402
42.	2005	+	Enf.	23	1	4	§101
44.	2004	+	Cit. S.	4, 23	2	2	§402
55.	2002	–	Cit. S.	3, 7, 10, 22, 23	5	2	§402
57.	2002	+	Cit. S.	4, 23	2	2	§402
59.	2002	+	Cit. S.	22, 23	3	2	§402
60.	2002	–	Cit. S.	23	1	1	§402
63.	2001	+	Jud. Rev./ Enf.	2, 23	2	1	§404
66.	2000	–	Cit. S.	22, 23	2	5	§§402/404
74.	1997	+	Crim.	4, 22, 23	3	5	§402
82.	1994	+	Cit. S.	4, 7, 22, 23	4	7	§402
84.	1993	+	Cit. S.	22, 23	2	2	§402
85.	1993	–	Crim.	1, 8, 10, 16, 17, 22, 23	7	7	§402
93.	1991	+	Cit. S.	4, 10, 23	3	3	§402
94.	1991	+	Cit. S.	22	1	–	§402
103.	1984	+	Jud. Rev.	16, 22, 23, 24	4	2	§402
108.	1983	+	Cit. S.	22, 23	2	2	§402
110.	1982	–	Cit. S.	2, 4, 7, 16, 22, 23, 24	7	1	§402
111.	1982	–	Cit. S.	22	1	–	§402
113.	1980	+	Cit. S.	22, 23	2	2	§402
116.	1979	+	Enf.	4, 11, 16, 22, 23, 24	6	2	§402
119.	1977	+	Jud. Rev.	6, 16, 18, 23, 24	5	1	§402
123.	1976	–	Jud. Rev.	22, 24	2	–	§402
U.S. District Court Decisions							
138.	2010	+	Cit. S.	22	1	–	§402

A	B	C	D	E	F	G	H
140.	2010	+	Jud. Rev.	1, 23, 24	3	8	§404
141.	2010	+	Enf.	22, 23	2	1	§404
148.	2009	–	Cit. S.	7, 23	2	1	§402
149.	2009	+	Cit. S.	22, 23	2	2	§404
150.	2009	+	Cit. S.	4, 7, 10, 21, 23	6	2	§402
151.	2009	+	Enf.	23	1	5	§404
152.	2008	–	Cit. S.	23	1	9	§402
160.	2007	+	Cit. S.	4, 22, 23	3	7	§402
171.	2003	+	Cit. S.	22, 23	2	5	§§402/404
172.	2003	+	Cit. S.	4, 7, 23	3	3	§402
176.	2002	+	Cit. S.	1, 2, 4, 5, 23	5	8	§402
182.	2002	+	Cit. S.	4, 7, 22, 23	4	3	§402
184.	2001	–	Cit. S.	8, 13, 21, 23	4	5	§402
192.	1999	+	Cit. S.	22, 23	2	1	§402
195.	1998	–	Cit. S.	23, 24	2	2	§§402/404
198.	1997	+	Cit. S.	1, 23	2	3	§402
202.	1996	+	Jud. Rev.	4, 23, 24	3	4	§402
203.	1996	+	Crim.	23	1	1	§404
204.	1996	–	Cit. S.	13, 23	2	2	§402
205.	1996	+	Cit. S.	4, 23	2	8	§402
206.	1995	+	Cit. S.	4, 7, 23	3	2	§402
208.	1995	+	Cit. S.	22, 23	2	1	§402
210.	1994	+	Cit. S.	4, 7, 23	3	8	§402
215.	1991	+	Crim.	1, 4, 7, 8, 13, 16, 22, 23, 24	9	5	§402
217.	1990	+	Enf.	23	1	3	§404
220.	1990	–	Cit. S.	1, 4, 7, 23	4	3	§402
227.	1987	+	Cit. S.	22, 23	2	2	§402
228.	1986	+	Cit. S.	4, 23	2	4	§402
240.	1981	+	Cit. S.	22, 23	2	3	§402
241.	1980	+	Enf.	22, 23	2	1	§404
242.	1980	+	Crim.	22, 23, 24	3	1	§402
244.	1979	+	Cit. S.	22, 23	1	2	§402
247.	1978	+	Cit. S.	22, 23	2	1	§402
251.	1976	+	Cit. S.	6, 24	2	–	NEPA
259.	1974	+	Enf.	22	1	–	§402

Appendix I
EPA's Regulatory Definition of "Navigable Waters"
40 C.F.R. §122.2

Note to reader: This appendix captures EPA's regulatory definition as it existed prior to and after EPA's amendment of the definition on June 29, 2015. The text that is common to both the earlier and the amended versions is *italicized*. The text that existed prior to the amendment, but deleted by the amendment is underlined. The text that was added by the amendment is in plain text. As this book was in production, EPA and the Corps proposed to rescind the 2015 rule, 82 Fed. Reg. 34899 (July 27, 2017), making the appendix more useful than originally expected at the time of its writing.

Waters of the United States or waters of the U.S. means:

(a 1*) For purposes of* all sections of *the Clean Water Act, 33 U.S.C. 1251 et seq. and its implementing regulations, subject to the exclusions in paragraph (*b 2*) of this definition, the term "waters of the United States" means:*

(a i *) All waters which are currently used, were used in the past, or may be susceptible to use in interstate or foreign commerce, including all waters which are subject to the ebb and flow of the tide;*

*(*b ii *) All interstate waters, including interstate wetlands;*

*(*f iii *) The territorial seas;* and

(c) All other waters such as intra state lakes, rivers, streams (including intermittent streams), mudflats, sandflats, "wetlands," sloughs, prairie potholes, wet meadows, playa lakes, or natural ponds the use, degradation, or destruction of which would affect or could affect interstate or foreign commerce including any such waters

(1) Which are or could be used by interstate or foreign travelers for recreational or other purposes;

(2) From which fish or shellfish are or could be taken and sold in interstate or foreign commerce; or

(3) Which are or could be used for industrial purposes by industries in interstate commerce;

(d iv) *All impoundments of waters otherwise identified as waters of the United States under this section*;

(e v) All *tributaries*, as defined in paragraph (3)(iii) of this section, of waters identified in paragraphs (1)(i) through (iii) of this section *of waters identified in paragraphs (a) through (d) of this section;*

(g) "Wetlands" adjacent to waters (other than waters that are themselves wetlands) idenitified in paragraphs (a) through (f) of this definition.

(vi) All waters adjacent to a water identified in paragraphs (1)(i) through (v) of this definition, including wetlands, ponds, lakes, oxbows, impoundments, and similar waters;

(vii) All waters in paragraphs (1)(vii)(A) through (E) of this definition where they are determined, on a case-specific basis, to have a significant nexus to a water identified in paragraphs (1)(i) through (iii) of this definition. The waters identified in each of paragraphs (1)(vii)(A) through (E) of this definition are similarly situated and shall be combined for purposes of a significant nexus analysis, in the watershed that drains to the nearest water identified in paragraphs (1)(i) through (iii) in this definition. Waters identified in this paragraph shall not be combined with waters identified in paragraph (1)(vi) of this definition when performing a significant nexus analysis. If waters identified in this paragraph are also an adjacent water under paragraph (1)(vi), they are an adjacent waster and no case-specific significant nexus analysis is required.

(A) Prairie potholes. Prairie potholes are a complex of glacially formed wetlands, usually occurring in depressions that lack permanent natural outlets, located in the upper Midwest.

(B) Carolina bays and Delmarva bays. Carolina bays and Delmarva bays are ponded, depressional wetlands that occur along the Atlantic coastal plain.

(C) Pocosins. Pocosins are evergreen shrub and tree dominated wetlands found predominately along the Central Atlantic coastal plain.

(D) Western vernal pools. Western vernal pools are seasonal wetlands located in parts of California and associated with topographical depression, soils with poor drainage, mild, wet winters and hot, dry summers.

(E) Texas coastal prairie wetlands. Texas coastal prairie wetlands are freshwater wetlands that occur as a mosaic of depressions, ridges, intermound flats, and mima mound wetlands located along the Texas Gulf Coast.

(viii) All waters located within the 100-year floodplain of a water identified in paragraph*s* (1)(i) through (iii) of this definition and all waters located within 4,000 feet of the high tide line or ordinary high water mark of a water identified in paragraphs (1)(i) through (v) of this definition where they are determined on a case-specific basis to have a significant nexus to a water identified in paragraphs (1)(i) through (v) of this definition. For waters determined to have a significant nexus, the entire water is a water of the United States if a portion is located within the 100-year floodplain of a water identified in (1)(i) through (iii) of this definition or within 4,000 feet of the high tide line or ordinary high water mark. Waters identified in this paragraph shall not be combined with waters identified in paragraph (1)(vi) of this definition when performing a significant nexus analysis. If waters identified in this paragraph are also an adjacent water under paragraph (1)(vi), they are adjacent water and no case-specific significant nexus is required.

(2) The following are not "waters of the United States" even where they otherwise meets the terms of paragraphs (1)(iv) through (viii) of this definition.

(i) Waste treatment systems, including treatment ponds or lagoons designed to meet the requirements of the Clean Water Act. This exclusion applies only to manmade bodies of water which neither were originally created in waters of the United States (such as disposal area in wetlands) nor resulted from the impoundment of waters of the United States. [See Note 1 of this section.]

(ii) Prior converted cropland. Notwithstanding the determination of an area's status as prior converted cropland by any other Federal agency, for the purposes of the Clean Water Act, the final authority regarding Clean Water Act jurisdiction remains with EPA.

(iii) The following ditches:

(A) Ditches with ephemeral flow that are not a relocated tributary or excavated in a tributary.

(B) Ditches with intermittent flow that are not a relocated tributary, excavated in a tributary, or drain wetlands.

(C) Ditches that do not flow, either directly or through another water, into a water identified in paragraphs (1)(i) through (iii) of this definition.

(iv) The following features:

(A) Artificially irrigated areas that would revert to dry land should application of water in that area cease;

(B) Artificial, constructed lakes and ponds created in dry land such as farm and stock watering ponds, irrigation ponds, settling basins, fields flooded for rice growing, log cleaning ponds or cooling ponds;

(C) Artificial reflecting pools or swimming pools created in dry land;

(D) Small ornamental waters created in dry land;

(E) Water-filled depressions created in dry land incidental to mining or construction activity, including pits excavated for obtaining fill, sand, or gravel that fill with water;

(F) Erosional features, including gullies, rills, and other ephemeral features that do not meet the definition of tributary, non-wetland swales, and lawfully constructed grassed waterways; and

(G) Puddles.

(v) Groundwater, including groundwater drained through subsurface drainage systems.

(vi) Stormwater control features constructed to convey, treat, or store stormwater that are created in dry land.

(vii) Wastewater recycling structures constructed in dry land; detention and retention basins built for wastewater recycling; groundwater recharge basins; percolation ponds built for wastewater recycling; and water distribution structures built for wastewater recycling.

(3) In this definition, the following terms apply:

(i) Adjacent. The term adjacent means bordering, contiguous, or neighboring a water identified in paragraph (1)(i) through (v) of this definition, including waters separated by constructed dykes or barriers, natural river berms, beach dunes, and the like. For purposes of adjacency, an open water such as a pond or lake includes any wetlands within or abutting its ordinary high water mark. Adjacency is not limited to waters located laterally to a water identified in paragraphs (1)(i) through (v) of this definition. Adjacent waters also include all waters that connect segments of a water identified in paragraphs (1)(i) through (v) or are located at the head of a water identified in paragraph(1)(i) through (v) of this definition and are bordering, contiguous, or neighboring such water. Waters being used for established normal farming, ranching or silvaculture activities (33 U.S.C. 1344(f) are not adjacent.

(ii) Neighboring. The term neighboring means:

(A) All waters located within 100 feet of the ordinary high water mark of a water identified in paragraphs (1)(i) through (v) of this definition. The entire water is neighboring if a portion is located within 100 feet of the ordinary high water mark;

(B) All waters located within the 100-year floodplain of a water identified in paragraphs (1)(i) through (v) of this definition and not more than 1,500 feet from the ordinary high water mark of such water. The entire water is neighboring if a portion is located within 1,500 feet of the ordinary high water mark and within the 100-year floodplain;

(C) All waters located within 1,500 feet of the high tide line of a water identified in paragraphs (1)(i) through (v) of this definition, and all waters within 1,500 feet of the ordinary high water mark of the Great Lakes. The entire water is neighboring if a portion is located within 1,500 feet of the high tide line or within 1,500 feet of the ordinary high water mark of the Great Lakes.

(iii) Tributary and tributaries. The terms tributary and tributaries each mean a water that contributes flow, either directly or through another water (including an impoundment identified in paragraph (1)(iv) of this definition), to a water identified in paragraphs (1)(i) through (iii) of this definition that is characterized by the presence of the physical indicators of a bed and banks and an ordinary high water mark. These physical indicators demonstrate there is a volume, frequency, and duration of flow sufficient to create a bed and banks and an ordinary high water mark, and thus to qualify as a tributary. A tributary can be natural, man-altered, or man-made water and includes waters such as rivers, streams, canals, and ditches not excluded under paragraph (2) of this definition. A water that otherwise qualifies as a tributary under this definition does not lose its status as a tributary if, for any length, there are one or more constructed breaks (such as bridges, culverts, pipes, or dams), or one or more natural breaks (such as wetlands along the run of a stream, debris piles, boulder fields, or a stream that flows underground) so long as a bed and banks and an ordinary high water mark can be identified upstream of the break. A water that otherwise qualified under this definition does not lose its status as a tributary if it contributes flow through a water of the United States that does not meet the definition of tributary or through a non-jurisdictional water to a water identified in paragraphs (1)(i) through(iii) of this definition.

(iv) Wetlands. The term wetlands means those areas that are inundated or saturated by surface or groundwater at a frequency and duration sufficient to support, and that under normal circumstances support, a prevalence of vegetation typically adopted for life in saturated soil conditions. Wetlands generally include swamps, marshes, bogs, and similar areas.

(v) Significant nexus. The term significant nexus means that a water, including wetlands, either alone or in combination with similarly situated waters in the region, significantly affects the chemical, physical, or biological integrity of a water identified in paragraphs (1)(i) through (iii) of this definition. The term "in the region" means the watershed that drains to the nearest water identified in paragraphs (1)(i) through (iii) of this definition. For an effect to be significant, it must be more than speculative or insubstantial.

Waters are similarly situated when they function alike and are sufficiently close to function together in affecting downstream waters. For purposes of determining whether or not a water has a significant nexus, the water's effect on downstream (1)(i) through (iii) waters shall be assessed by evaluating the aquatic functions identified in paragraphs (3)(v)(A) through (I) of this definition. A water has a significant nexus when any single function or combination of functions performed by the water, alone or together with similarly situated waters in the region, contributes significantly to the chemical, physical, or biological integrity of the nearest water identified in paragraphs (1)(i) through (iii) of this definition. Functions relevant to the significant nexus evaluation are the following:

(A) Sediment trapping.

(B) Nutrient recycling.

(C) Pollutant trapping, transformation, filtering, and transport.

(D) Retention and attenuation of flood waters.

(E) Runoff storage.

(F) Contribution of flow.

(G) Export of organic matter.

(H) Export of food, and

(I) Provision of life cycle dependent aquatic habitat (such as foraging, feeding, nesting, breeding, spawning, or use as a nursery area) for species located in a water identified in paragraphs (i) through (iii) of this definition.

(vi) Ordinary high water mark. (Omitted from this Appendix)

(vii) High tide line. (Omitted from this Appendix)

Notes

Notes

Notes

Notes

Notes